By the 1980s the Soviet science establishment had become the largest in the world, but very little of its history was known in the West. What has been needed for many years in order to fill that gap in our knowledge is a history of Russian and Soviet science written for the educated person who would like to read one book on the subject. This book has been written for that reader.

The main theme of the book is the shaping of scientific theories and institutions in Russia and the Soviet Union by social, economic, and political factors. Major sections include the tsarist period, the impact of the Russian Revolution, the relationship between science and Soviet society, and the strengths and weaknesses of individual scientific disciplines. The book also includes discussions of changes brought to science in Russia and other republics by the collapse of communism in the late 1980s and early 1990s.

Science in Russia and the Soviet Union

CAMBRIDGE HISTORY OF SCIENCE

Editors

GEORGE BASALLA
University of Delaware

OWEN HANNAWAY
Johns Hopkins University

Physical Science in the Middle Ages
EDWARD GRANT
Man and Nature in the Renaissance
ALLEN G. DEBUS
The Construction of Modern Science:
Mechanisms and Mechanics
RICHARD S. WESTFALL
Science and the Enlightenment
THOMAS L. HANKINS
Biology in the Nineteenth Century:
Problems of Form, Function, and Transformation
WILLIAM COLEMAN
Energy, Force, and Matter: The Conceptual Development of
Nineteenth-Century Physics
P. M. HARMAN
Life Science in the Twentieth Century
GARLAND E. ALLEN
The Evolution of Technology
GEORGE BASALLA
Science and Religion: Some Historical Perspectives
JOHN HEDLEY BROOKE
Science in Russia and the Soviet Union: A Short History
LOREN R. GRAHAM

SCIENCE IN RUSSIA AND THE SOVIET UNION
A Short History

LOREN R. GRAHAM

CAMBRIDGE
UNIVERSITY PRESS

Published by the Press Syndicate of the University of Cambridge
The Pitt Building, Trumpington Street, Cambridge CB2 1RP
40 West 20th Street, New York, NY 10011, USA
10 Stamford Road, Oakleigh, Melbourne 3166, Australia

First published 1993

Printed in the United States of America

Library of Congress Cataloging-in-Publication Data
Graham, Loren R.
Science in Russia and the Soviet Union : a short history / Loren
R. Graham.
p.   cm. – (Cambridge history of science)
Includes bibliographical references and index.
ISBN 0–521–24566–4
1. Science – Soviet Union – History. I. Title. II. Series.
Q127.S696G729   1992
509.47 – dc20   92-5087
                                                                   CIP

A catalog record for this book is available from the British Library.

ISBN 0–521–24566–4 hardback

TO MY STUDENTS

Contents

Preface

I HAVE written this book with a certain type of reader in mind: an educated person who knows little about the history of Russian and Soviet science and technology but would like to read one book on the subject. The book is not, therefore, aimed primarily at Russian area specialists or at that very small circle of scholars who specialize in Russian and Soviet science and technology. Many of the latter people are friends and colleagues of mine, and their names will often be found in the Bibliography and in the Notes. I have gained immensely from their insights. Much as I would like for them to think well of the book, I would be even more pleased if a nonspecialist would read it and feel that it had introduced him or her to the field.

Such an introduction is surely needed. By the 1980s there were more scientists and engineers in the Soviet Union than in any other country in the world, but the history and achievements of that scientific community are poorly known in the West. I have tried to present a short history of that community in a readable form. As I have written that history I have again and again been struck by its remarkable features.

The book is in four parts, the first three of which are in the text proper, and a fourth that appears as Appendix Chapter A: The physical and mathematical sciences, and Appendix Chapter B: The biological sciences, medicine, and technology. The two Appendix chapters are analyses of the strengths and weaknesses of Russian and Soviet science field by field. They are placed in the Appendix because the casual reader may not be interested in the amount of detail that they contain about specific fields of research and individual Russian and Soviet scientists. These chapters are not, however, mere lists or compilations of data but, instead, discursive historical interpretations of the core of Russian and Soviet science. They have been indexed along with the rest of the book so that the person seeking information about a specific field or a specific scientist will be able to find the appropriate section in the index.

I have dedicated this book to my students, both undergraduate and graduate, who at four different universities – Indiana, Columbia, MIT, and Harvard – have inspired me intellectually and sustained me personally. I would like to name a few of them here, but would inevitably leave out others equally deserving of citation. Some of my former students' names can be found in the Notes, because a few of them continued to work in the same field as I.

Throughout the text whenever I have given Russian language terms or names I have used the standard Library of Congress transliteration system, without diacritical marks, except in cases of words or names so well known in English according to different spellings that it is best to follow the current practice (Trotsky instead of Trotskii, *glasnost* instead of *glasnost'*). All translations are mine except when otherwise noted.

As in the case of all researchers, I am seriously in debt to generous institutions and individuals. Research on this book began at the Kennan Institute for Advanced Russian Studies at the Woodrow Wilson Center of the Smithsonian Institution. In intervening years I have relied on the services of the Program on Science, Technology, and Society at the Massachusetts Institute of Technology and the Russian Research Center and the Department of the History of Science at Harvard University. In 1991 I was pleased to receive support from the John D. and Catherine T. MacArthur Foundation for a series of joint U.S.–Russian seminars entitled "Science and Technology with a Human Face," which has increased my knowledge of Russian and Soviet science. In the Soviet Union I have worked at the Institute of Philosophy and the Institute of the History of Science and Technology of the Soviet Academy of Sciences, as well as the archives of the Academy and the Central Governmental Archives of the October Revolution. Among libraries the three that have provided me the most support are the Widener Library at Harvard University (truly one of the glories of American education), the Library of Congress, and the Lenin Library in Moscow.

This book draws upon several decades of teaching and research, and it would be impossible to name all the individuals who at one time or another have helped me. Let me name only two, who stand out: my wife, Patricia Albjerg Graham, whose name appears only this one time but who might be footnoted on practically every page, and Kenneth Keniston, who, as the director of the STS Program at MIT, provided an ideal environment for working on a book of this type.

Loren Graham
Grand Island, Lake Superior
October 1992

1. *Presidium building, Russian Academy of Sciences. This eighteenth-century mansion was originally built by a wealthy factory owner and later was owned by Tsar Nicholas I. Since 1934 it has been the headquarters of the Academy. It is located on commodious grounds just off the busy Lenin Prospect near downtown Moscow. Photo courtesy of Institute of the History of Science and Technology, Moscow.*

ЛОМОНОСОВЪ

2. M. V. Lomonosov (1711–1765), Russia's first significant scientist and founder of Moscow University (1755). Illustration courtesy of Institute of the History of Science and Technology, Moscow.

3. A model of Lomonosov's chemical laboratory in St. Petersburg, the first chemical laboratory in Russia, opened in 1748. Illustration courtesy of Institute of the History of Science and Technology, Moscow.

4. *N. I. Lobachevskii (1792–1856), the mathematician who first published a non-Euclidean geometry (1829–1830). Illustration courtesy of the Institute of the History of Science and Technology, Moscow.*

5. D. I. Mendeleev (1834–1907), the father of the periodic table of chemical elements (1869). Photo courtesy of the Institute of the History of Science and Technology, Moscow.

6. *V. V. Dokuchaev (1849–1903), international leader in soil science.*
Photo courtesy of the Institute of the History of Science and Technology,
Moscow.

7. I. P. Pavlov (1849–1936), noted physiologist who developed the concept of conditioned reflexes and the first Russian to receive the Nobel Prize (1904). Photo courtesy of the Institute of the History of Science and Technology, Moscow.

8. A. F. Ioffe (1880–1960), founder of the Leningrad Physico-Technical Institute, often called the "cradle of Soviet physics." Photo courtesy of the Institute of the History of Science and Technology, Moscow.

9. N. N. Semenov (1896–1986), winner of the 1956 Nobel Prize in chemistry.

10. T. D. Lysenko (1898–1976), promoter of a theory of genetics based on the inheritance of acquired characteristics and tyrant of Soviet biology from 1948 to 1965. Photo courtesy of the Institute of the History of Science and Technology, Moscow.

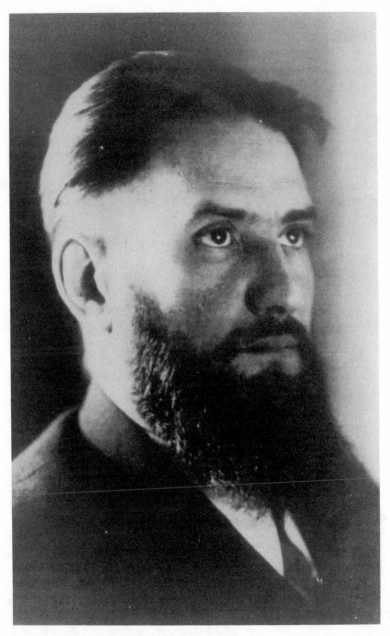

11. *I. V. Kurchatov (1903–1960), physicist who was the leader of the Soviet atomic bomb project. Photo courtesy of the Institute of the History of Science and Technology, Moscow.*

12. *A. N. Kolmogorov (1903–1987), outstanding Soviet mathematician. Photo courtesy of the Institute of the History of Science and Technology, Moscow.*

13. *N. I. Vavilov (1887–1943), Soviet biologist who died in prison, a victim of Lysenkoism. Photo courtesy of the Institute of the History of Science and Technology, Moscow.*

14. *P. L. Kapitsa (1894–1984), Nobel Prize-winning Soviet physicist who worked with Ernest Rutherford in England from 1924 to 1934 but who was detained by Stalin when visiting the Soviet Union in 1934 and was forced to move his research laboratory from Cambridge to Moscow. Photo courtesy of the Institute of the History of Science and Technology, Moscow.*

15. BESM-6, Soviet second-generation computer produced in the 1960s. Photo courtesy of Gregory Crowe.

16. Elbrus-II, a leading Soviet computer of the early 1990s. Photo courtesy of Peter Wolcott.

17. *A. D. Sakharov (1921–1989), a leading physicist in the construction of the Soviet hydrogen bomb, later famous dissident and critic of the Soviet regime. Photo taken during his 1988 visit to the United States. AP/Wide World Photos.*

Introduction

THE main theme of this book is the shaping of science and scientific institutions in Russia and the Soviet Union by social, economic, and political factors. If it is true, as many modern historians of science believe, that such factors are major influences on science and its development, surely those influences will manifest themselves in the case of Russia and the Soviet Union. No one will deny that Russian society and culture have in the thousand years of Russian history differed from society and culture in Western Europe, where modern science was born. Russia has followed a different economic path from that of Western Europe, and it has religious, political, and cultural traditions quite unlike those of its Western neighbors.

The most fruitful comparisons are made, however, not between entities that are totally different, but between those that are similar enough for some common elements to be found but diverge sufficiently that the variations can be studied. Russian science fits these criteria well. Imported initially from Western Europe, it took root and developed in distinct ways. One of the theses of this book is that the variations that arose were not only organizational and economic, but cognitive. The intellectual pathways of many areas of Russian and Soviet science are dissimilar from those in Western Europe and America.

Studying how those differences arose can reveal a great deal about the ways in which science develops in distinct environments. Why, for example, did most biologists in nineteenth-century Russia accept Darwin's theory of evolution enthusiastically yet reject his term "struggle for existence"? Why has Russia had such a strong tradition in mathematics and astronomy, but weak ones in experimental science? Why did Soviet physicists for thirty years refuse to use Niels Bohr's term "complementarity"? Why did Soviet scientists lag in the development of plate tectonics in geology but lead in the promotion of magnetohydrodynamics? Why have Soviet astrophysicists been enthusiastic pioneers in developing "inflationary theories" of the development of the universe and critics of "big bang" theories? What factors account for the Soviet

1

Union being the first country in the world to build an atomic power station and the first to launch an artificial satellite?

And then there are the differences that are rooted in the authoritarian and ideologically troubled history of Russian and Soviet science. Why did so many Russian scientists get into political difficulties, both before and after the 1917 Revolution? Why did Lysenkoism, a biological doctrine that denied the achievements of modern genetics being pursued elsewhere in the world, reign in the Soviet Union for several decades? Why did Soviet authorities purge thousands of its best scientists and engineers exactly at the moment when it needed them most? And how did Soviet science not only survive these terrible losses but even, in some instances, flourish under conditions of tyranny? How do we explain the fact that Russian and Soviet science seemed to do best when the political situation was worst, and, with the emergence of democratic reforms in the nineties, seemed to go into decline?

These questions point to some of the ways in which Russia and the Soviet Union offer a unique opportunity for the study of science under conditions different from those in the industrialized West. In the chapters that follow, I have tried to answer these questions while, at the same time, giving an overview history of Russian and Soviet science.

The history offered here is not, of course, a comprehensive coverage of Russian and Soviet science; such a treatment would occupy many volumes. I have tried to analyze the most essential points. Where I have thought that more detail is necessary even though all readers may not be equally interested, I have placed that material in the appendix chapters under the individual scientific disciplines discussed there.

The main text of the book analyzes different aspects of the shaping of science in Russia by the social environment. The first chapter contains an analysis of the receptivity of early Russian culture to science, and the importation of science by Peter the Great in the early eighteenth century. In this first stage of the history of Russian science some issues emerge that remain constant in the later history: the authoritarian method of promoting science, the effort by Russia's leaders to catch up with Western science and technology without losing their own cultural and political identities, and the greater success of Russia in mounting large centralized projects such as explorations of the Arctic and Siberia than in fostering scientific creativity in a broad spectrum of official and unofficial organizations.

The next chapter, concentrating on the nineteenth century, describes the remarkable appearance of two scientists in Russia who left permanent marks on the history of world science, the mathematician Nikolai Lobachevskii and the chemist Dmitrii Mendeleev. Their improbable social origins and remarkable achievements are consonant in striking ways, and even the topics on which they worked are linked to the situations in which they found themselves. During their lives the Russian university system grew impressively, even though it often suffered from political oppression and financial deprivation. The tsarist government gradually came to realize that the promotion of science and technology was in its interest. By the end of the century, science had become an important part of Russian culture, and in some areas – mathematics, physiological psychology, soil science, animal and plant ecology – Russia was emerging as a leader.

Chapter 3 concentrates on the ways in which the Russian socio-economic environment affected one area of science: the reception of Darwinian evolution. By the late nineteenth century radical political views, of several different types, were beginning to sweep across Russia. Politics affected attitudes toward evolution, resulting in debates that differed in some ways from those going on at the same time in Western Europe and America. A number of Russian scientists and analysts of evolution maintained that Charles Darwin himself had been influenced by the politics of nineteenth-century England while developing his great theory, and they tried to make revisions to Darwinism that would make it more palatable both to them and their Russian audience.

The second part of the book, opening with Chapter 4, analyzes the impact of the Russian Revolution of 1917 and the early Soviet regime on the development of science. The relationship of science to that revolution was far more intimate than a casual observer might think. The victorious Marxist leaders of the Revolution considered their view of history and politics to be scientific, and they had begun developing even before 1917 a Marxist philosophy of science. They wanted to promote science and technology rapidly, but they distrusted many of the scientists and engineers who had been educated before the Revolution, most of whom had little sympathy with the Bolsheviks. Gradually, however, a number of the scientists and especially the engineers began to work out a modus vivendi with the new government. The 1920s was a time when the future of the Soviet Union was still undetermined, and some technical specialists became intrigued with the vision of a scientifically planned economy. On the other hand, the most militant of the revolutionaries continued to be suspicious of the technical specialists. The scene was gradually set for a violent confrontation, which was promoted by Stalin in 1929.

Chapter 5 explores the ways in which several distinguished Soviet scientists were influenced by Marxism in the development of innovative theories about nature. This is an aspect of the history of Soviet science that is little known, and which, with the fall of communism in the Soviet Union in the 1990s, is now often denied, both in the former USSR and abroad. Most people now assume that all influence of Marxism on Soviet science was deleterious. On the contrary, in the works of scientists such as L. S. Vygotsky, A. I. Oparin, V. A. Fock, O. Iu. Schmidt, and

A. N. Kolmogorov, the influence of Marxism was subtle and authentic. It shows up in the cognitive core of their work.

Better known, although often misunderstood, is the catastrophe of Soviet biology because of Lysenkoism, the subject of Chapter 6. Many foreign commentators on Lysenkoism have assumed that its origins can be found in the hope of creating a "New Soviet Man" through the inheritance of acquired characteristics by human beings. On the contrary, Lysenko rejected such arguments and based his "Michurinist biology" on an effort to increase yields from agricultural crops and livestock. His rise was intimately connected with the crisis of Soviet agriculture stemming from the disastrous forced collectivization program. His overthrow was an incredibly protracted affair that involved the efforts of dozens of heroic Soviet scientists and intellectuals. The Lysenko affair was only the most dramatic of many political intrusions in Soviet intellectual life under Stalin and his immediate successors that resulted in the deaths of tens of thousands of Soviet scientists and engineers.

The last part of the book contains three chapters analyzing the ways in which Soviet society and the Soviet government over seventy-four years from 1917 to 1991 formed attitudes toward science and technology and the organizational framework of research and development. Chapter 7 shows that the Soviet Union was a pioneer in the 1920s in promoting the historical and social study of science and technology but that this effort soon was stymied by major political troubles. Chapter 8 examines the complicated relationship between the technical intelligentsia and the leaders of the Soviet state, two groups who simultaneously needed and yet distrusted each other. In the life of Andrei Sakharov, the famous physicist and dissident, we see a poignant illustration of this ambiguous relationship. Chapter 9 examines the history of the organization of science in the Soviet Union. This history is particularly striking, for it contains an ambitious effort by the Soviet Union to establish an alternative to the organization of science in the West that would, Soviet leaders thought, produce the finest and most productive science establishment in the world. The failure of that utopian scheme and the effort in the nineties to return to Western models of research are the subjects of the last section of this chapter.

The two appendix chapters, which examine each field of Russian and Soviet science and technology separately, describe the strong and weak aspects of research and then attempt to explain those strengths and weaknesses in terms of the social, economic, and political factors discussed in the earlier chapters. The history of Soviet science and technology contains impressive achievements, many of them little known in the West. How many people in the West know, for example, that the term "gene pool" is of Russian origin? Or that several of the pioneers of population genetics worked in Russia in the 1920s? Or that the basic terminology of soil science is Russian? Or that the Soviet Union held many aviation records before World War II? Better known are the spectacular early achievements of the Soviet space program. The Soviet Union was the first country in the world to launch an artificial satellite and the first to place a human being in orbit. Western physicists know that the prevalent design for nuclear fusion, the Tokamak reactor, is also of Soviet origin.

Despite these achievements, the overall record of Soviet science and technology, considering the enormous size of the Soviet science establishment, was disappointing, especially to Soviet leaders. Fields with tens of thousands of researchers, such as chemistry, failed to match the level of outputs of countries with many fewer specialists in the field. In computers, after an early rather promising start, the Soviet Union fell behind badly. Medicine and public health, fields the Soviet Union held up for emulation by other developing countries in past decades, slipped catastrophically, as evidenced by deteriorating life expectancy and infant mortality statistics. Soviet biology never attained the excellence that it possessed before the calamity of Lysenkoism. Throughout the Soviet science establishment creativity fell to low levels, even in fields such as physics where a strong tradition of excellence existed.

These weaknesses have organizational, political, and social roots. Soviet science was not traditionally organized on the basis of peer review and research grants, but instead on block funding of entire institutes. This funding system is one of the many factors analyzed that contributed to the low productivity in Soviet science.

The end of the Soviet Union in late 1991 means that "Soviet science" as such has disappeared. "Russian science," however, will continue, and it will be one of the largest science establishments in the world, even if diminished from its size of a few years ago. The biggest scientific centers of the old Soviet Union – in Moscow, Leningrad (now again St. Petersburg), Novosibirsk – are in Russia. The new Russian Academy of Sciences, created in 1991, inherited most of the institutions and personnel of the old Soviet Academy of Sciences. Throughout the former Soviet Union the center of gravity shifted to the newly independent republics, and Russia, while dominant in science, is not the only one with a viable scientific establishment; Ukraine, in particular, has an impressive scientific community.

In the coming era, the influence of scientific institutions and modes of management formed during the Soviet period will long exercise influence in the successor states to the Soviet Union. A great debate is currently going on there about how much these institutions and their modes of governance should be changed in the post-Soviet era, a debate that I witnessed and even participated in while in Moscow in the last weeks of 1991, as the Soviet Union collapsed around us. I have described these discussions in the last part of Chapter 9.

In short, I began writing this book when one could speak of Soviet science as currently organized and practiced and Russian science as historically organized and practiced. By the time this book was completed the situation was reversed. Soviet science is now historical and Russian science is both historical and current.

PART I

The tsarist period

1

Russian science before 1800

WHY DIDN'T A SCIENTIFIC TRADITION DEVELOP IN OLD RUSSIAN CULTURE?

The study of the history of science in Russia and the Soviet Union raises basic questions about both Russian history and the nature of science. Some historians of Russia have argued that old Russia was particularly backward compared to Western Europe in the area of rational and naturalistic thought, and that the cause of this backwardness was an irrational or mystical characteristic in Russian Orthodoxy or even in the Slavic personality.[1] A study of the history of science in Russia will allow us to examine this viewpoint critically. This issue is elevated from a historical to a contemporary one by virtue of the fact that the Soviet Union became a major scientific force with an enormous scientific and technical community.[2]

Although I will concentrate in this book on the Soviet period, I will consider in the first four chapters the history of science before 1917, when the Soviet system of government was established. It is important, therefore, to recall the major divisions of Russian history: the first state, known as that of the Kievan Rus', from the ninth century A.D. to its conquest by the Mongols in approximately 1240; the period of Mongol rule from 1240 to 1480; the time of Muscovite primacy from 1480 to 1700; the Imperial period centered on St. Petersburg from 1700 to the Revolutions of 1917; and the Soviet period from 1917 to 1991. A new period commenced with the end of Communist Party rule in the latter year. The systematic study of nature according to the methods of Western science is almost entirely a phenomenon beginning in the early eighteenth century, but the cultural matrix in which this science arose in Russia had deep roots in the earlier periods.

The Kievan era was the formative period of Russian culture, the time when the Rus' were Christianized (988), and when they turned to Byzantium for religious, ideological, and literary models. It was also a period of remarkable cultural and economic development. Kiev at the end of the tenth and the beginning of the eleventh centuries was one of the largest cities in Europe, possessing, according to foreign accounts, some 400 churches and eight markets.[3] In architecture and some of the decorative arts the legacy of Kiev remains a world treasure to this day.

If one considers the place of Kiev in terms of its potential in the history of science, examining the issue at first from a speculative point of view, one might think that the culture of the Rus' was advantageously placed for scientific development. Historians of science in Western Europe often observe that two significant ingredients of the Scientific Revolution of the seventeenth century were ancient Greek thought and Arabic science. In geographic and cultural terms, the Kievan Rus' seem, at first sight, to have been well positioned. Kiev was a part of the Byzantine intellectual world with its ancient Greek sources of knowledge; furthermore, it had trading and even dynastic ties with the medieval cities of Europe, the Islamic centers of Asia, and with Constantinople itself. At the very beginning of Russian history, then, we face a riddle in the history of science: Why did Kiev miss this apparent opportunity to develop into a center of scientific studies?

Of the two important possible influences, the Arabic and the Byzantine, the Arabic is the easier to analyze. The possibilities of the Kievan Rus' for making contact with Arabic science were much more apparent than real. The most productive Islamic centers of scientific learning were at the Western end of the tier of Islamic lands – Moorish Spain (Toledo), Morocco, Southern Italy (Salerno), Sicily – and were therefore far from Kiev geographically. The Salerno school of medicine was very significant in the twelfth and thirteenth centuries. Recent research has shown that intricate mechanical clocks and devices were developed in Morocco in the same period. The Eastern tier of Muslim states, with which Kiev was in the closest proximity, seems to have been somewhat less creative in the natural sciences. Important exceptions exist, to be sure – Avicenna, or Ibn Sina, the greatest of the Arabic philosophers of the East – lived in Bukhara (later a Soviet city) exactly at the time of Kiev's flowering. Throughout this period, however, the Rus' were cut off from the enlightened Persian dynasty of the Saminids by fierce nomadic tribes such as the Patzinaks and the Cumans. The Rus' did have contact with the Volga Bulgars, Muslims who traded with the Arabs, but who were not interested in natural philosophy. Still, some knowledge penetrated to Kiev, especially in medicine, a field in which the Rus' became strong, and where Armenian, Syrian, and Arabic influences were important. The Bulgars also supplied silver to the Rus' and contributed to the development of a sophisticated art of the working of precious metals.

The failure of the Kievan Rus' to partake deeply of the Greek sources in science available in Byzantium is somewhat more difficult to explain. To the Kievans, Byzantium was not just one more potential influence or source of knowledge, it was *the* preeminent cultural force, sometimes disliked or envied, to be sure, but never disregarded. In liturgy, theology, political ideology, and art, Byzantine influences were dominant in early Kievan culture. Yet, oddly enough, the Rus' and other Orthodox Slavs were not as interested in Byzantine science as members of some other cultures more peripheral to the Byzantine orbit. As Ihor Sevcenko, who studied the issue, remarked, the "Orthodox Slavs as a whole were certainly more influenced in their culture and literature by Byzantium than were Islam or Western Christianity, and even perhaps the Syrian Christians. With these Slavs, translated literature by far outranked original works in bulk, prestige and popularity. Yet, the Orthodox Slavs translated fewer of the scientific and philosophical works available in Byzantium than did the Syrians, Arabs or Latins."[4]

In fact, during the middle ages the Orthodox Slavs translated no complete major work of ancient Greek science.[5] And yet in this same period they translated many Greek manuscripts on other subjects into Church Slavonic, a language common to the Eastern Slavs. The most active translators were the Bulgarians, who created a literature in translation upon which the Kievan Rus' drew.

At the time of the greatest translation activity, the tenth century, Byzantium was actually enjoying a modest scientific revival. In the ninth century Byzantine scholars produced standard texts of Ptolemy and Euclid, and one of them, Leo the Mathematician, was familiar with the works of Archimedes and Proclus as well. None of these texts showed up in Kiev, however, a riddle that appears even deeper when one notices that Leo was described as the teacher of the Apostle of the Slavs, Cyril.

Despite the relative lack of interest in the Eastern Slavs in Byzantine science, Kiev did not go uninfluenced by Greek ideas about the natural world. Translations of Euclid or Ptolemy were not necessary in order for Greek terms and rudimentary ideas of nature to penetrate Kievan culture. Some of the Kievan monks and literati could read Greek, and even for those who did not, the translations of nonscientific texts from Byzantium often provided glimpses of Greek ideas about nature. According to these sources, fire, for example, was described as one of the four elements, and not as a pagan principle, as it had been regarded by the pre-Christian Rus'. The concept of the sphericity of the earth was also contained in some of the translated texts, as were tidbits of knowledge about biology that can be traced back to Aristotle. Educated Kievans encountered the names, and some of the sayings (often incorrect or distorted), of Thales, Parmenides, Democritus, Pythagoras, Socrates, Plato, and Aristotle. Words such as "planet" (*planida, planeta*), the translations of the signs of the zodiac, and technical terms crept into Kievan writings from Byzantine sources. But, considering the possibilities, the total impact of Greek and Byzantine science upon the Kievan Rus' was still quite small.

Some writers have attempted to account for the inattention of the

Kievan Rus' to science by pointing to the ascetic traditions of early Russian Christianity, often giving as an illustration the Monastery of the Caves in Kiev, where monks imprisoned themselves in underground cells and forsook all secular contacts. Such a tradition, the analysts have said, led Kievans to regard book learning as a deviation from piety, for books might weaken the faith of believers by describing natural events in ways that diminished the role of God.

One should be cautious, however, in concluding that Russian Orthodoxy was uniquely antithetical to scholarship and book learning. In the primary chronicles (the official records of the reigns of the princes) of the Kievan period a rejection of learning can not be found. Indeed, one chronicler described the cultural activities of the eleventh-century Kievan ruler Prince Iaroslav in the following way:

> He applied himself to books, and read them continually day and night. He assembled many scribes, and translated from Greek into Slavic. He wrote and collected many books through which true believers are instructed and enjoy religious education. . . . For great is the profit from book-learning. Through the medium of books, we are shown and taught the way of repentance, for we gain wisdom and continence from the written word. Books are like rivers that water the whole earth; they are the springs of wisdom.[6]

By 1076 learned people in Kiev were reading praise of books.[7]

Additional information about the attitudes of the early Rus' toward learning can be gleaned from hagiography, from the lives of saints. The biographies of the saints do not provide us with reliable factual material, because they were written more to glorify their subjects than to record data, but this very goal of sanctification informs us indirectly about the values of Kievan society. G. P. Fedotov tells us that on the subject of attitudes toward scholarship there were two standard formulas in the Greco-Byzantine tradition for the writing of saintly biographies: (1) The saint as a child was precocious and learned very rapidly. (2) The saint as a child had an ascetic aversion to learning and school.[8] Evidently, either path could lead to piety. In the Kievan tradition the first type of saint is amply present.

Nonetheless, the Orthodox Christian faith of the Kievan Rus' contained significant resistance to secular learning. Notice that in the quotation from the primary chronicle just given, book learning was not praised for its value in general, but because it taught repentance and religious principles. The praise given to at least some saints for avoiding learning as children is further evidence of this resistance. However, these same characteristics were common in Western Christianity of the time as well, and therefore provide an insecure basis for concluding that Orthodox Christianity was uniquely hostile to learning. Thomas à

Kempis, an Augustinian monk who lived in Utrecht in the fifteenth century, wrote in his *Imitation of Christ:*

> Rest from too great a desire to know, because therein is found great discord and delusion. Learned men are very eager to appear, and to be called learned. There is much which it profits the soul little or nothing to know. And foolish indeed is he who gives his attention to other things than those which make for his salvation.[9]

The *Imitation of Christ* had an enormous influence in Western Christianity and is said to have been translated into more languages than any other book than the Bible itself.

The practice of shutting oneself off from the secular world was also common in Western Europe, as the term "anchorite" still reminds us. Even being enclosed in cells, similar to those at the Monastery of the Caves in Kiev, was prevalent in the West. Often priests, nuns, or laypeople were walled up until death in a cell attached to a church, with a window through which to receive food and the holy sacraments. When in 1329 a young woman, Christine Carpenter, was enclosed at her own request in a cell in Surrey, England, the Bishop of Winchester observed that she would be able to "keep her heart undefiled by this world."[10] It is thus dubious to ascribe to the Orthodox Christianity of the Kievan Rus' a resistance to secularism or learning that did not exist in Western Christianity.

In order to explain the lack of attention by the Kievan Rus' to science, we must delve deeply into Kievan history and culture, identifying political and social reasons why interest in learning was rare. The question of early Kievan receptivity to Greek science is an issue of the diffusion of cultures. For cultural ideas to be readily transmitted from one culture to another, more is needed than merely for the first culture (in this case, the Byzantine) to possess the ideas in a readily accessible form. The second culture (the Kievan) must have the equipment and the needs for the particular cultural ideas to be absorbed and propagated.[11] Several reasons can be cited why the Kievan Rus' were not ready to study the Greek treatises on the exact sciences located in Constantinople. Kiev had only very recently become Christian, and the buttressing of that faith was a primary concern both of the Kievan princes and the patriarch in Constantinople, who during the first several centuries of the existence of Christianity in Kiev continued to name the high officials of the church. Furthermore, Christianity was initially seen as something imposed by the princes, the early ones of whom were of Viking (Varangian) stock. Paganism still lurked in the hearts of many people; several pagan revivals were even attempted. As Alexander Schmemann has written: "Christianity was long a foreign religion – even doubly foreign, being Greek and coming from the prince as well, which meant its support by the Varangian *druzhina*, the ruling clique in Russia."[12]

In these conditions, the first goal of the rulers of Kiev was to reinforce piety and allegiance, with book learning used for this purpose. The Kievan monks who sought knowledge, mostly in Slavic translations of Byzantine texts, were serving both the Kievan church and the Kievan prince; at a time of continuing internal and external threats to their ideology, they were not very interested in knowledge that did not strengthen the ideological fabric of society.

In Western Europe in these same centuries, most monks who were learned were similarly interested in the relationship of knowledge to faith. But by the eleventh and twelfth centuries a few centers of learning existed in Western Europe, especially in Italy (Salerno, Bologna), France (Paris), and in England (Oxford), where embryonic professional study began (law, medicine, theology) and where the seven liberal arts (three of which – arithmetic, geometry, and astronomy – were scientific) took root. For all the glory of medieval Kievan civilization, these nutrients of scholarship were not yet present, and Byzantium did not possess a center of learning in the Western sense.

Yet another reason for the absence in the lands of the Kievan Rus' of centers of scientific study concerns the intermediaries through which most of its Byzantine sources came. As already noted, the transmission sources were largely other Eastern Slavs, especially the Bulgarians, who became Christians before the Rus'. When searching for Byzantine learning, the Kievan Rus' most frequently turned to the Church Slavonic translations already made by the Bulgarians or other Balkan Slavs, rather than to the Greek originals. The total repository of such translations was already rich by the time Kiev became Christian, and seemed quite adequate, and certainly more convenient, to the Rus'. Unfortunately, this literature was extremely poor in secular literature.

On reflection, the original question about why Kiev was not more stimulated and affected by Greek learning available in Constantinople seems somewhat misplaced; it ascribes too great a motivating power to ideas by themselves, without paying much attention to supporting elite groups, economic structures, and social and political needs. Surely one does not assume that even if Euclid or Ptolemy or Aristotle had been translated into Church Slavonic by the medieval Rus' such manuscripts would automatically have led to a flowering of science. Indeed, even in Western Europe, where such translations did lead to some activity in science as early as the fourteenth century (one thinks of the writings on physics at Oxford and Paris by scholars like Oresme and Buridan), no great burst of scientific creativity followed. The concerns of medieval monks everywhere in Europe (West and East) were primarily theological; the seizing upon Greek science and the advancement of it in Western

Europe through the Arabic intermediary had to await the development of a fertile institutional and economic matrix, which came several centuries later. To emphasize the "failure" of early Russian civilization to take advantage of its direct connection to Greek science in Constantinople is not only to exaggerate the significance of mere intellectual contact (as if it were a spark that automatically ignites), but also to overlook the fact that some of the same factors that obstructed science among the Kievan literati were present in Western Europe as well in the middle ages and diminished only with the development of modern economic relations and social institutions.

I have dwelt here at some length on the lack of development of a scientific tradition in old Russian culture even though I intend to concentrate in this book on the late nineteenth and twentieth centuries; the reason is that moments of birth are always important, and in this case even more so, since the problems of retardation and knowledge transfer in early Russian history remained in the later period.

The invasion of the Mongols in the thirteenth century and their occupation of the lands of the Kievan Rus' and their immediate neighbors separated the area even more from Western Europe than its earlier connection with Byzantium had done. For two and a half centuries, while Western Europe grew in political strength and cultural achievement, the Russian principalities were under foreign rule. The Mongols were interested primarily in political submission and taxation; after submission was established, they allowed the Russians to maintain the Orthodox church and their princely order because this system facilitated administration and payment of taxes. A side effect of the system, however, was to increase political and intellectual authoritarianism (already strong even before the arrival of the Mongols) throughout Russia, and to reduce its contacts with Western Europe. The simultaneous decline of Constantinople, its eventual fall, and the Muslim subjugation of the Balkan Slavs – the early mentors of the Kievan Rus' – threw Russia back on its own still meager intellectual resources while increasing its sense of ideological isolation.

The early modern period of Russian history, up to the beginning of the eighteenth century, was one in which science, or "natural philosophy" as it was known in Western Europe, played almost no role. To emphasize this lack of rational knowledge about nature does not, of course, mean to deny the greatness of the achievements of Old Russia in areas such as religious art. Old Russian culture before Peter the Great produced much of interest and value in art, music, and architecture.[13] Science, however, came to Russia only in the eighteenth century. The Renaissance of the fifteenth and sixteenth centuries and the Scientific Revolution of the seventeenth were momentous episodes in the rise of Western Europe in which Russia did not participate. It is true that historians have found a few faint echoes in Russia of these Western developments. For example, in the late fifteenth and early sixteenth centuries there was an intellectual current in Novgorod known as the Judaizer heresy whose adherents were familiar with some of the astronomical ideas of Aristotle and Ptolemy.[14] In Moscow in the sixteenth century some Western ideas were spread by a Greek foreigner of Byzantine culture, known as Maxim the Greek, who had been exposed to cultural currents in Italy at the height of the Renaissance; his interests, though, were mainly moral and literary rather than the study of nature.[15] Perhaps the most obvious influence of Renaissance currents on Muscovy was in architecture, as the buildings and fortifications of the Kremlin still attest. When the Poles briefly occupied Moscow in the early seventeenth century, one of the Polish dukes imported Galileo's writings and actually corresponded with Galileo.[16] In 1632 Andrew Vinius, a Dutchman, established the Tula armament works, which have continued to the present day. Throughout the middle ages and early modern period Russian practical technology and arts continued to develop, a tradition that historians of technology have only begun to study with modern methods. In the last half of the seventeenth century, during the reign of Peter the Great's father Alexei Mikhailovich (1645–1676), a group of "Westernizers" around the throne exercised some influence in adopting Western customs passed via Poland and Ukraine. None of these developments, however, came close to the introduction of science. Arabic numerals did not become widespread in Russia until the eighteenth century, and the astronomical ideas of Copernicus were not fully described in a Russian source until 1717, over 170 years after they first became available in print in neighboring Poland. The introduction of science to Russia was a part of the early eighteenth-century reforms of Peter the Great, and even then the process was very difficult.

SCIENCE UNDER PETER THE GREAT

The reign of Peter the Great (1689–1725) was clearly momentous for Russian history, most notably in the introduction of West European science and technology. The revolutionary nature of Peter's reign is obvious, even though certain Russian historians, such as Solovev and Kliuchevskii, have pointed out that many of Peter's reforms were prefigured in earlier years, especially under Alexis, Peter's father. It is true that Moscow already had a foreign settlement before Peter's time, and that West European crafts had already achieved a foothold there. But if we define becoming "Europeanized" to mean not only using certain techniques but also becoming progress-oriented, or science-oriented, then the process of becoming European began in Russia with Peter the Great.[17] And if we look not just at technology, but also at science, we see that no real science was being pursued in Russia before the eighteenth century. Even after Peter it would be several decades before native Russians, as opposed to imported Western Europeans, began to pursue independent scientific research of quality. It was Peter, though, who created the first institutions making such work possible.

At the end of the seventeenth century Russia possessed no scientific organizations, although theological academies in Kiev and Moscow supplied important instruction in languages, as well as religious teaching. Mikhail Lomonosov, Russia's first great scientist, received his education in one of these academies. In 1701 Peter established a navigation school in Moscow, and in 1715 a Naval Academy in St. Petersburg. An artillery school, an engineering school, and a medical school were also established in his reign, and an ambitious and unsuccessful start was even made in primary education.

Peter was an unsophisticated man, and many of his educational enterprises had narrow utilitarian goals, especially improving Russia's military and naval power. Furthermore, many of his projects failed entirely. Nonetheless, Peter gradually gained some appreciation of science, at least for its contribution to national prestige. During his visits abroad he talked not only to military and naval experts, but also to leading scientists, including, we are told, Isaac Newton.[18]

Peter created an atmosphere that fostered the penetration into Russia of European culture and science, in contrast to many of his predecessors. Newton's *Principia*, for example, was accepted earlier by some of Peter's advisors in Russia than it was in colonial America.[19] From Peter's time forward a small portion of Russia's population, primarily some of the nobility and a few academicians and literati, looked upon education and Western learning as desirable aims. Such intellectuals as Feofan Prokopovich, Antiokh Kantemir, and Vasilii Tatishchev, while not scientists, defended and pursued scholarship. Most of the young nobles were not enamored with Peter's insistence that they learn arithmetic and geometry (which he saw as the key to military prowess in areas such as artillery), preferring to be "educated gentlemen" conversant in literature and the arts, but during the eighteenth century following Peter's reforms the Russian nobility gradually overcame its historic illiteracy.

That Peter is not adequately described as a simple utilitarian is revealed by his effort to create museums, theaters, and, most important for our study, an Academy of Sciences. Despite his own poor education, Peter seemed to understand that for Russia to become competitive in European politics it needed to do more than just import technicians and copy West European arms. Peter visited the academies of science of France, England, and Prussia, the observatory at Greenwich, and other places in Europe where research was being performed. He decided that Russia must also create such centers.

THE FOUNDATION OF THE ACADEMY OF SCIENCES

In their conception, goals, method of financing, regulations, and organization the numerous scientific societies that sprang up in Europe in the seventeenth and early eighteenth centuries were surprisingly diverse. Some were founded privately, by individuals, and only later recognized by the state (an example is the Royal Society in England); others were from the beginning wards of the ruler (for example, the Prussian Academy and the Accademia del Cimento of Florence, created largely through the efforts of the Hohenzollern and Medici families, respectively). What was common to these various societies and academies in Western Europe was that they were established in opposition to the universities, which had already existed in their respective countries for centuries, and which the members of the academies considered bastions of arid scholasticism. The academies represented the "new science" of the seventeenth century. Since the academies of Western Europe followed the universities chronologically, their eventual fates were linked to the reaction to the rise of science by the universities. So long as the universities continued to ignore science, the academies had a special role to play.

In Russia, in contrast to the countries it was emulating, there were no universities at the time of the foundation of the Academy of Sciences, a fact that would long continue to contribute to the unusual role and prestige of the Academy. We will see that at the end of the nineteenth century the Russian universities overtook the Academy of Sciences, and the trend already clearly manifested in Western Europe – for the academies to diminish from actual foci of research to mere honorific organizations – was also evident in Russia. That transition was not yet complete, however, at the time of the Revolution of 1917; the continued dominance in prestige of the Academy of Sciences over the universities, to be greatly enforced by the Soviet government after 1917, was therefore based on roots stretching back to the time of Peter the Great.

When Peter decided to create a scientific society he had a number of models from which to choose. The tsar communicated with other academies in Europe, and he personally inspected the activities of several of them. He conducted a long correspondence with Leibniz, founder and president of the Berlin Academy. He also consulted with Christian Wolff, professor at the University of Halle and, later, the University of Marburg.

Each of the European academies reflected the characteristics of the social milieu in which it had arisen as well as the philosophy of its founders and subsequent leaders. Peter's choice would be a wedding of borrowed foreign models with innovations required by the characteristics of the Russian environment.

The subject of discussion for over twenty years before its actual appearance, the Academy did not meet for the first time until 1725, after Peter's death.[20] Peter's court physician, Lavrentii Blumentrost, drew up the final plan for the organization. Blumentrost's statute for the Academy, approved by Peter before his death, was clearly the result of a diligent comparative study of foreign academies and an appraisal of Russia's scientific potential.

The goal of the academy project – and one of the characteristics that distinguished it from the previous academies of Western Europe – was the cultivation of native Russian science from imported seeds. The entire organization would consist of three tiers, with the top layer entirely foreign and the bottom mostly Russian. The uppermost level would be composed of academicians who would investigate the sciences in their highest development and also serve as professors of their special fields. The academicians would bring university students with them from Western Europe who, together with Russian students, would form the second level, or university. The university students (adjuncts) would in turn act as teachers in the third level, the gymnasium, all of whose students would be Russian. Thus, one institution, the Academy of Arts and Sciences, would assume responsibilities that in other states three different institutions undertook. The aim of the scheme was the gradual rise from the bottom to the top of the Russian elements.

Unfortunately, only the top level of the institution enjoyed a permanent existence. The gymnasium at first prospered but in later years lost much of its importance. The university did not come into existence until 1747 and finally expired at an unknown date, sometime near the end of the century (Moscow University, the oldest Russian university today, was established in 1755). Long before the disappearance of the gymnasium and the university, however, the official goal of cultivating ethnically Russian science was abandoned. The Academy was largely an organization of foreigners, most of them German speaking. The first Russian academician was not elected until twenty years after the birth of the society; a century and a half was to pass before the ethnic Russians won control of the Academy.[21]

The Academy was from the beginning treated as a branch of the government and subject to imperial command. The original project drawn up by Blumentrost and approved by Peter granted the Academy the privileges of self-government, including the right to elect its own members and president. However, this provision was violated from the start. The entire 1725 project, carrying Peter's signature, was hidden from the academicians by the court librarian, J. D. Schumacher, who acquired personal control of the Academy, establishing the precedent of rule by court favorites, which became a feature of the Academy's history. The promised right of electing their own presidents was withheld from the academicians until the advent of the provisional government in 1917.

Instead, the presidents were appointed by the crown and they in turn confirmed the membership of new academicians.

Despite the difficulties of its birth, the academy in its first years was a competent institution. The foreign scholars who came to St. Petersburg included several distinguished scientists, and they helped a few Russians to become familiar with European science.

RUSSIA'S FIRST GREAT SCIENTIST: MIKHAIL LOMONOSOV

Mikhail Vasilievich Lomonosov (1711–1765) was Russia's first eminent scientist. Lomonosov was versed in many fields, including chemistry, physics, mineralogy, mining, metallurgy, and optics. In addition, he was interested in history, and composed a great deal of verse. Proud of Russian accomplishments, he promoted the advancement of learning in his native land. The emergence of such an unprecedented person naturally attracts our attention, and we will want not only to consider some of his major accomplishments but also to understand the way in which Lomonosov, with both his strengths and weaknesses, was connected to the characteristics of his aspiring but still backward homeland.[22]

Lomonosov was born in the small village of Mishaninskaia near Kholmogory in the far north of European Russia, on the White Sea. At first glance this remote arctic location seems improbable for the emergence, childhood education, and adolescence of a scholar who would later try to lead Russian science onto the European stage. Yet the site of his birth did not bring only disadvantages to Lomonosov; in the seventeenth and early eighteenth centuries the area was a major channel for imports by sea from Western Europe, through ports such as Arkhangelsk and Kholmogory. Foreign influences were widespread. Furthermore, the area was so far north that it had been spared both Mongol rule and the blight of serfdom. Although Lomonosov and his family were, legally speaking, peasants, they enjoyed a freedom not often known to peasants on the estates of central Russia. In fact, Lomonosov's father, an active merchant, owned several fishing and cargo ships. His mother was the daughter of a deacon, and Lomonosov learned as a child to read and write, in both Russian and Church Slavonic.

Lomonosov's childhood was not one of privilege, however. The family circumstances were decidedly modest, despite the father's local prominence. Furthermore, Lomonosov's mother died while he was a child, and the several succeeding stepmothers evidently did not appreciate his interest in reading and study, considering such pastimes idle diversions from the practical tasks that surrounded the Lomonosov family. Lomonosov had no opportunity in his youth to learn Latin or the other European languages important for science, a disadvantage that, unless repaired, would have rendered a significant scientific career impossible.

Desiring further education, Lomonosov in 1730 obtained the permission of local authorities to go to Moscow to apply for admittance to the Slavic-Greek-Latin Academy, the best higher educational institution in Russia at that time, although devoted to theology and the preparation of clergy. A thoroughly Orthodox institution, the Academy nonetheless had roots in two of Russia's intellectual traditions, the Greek school of old Muscovy and the Catholic-influenced Latin tradition transmitted through Kiev and Poland. Peasants were not admitted to the Academy, and Lomonosov therefore concealed his origins, telling the Academy authorities that he was the son of a priest, as his knowledge of Church Slavonic supported. By the time his superiors learned of the falsehood Lomonosov had so impressed them by catching up with his classmates (most of whom were younger than he) in Latin and surpassing them in most other subjects that they permitted him to stay.

At this point in his life, a fortunate event occurred that opened up the possibility for Lomonosov to become a scientist. At first glance an accident, it should instead be seen as a symbol for what was happening to Russia in these years, gradual secularization and Westernization. The Academy of Sciences was by 1735 in full operation, but only on the basis of academicians imported from Europe; the university that was supposed to be a component of the Academy, attended by Russians, lacked students. Therefore, the director of the Academy, Baron Korf, requested that monasteries and ecclesiastical academies send students to the Academy's university to study under the foreign academicians. Lomonosov was therefore sent, along with eleven others, to the Academy in St. Petersburg, where he began to study mathematics and physics.

Again, larger events intervened. The Academy was at this time organizing expeditions for the exploration of Siberia and the Arctic, and it needed a chemist with mining experience to seek valuable minerals. Finding no such person in St. Petersburg, the Academy authorities decided to prepare for similar needs in the future by sending Russian students to study chemistry and mining in the Universities of Marburg and Freiburg. Thus Lomonosov was sent to Western Europe in 1736, where he stayed for almost five years. During this time he managed to learn a great deal of European science while at the same time leading the disordered and argumentative life-style for which he became famous. The European scholars who made the deepest impression upon Lomonosov were Christian Wolff at Marburg and Johann Friedrich Henkel at Freiburg.

Soon after returning to St. Petersburg Lomonosov was made an adjunct of the Academy of Sciences in physics, and later professor of chemistry. In 1748 he opened the first scientific chemical laboratory in

Russia, one equipped with balances and other equipment similar to that he had seen in Europe. In later years he was made head of the geographical department of the Academy of Sciences. Foreign honors included honorary memberships in the Swedish Academy of Sciences and the Bologna Academy of Sciences.

Lomonosov's scientific activity can be divided, for convenience, into three phases. From 1740 to 1748 he was occupied primarily with theoretical physics. In these years he wrote and taught about corpuscular philosophy, heat and cold, and the elasticity of air. He compiled a syllabus in physics and in 1746 he delivered the first public lecture on physics ever presented in the Russian language. Most of his writings, however, were in Latin. From 1748 to 1757, after the construction of his chemical laboratory, Lomonosov was active in chemistry, working on the characteristics of saltpeter, the nature of chemical affinity, the production of glass and mosaics, the freezing of liquids, and the nature of mixed bodies. From 1757 to his death in 1765, Lomonosov was heavily involved in scientific administration, exploration, mining and metallurgy, and navigation. Throughout all these periods he also wrote poetry and promoted the Russian language and Russian history.

In the area of scientific theory, Lomonosov's most significant work was his extension of the corpuscularian or mechanical philosophy common in the seventeenth and early eighteenth centuries to a wide variety of phenomena. This philosophy was based on the earlier work of Gassendi, Descartes, and Boyle, and had come to Lomonosov not only through his knowledge of Boyle's works, in particular, but also by way of Christian Wolff in Marburg, Lomonosov's teacher in philosophy. Through Wolff Lomonosov learned of the Leibnizian principles of knowledge and reason, as well as the concept of monads as indestructible units of the world. Lomonosov was impressed by this general approach to natural phenomena, but he developed it in his own way. In particular, he converted the Leibnizian monads from abstract, dimensionless force units into concrete physical realities.[23] In his earliest writings Lomonosov called the most elementary units of matter "physical monads," but later he spoke of "elements" and "particles," and his view of nature became a thoroughly mechanical one. His favorite way of describing nature was in terms of concrete pictures and mechanical models, often reasoning by analogy from such models.

This approach, applied in a rather literal fashion, sometimes led Lomonosov to views that seem prescient to modern readers and sometimes to conclusions that later science contradicted. Of course, we should not judge Lomonosov by later science, but in terms of the science of his day. Seen in that way, Lomonosov was clearly a distinguished scholar. However, much Soviet literature on Lomonosov, particularly that published before about 1960, contains exaggerated claims about his achievements. He has been credited with being the "founder of physical chemistry," "the first scientist to disprove the phlogiston theory," "the discoverer of the principles of conservation of matter and the conservation of energy," "the first scientist to describe absolute zero," and "the definer of the difference of the molecule and the atom."[24] Later Soviet historians of science ceased making many of these claims, as evidenced by the excellent article on Lomonosov by B. M. Kedrov in the *Dictionary of Scientific Biography* (although Kedrov's emphasis on the originality of Lomonosov's formulation of the law of conservation of matter needs to be further examined).[25]

Examples of the ways in which Lomonosov used the corpuscular philosophy can be seen in his views, respectively, on gravitation and on heat. Lomonosov, like many followers of mechanical philosophy of the time, could not accept Newton's concept of gravitation, since its apparent reliance on action at a distance seemed to Lomonosov to be a return to occult qualities similar to those from which science had just escaped in the preceding century. Lomonosov believed that forces could be transmitted only by contact. Therefore, he postulated the existence of a corpuscular gravity material that would account for gravitational phenomena.

In describing heat, Lomonosov was similarly loyal to mechanical principles. H. M. Leicester, a historian of chemistry, observed that Lomonosov employed the corpuscular viewpoint "more completely and more consistently" than any of its other adherents. To Lomonosov, heat was internal movement, in fact, "rotational" internal movement of the particles in a body. This kinetic viewpoint led him logically to the conclusion that there is no upper limit to temperature, since the internal particles can always move more rapidly, but that a lowest possible temperature would be the point at which the corpuscles are motionless. Lomonosov's clarity on this issue deserves quotation from his works: "There must exist a greatest and final degree of cold which should consist in absolute rest from the rotary motion of the particles."[26] Lomonosov could not agree with the increasingly popular view that heat is a caloric fluid, since he could not conceive of this fluid in mechanical terms.

Lomonosov did not refute the theory of phlogiston, as sometimes is claimed.[27] In fact, he relied upon the theory in explaining several phenomena, including the luster and ductility of metals. He believed that the greater the phlogiston content, the more noble a metal. However, because of his mechanical views, he could not conceive of phlogiston as a weightless fluid, or a substance possessing negative weight, as a few of its supporters did. Furthermore, Lomonosov correctly sensed something wrong with Robert Boyle's conclusion that the gain in weight of calcined (in modern terms, "oxidized") metals comes from "the matter of fire" that penetrates the walls of the glass retort and unites with the metal. In 1756 Lomonosov reported to the Academy that "we made experiments in tightly sealed glass vessels so as to study whether the weight of the metals increased from pure fire; by these experiments it was found that the opinion of the famous Robert Boyle was false, for without entry of the internal air, the weight of the burned metal remained at the same measure."[28] In approach, but not in rigor, these experiments anticipated the work of Lavoisier.

Lomonosov never published detailed accounts of these experiments, and he gave several conflicting explanations of them. At one point, he said the increase in weight of calcined metals could come from "particles flying around in the air." At another, he said that calcination might cause the surfaces of the metallic corpuscles to be more freely exposed to the action of gravity particles, and, hence, heavier.[29] Lomonosov's views on combustion, then, were provocative but somewhat unclear and certainly not conclusive. His criticism of Boyle's work on calcination may actually have come to the attention of Lavoisier; we know that Lavoisier read portions of the volume of the memoirs of the St. Petersburg Academy in which it appeared, but the further conclusion that Lomonosov was a stimulus to Lavoisier, as proposed by Dorfman, is highly speculative. After analyzing the claims to this effect, Leicester concluded that Lavoisier "could not have been expected to acknowledge a debt to Lomonosov, since he did not owe one."[30]

Although a thorough evaluation of the place of Lomonosov in the general history of science is yet to come, his prominent place in the history of Russian science cannot be questioned. Lomonosov was a symbol of emerging Russian scholarship, talented but still only partially developed. He had brilliant ideas, but lacked discipline, was somewhat unsure of mathematics, and scattered his efforts over a broad front. Nonetheless, he was an unprecedented phenomenon, a native Russian champion of science and learning who was, in important areas, at the forefront of knowledge. He would serve as a model for young Russian scholars for generations. It was, after all, Lomonosov who promoted the establishment of Moscow University, and who called upon the youth of his day "to show that the Russian land can give birth to its own Platos and quick-witted Newtons." To issue such a summons, to create institutions that may help serve it, and to show by example that the goal was not unrealistic were profoundly significant services to Russian science.

The role of geographical explorations. One of the strongest areas of Russian science from the earliest days to the present has been that of geographical explorations. The presence of a large virgin country inhabited sparsely by native tribes was a strong incentive for such explorations, just as was also the case in America in the eighteenth and early nineteenth centuries. The Russian forays before 1700 were often carried out by Cossack explorers, intrepid but usually illiterate, and the information they brought back was not very useful to science. In the sixteenth century the Cossack leader Yermak invaded Siberia and conquered the Western portion of that area. The city of Tomsk, situated in mid-Siberia, was founded in 1604. Later in the century Russians penetrated to the Pacific, although the geography of the area was still very poorly known.

By the beginning of the eighteenth century cartography and mapmaking had become sciences, based on mathematics and exact computation. Expeditionary work was also increasingly seen as a contribution to knowledge of nature, and European countries possessed systematic collections of fauna and flora. As this knowledge penetrated Russia, foreign explorers, many of them employed by Peter the Great, began to replace the hardy Cossacks as the leaders of expeditions. Peter decreed that Russians also learn geodesy and navigation, with the result that native Russians eventually became trained explorers and naturalists. Perhaps the best known of the early Russians was Stepan Krasheninnikov, author of the classic *Description of the Land of Kamchatka*. Lomonosov was also deeply interested in the Arctic and Siberia.

The eighteenth-century expeditions involved hundreds of men organized in whole constellations of activities. The best known were the First Kamchatka Expedition (1725–9) under the leadership of the Dane Vitus Bering; the Second Kamchatka, or Great Northern Expedition (1734–43), directed by the Admiralty; and the academic expeditions of 1768–74 guided primarily by Peter Simon Pallas, the great German naturalist who became a member of the St. Petersburg Academy and spent many years in Russia.[31] The Great Northern Expedition included many well-qualified scientists, including the astronomer Louis Delisle de la Croyere, the naturalists Johann Gmelin and Georg Steller, the historian Gerhard Müller, and the young Russian naturalist Krasheninnikov. The scientific output of these expeditions was enormous, including important new maps, dozens of volumes of descriptions of fauna and flora, hundreds of boxes of specimens that formed the basis of the great collections in St. Petersburg and Leningrad, and archival collections of Siberian history. Popular accounts of these expeditions usually emphasize Bering's discovery of the strait now named for him, but the most important result for Russia was the creation of its impressive tradition in the earth and life sciences.

The scientific nature of these expeditions is further illustrated by the fact that when he participated in the Great Northern Expedition Academician de la Croyere took to Siberia nine wagonloads of scientific instruments, including telescopes thirteen and fifteen feet long, as well as a scientific library of several hundred volumes. His colleagues were similarly equipped. When Müller and Gmelin saw Georg Steller off on the Bering trip, which was a part of the Great Northern Expedition, they gave him copies of the natural history works of such classic authors as

Gaspard Bauhin, John Ray, Joseph Pitton de Tournefort, and Thomas Willis.

Steller used these books as references on a long trip that took him to Siberia and Alaska, which he was the first natural historian ever to visit. The incredible voyage involved exposure to Arctic winters, shipwreck, and death from scurvy and starvation of nearly half the crew, including Bering himself. In the winter of 1741–2, living in an underground hut at the worst moment in the travails of this expedition, Steller made observations and collected data for his great work *De bestiis marinis* (published in 1751) which, along with his description of the now extinct Steller's Sea Cow, made him immortal in natural history.

Although the Russian explorations of the eighteenth century were important to science, they were not organized solely for the purposes of science. The Russian Empire was competing with the monarchies of Europe economically and militarily; the Pacific coasts of Siberia and America were the last promising regions of the northern hemisphere to be securely claimed by the maritime powers. Peter the Great was aware of this competition, and he and his successors masked the ultimate goals of their expeditions behind the stated aim of scientific research. Many standard accounts of Bering's voyages, both Russian and foreign, have accepted without question that their goal was the stated scientific one of ascertaining whether Russia and America are joined by land.

Scholars in the former Soviet Union and in the United States have recently begun to illustrate the weakness of this thesis. Most of these scholars have concluded, after looking at the archival materials, that Peter had a political goal in dispatching the first Bering expedition, but the scholars differ on the exact nature of that goal. Three different interpretations have been advanced: (1) Peter's ultimate goal was trade with Spanish colonies creeping up the West Coast of America; (2) Peter wanted the unclaimed parts of North America, as well as Siberia, in order to increase the lucrative fur trade; (3) Peter wanted to make Russia's eastern frontier secure from attack. Raymond Fisher examined these interpretations and the supporting evidence carefully, and concluded that the second thesis is the most plausible. The archival evidence indicates that Peter wanted to know more than merely whether the two continents were joined; he wanted to determine who and where his nearest European neighbors in America were and to assert Russian claims to the lands up to that point.[32]

However, there is no necessary contradiction between political goals and important scientific results. Need we remind ourselves that Darwin was sent as naturalist on the Beagle with instructions to look for commercially exploitable resources? Or that the United States used the International Geophysical Year of 1957 in order to help establish the principle that scientific satellites have the legal right to pass over the territory of other nations (and therefore, by implication, commercial and military satellites have similar rights)?[33] The fact that Russia's early brilliant achievements in scientific explorations were prompted by a mixture of political and scientific motivations simply illustrates what already should be clear, that science never proceeds in a political and economic vacuum. The introduction of science to Russia in the early eighteenth century was a part of Westernization, and Russia adopted the motives as well as the science of its powerful neighbors.

RUSSIAN EDUCATION IN THE EIGHTEENTH CENTURY

For decades Russian schools were unable to supply adequate numbers of educated scholars to replace the foreigners who dominated Russian science. Indeed, during the entire eighteenth century the Russian Empire lacked a school system, although several attempts to create one were made. The great majority of the population consisted of peasants, and serfdom remained in force. Under these conditions, education was restricted in practice to the privileged nobles, of whom few were interested in science in more than a dilettantish way. Nonetheless, the eighteenth century was a watershed in the history of Russian learning, for the principle was established, largely in the reigns of Peter I and Catherine II, that education was a proper concern of the state.

Peter was a practical man who enjoyed working with his hands (whether building boats or executing condemned prisoners), and the schools which he established for the nobility emphasized practical skills, especially military ones. He also established a system of state service for the nobles that required them to spend most of their careers in the civil or military bureaucracies. The nobles gradually adjusted to this system, and although they came to support the military and naval schools supplying their higher education, they favored a broad, humanistic curriculum rather than a narrow, technical one. They sought not mere "training" (*obuchenie*) but "enlightenment" (*prosveshchenie*). Enamored with European culture, they imitated West European nobles and began to speak languages other than Russian (usually French) at home. As a result, cultural divisions between the nobility, or *dvorianstvo*, and the peasants widened.

The better educated and well-traveled nobles often despaired of the cultural position of Russia relative to Western Europe. As they began to study European learning it was only natural that they were attracted to those theories that accounted for Russia's backwardness and simultaneously offered some hope that a favored few could escape from it. Educated Russians who compared the theories of learning of Descartes and Locke, for example, often thought they found implications for their own country. Descartes's emphasis on innate ideas was much less attractive to them than Locke's belief that environmental influences form the mind; an environmental account could explain Russia's intellectual backwardness and simultaneously extend the promise that the fortunate Russian who was raised in a superior environment could excel. Similarly, to educated Russians Rousseau's observation in *Emile* that "there is no original sin in the human heart, the how and why of the entrance of every vice can be traced" seemed to suggest that the flaws of Russian society could be blamed on the system of serfdom and the lack of culture among the peasantry.

Several of the most interesting educational projects in eighteenth-century Russia were based upon such environmental assumptions, often consciously drawn from those European models that seemed most appropriate. Ivan Betskoi, an illegitimate son of a prominent Russian noble, and a man who was educated in Western Europe and who knew several of the Encyclopedists including Rousseau, served as Catherine's advisor on education. Betskoi urged the creation of an educational system based on the philosophy of Rousseau. Since children should be protected from the pernicious influence of society, boarding schools were superior to other institutions of education. A number of schools were created in accordance with Betskoi's stipulations, including some open to students from a variety of social and economic backgrounds (except serfs). Special schools for women were also established, which was quite a radical departure, but in line with Rousseau's views and a vindication of Catherine's own place at the pinnacle of Russian society.

Catherine saw herself as a participant in the Enlightenment and (until frightened by peasant revolts in Russia and revolution in France) a supporter of political reform. She made a genuine effort to educate herself not only in the arts and literature but in natural philosophy. Catherine never, of course, became a scientist, but her positive attitude toward science and learning was significant in furthering their spread. Before she ascended the throne her tutor in science was the distinguished German physicist Franz Aepinus, who came to Russia as a member of the St. Petersburg Academy. Catherine even asked Aepinus to write a memorandum for her edification on the "system of the world"; Aepinus dutifully fulfilled his task by writing a treatise glorifying the Newtonian vision of the universe, with much emphasis on comets, a phenomenon just then very much in the public eye.[34]

The advancement of scientific education in Russia that occurred in the eighteenth century was more attitudinal than substantial. The upper levels of society gained a better appreciation of science, but little of substance was done that enabled Russian youth, particularly those of modest background, to follow scientific careers. The failure was most evident at the primary and secondary levels of education. Catherine's ambitious plan for a system of schools throughout the Empire was undermined by her failure to provide a proper means of financing. The situation was slightly better at the higher levels, with the Academy of Sciences existing after 1725, and Moscow University after 1755. The university, unfortunately, provided very little instruction in science. The Academy, especially in its early years, was a much more distinguished institution, but was a somewhat exotic, even artificial, focus of scholarship in a country where the overwhelming majority of the population was still illiterate. Nonetheless, the ground had been broken.

AN EVALUATION OF EIGHTEENTH-CENTURY RUSSIAN SCIENCE

The two greatest traditions of Russian science established in the eighteenth century were in mathematics and in the study of natural resources. The strength of these traditions is discernible even today. In both, but particularly the first, foreigners predominated, but also in both by the end of the eighteenth century native Russians were beginning to show genuine talent. On foreign roots Russian stock was grafted, painfully and with difficulty, and the nineteenth century would illustrate the vitality of the combination.

The most illustrious of the foreign academicians brought to St. Petersburg were in the mathematical and physical sciences. They included the great mathematicians Leonhard Euler (who arrived in 1727 as an adjunct in physiology), and Daniel and Nicolaus Bernouilli. Several others of the foreign academicians were thoroughly competent in mathematics and physics. As Alexander Vucinich observed, "Nearly half of the sixteen original members were skilled mathematicians who themselves made real contributions to the field."[35] They included the astronomer Joseph Delisle, the physicists Georg Bilfinger and Christian Martini, and the mathematicians Friedrich Mayer and Christian Goldbach.

The contributions of Euler and the Bernouillis alone would have given the St. Petersburg Academy an honored place in the history of mathematics. A native of Basel, Euler spent two long and productive periods in St. Petersburg, from 1727 to 1741, and from 1766 to his death in 1783. Several of his major works, including his *Mechanica* (1736), his *Institutiones calculi integralis* (1768–70), his *Anleitung zur Algebra* (1770), and his *Theoria motuum lunae* (1772) were written, partially or entirely, while he was in St. Petersburg. Euler arrived in Russia as a youth of twenty and spent his formative professional years there. And even when he was not in residence in Russia, he often sent his work there for publication. He once observed in a letter, "I and all others who had the good fortune to be for some time with the Russian Imperial Academy cannot but acknowledge that we owe everything that we are and possess to the favorable conditions that we had there."[36] That one of the founders of modern mathematical analysis should have gained scientific sustenance in a country where Arabic numerals had taken the place of Old Slavic letter-numerals only a few decades earlier ably illustrates how different levels of development can exist simultaneously in one country.

Russia was less important to the creativity of the Bernouilli brothers than to Euler. Nicolaus's promising career was cut off by his death at thirty-one, and his brother Daniel spent only seven or eight years in St. Petersburg before returning to Basel. Nonetheless, the example of the Bernouillis, along with that of Euler, would inspire Russian scholars for decades. Nicolaus before his premature death named a problem in the theory of probability "the St. Petersburg problem." Publication by members of the Bernouilli family (including father Johann) of articles in the first and subsequent issues of the St. Petersburg Academy's proceedings, *Commentarii Academiae scientiarum imperialis petropolitanae*, greatly increased the interest of scientists everywhere in that Russian organization and its publications. Daniel Bernouilli drafted his most important work, *Hydrodynamics*, while still in Russia.

Critics of eighteenth-century Russian science have often commented that these contributions are hardly best described as Russian, since they were made by Western Europeans, many of whom returned home after a few years. Furthermore, it is often said, the distinguished work in advanced mathematics was so distant from the needs of poorly educated Russia that it amounted to a distortion of priorities of research. These are correct but clearly incomplete criticisms, for they fail to notice how important these early achievements were in establishing scientific traditions in Russia. Euler, in particular, did not work alone in St. Petersburg, but created a school of younger mathematicians who attempted to carry on his work. To be sure, they could not match their master's achievements, but their emulation of his interests and methods had a significance all of its own. These disciples included the able Russians Stepan Rumovskii and S. K. Kotel'nikov, who by their teaching and their writing helped pull Russia into the world of contemporary mathematics, a field where Russians have excelled to the present day.

A second great legacy of the eighteenth century in Russian science was the study of natural resources, often the result of expeditionary work. The first and second Kamchatkan expeditions, which have already been briefly described, were only early prominent representatives of a type of descriptive work in the earth and life sciences for which Russia became famous. The American historian of Russian science Alexander Vucinich counted 161 published studies emanating from expeditionary work in Russia between 1742 and 1822.[37] Peter Simon Pallas (1741–1811), the eminent German naturalist, became a member of the St. Petersburg Academy and spent almost his entire adult life in Russia, arriving when he was twenty-six and leaving at sixty-nine. His Russian assistant V. F. Zuev eventually became a member of the St. Petersburg Academy himself and carried on his mentor's work. Other distinguished eighteenth-century researchers, both foreign and Russian, in the related fields of geology, geography, and biology included J. Gmelin, G. Steller, S. P. Krasheninnikov, M. V. Lomonosov, V. M. Severgin, A. I. Chirikov, I. I. Lepekhin, I. A. Güldenstädt, and N. Ia. Ozeretskovskii. The work of the noted embryologist and proponent of epigenesis Caspar Wolff, who lived in St. Petersburg from 1767 to 1794, was based on viewpoints he had worked out earlier in Berlin, but Wolff also took advantage of the rich specimen holdings in St. Petersburg, including the collections of monsters initiated by Peter the Great. Wolff found monsters useful for illustrating his belief that embryos are not preformed but develop differently in relation to their individual histories. (The same might be said about foreign scholars imported to the Russian Empire.)

Mathematics and natural history were the areas where eighteenth-century Russian science was most distinguished in terms of making contributions to general European science, but the St. Petersburg Academy was significant to Russian science in many other ways. It sponsored a large number of translations of European scientific works, it advised the government on technical questions, and, somewhat less successfully, it promoted education in science. Although they produced several distinguished scholars, the university and the gymnasium of the Academy were not permanent successes. Moscow University, founded in 1755, barely managed to exist during the latter half of the century, contributing very little to science in this period. Nonetheless, it provided an organizational base for an impressive development of university science in the next century.

Peter the Great at the beginning of the eighteenth century was impatient to bring science and technology from Western Europe to Russia, and he attempted to do it by starting from the top, with an academy of sciences. His critics doubted the wisdom of this method, maintaining, "There is no one to learn, for without secondary schools this academy will merely cost a great deal of money and yet be useless." Peter replied characteristically, "I have to harvest large shocks of grain, but I have no mill; and there is not enough water close by to build a water mill; but there is water enough at a distance; only I shall have no time to make a canal, for the length of my life is uncertain, and therefore I am building the mill first and have only given orders for the canal to be begun, which will force my successors to bring water to the completed mill."[38]

2

Science in nineteenth-century Russia

DURING the nineteenth century, Russian science and learning developed impressively. Progress occurred at all levels of education, from the primary schools to the universities. Most striking is the achievement in the universities and beyond, at the highest levels of advanced research. Universal primary education, on the other hand, remained a difficult problem that would not be solved until the twentieth century.

Russia produced in the nineteenth century a number of world leaders in science, of whom the names of Nikolai Lobachevskii, F. G. W. von Struve, Dmitrii Mendeleev, Pafnutii Chebyshev, Il'ia Mechnikov, and Ivan Pavlov are perhaps best known. In addition to these scientists of great fame, many others were at or near the same level in ability and in knowledge of international science. At the end of the century, Russia was economically and politically still a backward nation, but scientifically its promise was beginning to show. Russia in 1900 ranked well below Germany, France, and Great Britain in the range of brilliance of its scientific establishment, but, viewed against its position a century earlier, the achievement was spectacular.

This century of progress, however, was marked by long periods of academic reaction and intellectual retrogression resulting from the autocracy's continuing fear of political reform and Western ideological influence. In the area of educational policy, in particular, several reforms and liberalizing initiatives were followed by periods of political crackdowns that throttled creativity. The early reigns of Alexander I (1801–1825) and Alexander II (1855–1881), the two most innovative times, saw significant advances that were partially reversed in succeeding periods of reaction.

Nonetheless, significant advances in education for science and technology occurred even during the conservative eras. Indeed, the rulers of nineteenth-century Russia favored the development of science and technology if given the assurance that it would not undermine the existing political and social order. Their confidence in performing this difficult balancing act varied greatly with the political times, resulting in contradictory policies. In the latter half of this chapter, when we look at the lives and works of Nikolai Lobachevskii and Dmitrii Mendeleev, two of

Russia's greatest scientists of the century, we will see how they were affected both by periods of reform and periods of reaction. Their educations and even their research were intimately tied to the political, economic, and intellectual milieux of their times.

The young Alexander I turned to a group of personal friends, the so-called private committee, for advice in enacting reforms.[1] These men – Count Pavel Stroganov, Nikolai Novosil'tsev, Prince Adam Czartoryski, and Viktor Kochubei – were intrigued with French and Polish reforms stemming from the Enlightenment and the French Revolution. The first three had lived in France and Stroganov had even been a member of a Jacobin club. Czartoryski was the son of General Adam Czartoryski, who had been a member of the Polish Commission on National Education created during reforms in Poland between its first and second partitions. The Polish Commission is often called the first ministry of education in history.

Alexander and his advisors decided that the first step toward the solution of Russia's staggering educational problems was along the same path, the creation of a ministry of education. In 1802 the ministry was established, which lasted until the Russian Revolution of 1917, and within a few years it was charged with administering an impressive system of schools on four levels: elementary or parish schools, district schools, provincial schools, and universities. The Russian Empire contained only one city with a functioning university, Moscow, when Alexander ascended the throne, but Alexander added three more, at Kazan' and Khar'kov (1804), and St. Petersburg (1819). In addition, the Baltic Germans in Dorpat soon revived their earlier German-language university, and the Poles in Vilna transformed a local academy into a Polish-language university. Together with the existing Academy of Sciences, these six universities provided an important potential base for the development of science and learning.

Although this system was in some respects copied from the Polish precedent, it went beyond that model in its commitment to classless education. Here the most important inspiration was the scheme of public instruction introduced to the National Convention in 1792 by Condorcet, who believed that every child should be given an opportunity to develop to the limit of his or her innate potential. In the early phase of Alexander's reign, the tsar and his advisors adopted this characteristic of Condorcet's plan. All levels of the Russian educational system were, in principle, open to all social classes and even to both sexes. At first there was no tuition charge at any level, and state stipends were given to impecunious students.

This system never worked in practice the way that it appeared on paper. Not a single woman reached the university level during Alexander's reign. Social pressure was much more powerful than official regulations. The financial aid system for male students also worked imperfectly. Nonetheless, at one moment in Alexander's reign, when the total university enrollment was about 3,000, over 1,000 of the students were on state stipends.

Alexander's system of classless education did not have a chance of genuine success so long as the system of serfdom remained. The estate serfs were totally under the control of their masters, who wished to keep them as tillers of the soil. In 1806 Mikhail Speranskii, a more pragmatic advisor to the tsar than the members of the private commmittee, gave his ruler a report on education in which he said that the government could not emancipate the serfs until they are educated, but that it could not educate them until they are emancipated. While this observation exaggerates the intellectual obstacles to emancipation and underestimates the economic and political, it does point to Russia's dilemma in the first half of the nineteenth century: It wished to compete with Western Europe while retaining a feudal social order.

In the latter half of Alexander's reign the expenses of the Napoleonic Wars made it increasingly difficult for the government to provide the educational system with funds. At the same time the tsar became more conservative and mystical. No longer did the ideas of the French Revolution appeal to him or to his advisors. Russian nationalism grew in step with the defense of the motherland against the French invader.

In 1816 Prince Aleksandr Golitsyn, a religious crusader against "ungodly and revolutionary tendencies," was appointed minister of education. One of Golitsyn's officials, Mikhail Magnitskii, conducted a purge of Kazan' University. Over half of the faculty were dismissed, and the curriculum was thoroughly revised. Particularly suspect were West European doctrines of political economy, but natural science also suffered. Kazan' had early established a reputation in mathematics and chemistry, but during Magnitskii's administrative rule not a single major scientific work was produced there. The situation was similar at Khar'kov University, where the outstanding mathematician and rector of the university, T. F. Osipovskii, was dismissed because of an alleged lack of fervor in asserting that "God lives" during an oral examination of a graduate student.[2] At St. Petersburg University four professors were charged with ignoring Christian principles in their teaching of philosophy and history. Throughout Russia academic freedom suffered.

Nicholas I (1825–1855) was a conservative and militaristic ruler who kept Russian intellectual life under tight bureaucratic control. Paradoxically, this period witnessed a flourishing of literature heralded by such figures as Pushkin, Lermontov, Gogol, Dostoevskii, Nekrasov, and Turgenev. In science, a similar upsurge in creativity was promoted by Lobachevskii in mathematics, Struve in astronomy, von Baer in zoology, Hess in chemistry, Lenz in physics, among others.

The chief educational administrator of Nicholas's reign was Count S. S. Uvarov, a former school official of the St. Petersburg district. He has usually been described by historians, both Soviet and non-Soviet, as an idealistic, reform-minded young man who then turned into an extremely conservative bureaucrat who suppressed creativity. Recent research has shown a much more complex and interesting picture. Indeed, one American biographer of Uvarov, after studying in the archives in Leningrad, threw out her dissertation supporting the traditional interpretation and wrote a monograph refuting her own earlier thesis.[3] According to the new interpretation Uvarov remained loyal to his early ideals and was a dedicated supporter of science and learning of the highest quality; working under a conservative tsar, he had to serve these ideals in sophisticated, even devious, ways. On two different occasions, 1821 and 1849, Uvarov left office as an educational administrator because of the failure of the tsar to support his policies. But for many years he promoted the ideas of the early advisors to Alexander I, a time when Uvarov's own political and intellectual views were formed. And during those years Uvarov had many practical successes in building institutions and expanding learning.

Uvarov was not, of course, a radical or even a liberal in twentieth-century terms. He was a loyal supporter of the tsar who hoped to improve the existing system. A serf-owner himself, one who faced several revolts on his own estates, he freed only his household serfs and only in his will at his death. He opposed sudden changes in governmental principles, espousing the view that Russia was not yet "mature" enough to adopt a representative form of government, or even to emancipate the serfs. In the long view, however, he believed that Russia's future, after a long period of evolution, would be similar to that of the countries of Western Europe groping for political freedom. It was the task for another generation, thought Uvarov, to bring Russia both constitutional monarchy and emancipation of the serfs. The task of his generation was to promote education and prepare for the later day.

Uvarov promoted a renaissance of the Academy of Sciences and a new blossoming of scholarship in the universities. He created in Russia a tradition of excellence in Oriental studies and he strongly supported the existing one in mathematics. Under his administration, and with his strong support, Pulkovo Observatory became what one American scientist called "the astronomical capital of the world." He managed to retain the open-to-all-classes admission principle to the schools and universities against considerable opposition, a principle that was, admittedly, more important in theory than in practice. He promoted new buildings, supported the libraries, and recruited new and outstanding talent to the universities and the Academy of Sciences. Between 1833 and 1848 university enrollment more than doubled. As his biographer Cynthia Whittaker has observed, Uvarov's "insistence on high standards, coupled with his considerable administrative skills, accomplished one of the greater feats of the imperial era – to turn inconsequential and mediocre institutions into serious and first-rate centers of higher learning whose benefits redounded to the society at large."[4]

It is hardly surprising that the reforms promoted by the private committee under Alexander and by Uvarov under Nicholas were often frustrated. Various legislative acts and governmental pressures made it increasingly difficult for a student to rise above his class in the educational system, although no legislation absolutely forbade it. Imposition of increasingly stiff tuition as one ascended the educational ladder and segregation of classes along social lines had the effect of imposing a class system without formalizing it. At the same time, the autonomy of the students and faculty was restricted.

Not only did many conservatives in Russia oppose the reforms, but international events worked against them. The conservative reactions against the revolutions of 1830 and 1848 in Europe had damaging effects in Russia. In reponse to the first, Nicholas I closed Vilna University and generally clamped down on intellectuals. In response to the second, he tried to staunch the flow of Western ideas into Russia, and prohibited Russian scholars from studying in European universities, a serious blow to Russian science.

A second great period of reforms of Russian education and science came after the Crimean War, during the reign of Alexander II (1855–1881). The war had revealed how badly Russian military technology had slipped behind that of the West European states since the Napoleonic period, when Russia had possessed the mightiest armies of Europe. But Russia's problem was much larger than mere technological lag; its social and economic system made industrialization and modernization almost impossible.

The reforms of the 1860s and early 1870s could never have been drafted and pursued so rapidly had not a group of enlightened bureaucrats emerged earlier, in the 1840s and 1850s. These men, whose unofficial leader was Dmitrii Miliutin, came mostly from gentry backgrounds and received higher educations during the blossoming of the Russian universities. They believed that without the emancipation of the serfs, improvements in education, invigoration of agriculture and industry, and rationalization of the legal system, Russia had no chance of competing with other European states. They prepared the way for reforms by collecting data about the local economic conditions of the Empire, working within organizations like the Imperial Russian Geographical Society (founded in 1845), and cooperating closely with several members of the Academy of Sciences who were knowledgeable about statistics and economics.[5] They promoted slogans such as *glasnost'* (openness) and *za-*

konnost' (legality), and called for a transformation of Russian society along a path of moderate reform. These enlightened bureaucrats were too radical for conservative landowners and too moderate for a new left-wing intelligentsia that emerged in the late 1860s; furthermore, they could never be certain of the support of Alexander II, who vacillated in his enthusiasm for change and cooled toward many of his own reforms after an attempt on his life in 1866. Nonetheless, the enlightened bureaucrats had a moment of great influence during the early 1860s, when the reforms were being enacted. Miliutin himself managed to retain his office as minister of war until 1881.

The emancipation of the serfs was the major task of the day, but the reform movement advanced through most of the institutions of Russian society: the military, local government, the judicial system, and educational institutions. At no time since Peter the Great had the foundations of Russia been so stirred.

The reforms in education had the greatest effect on science. Admission to the universities was broadened, class privileges were reduced, the emphasis on applied sciences for commoners was softened, travel abroad for the purpose of study was again permitted, and the education of women on the secondary level was stimulated. A new university charter in 1863 eliminated most of the restrictive features of the 1835 charter.

Some reformers proposed that women be admitted to the universities, but this suggestion was not accepted. Instead, a few years later special "higher courses for women" were established separate from the universities, but taught by university faculty. In the 1860s and 1870s a remarkable generation of Russian women managed to acquire advanced educations, either through these women's courses or in Western Europe. They included the first women anywhere in the world to receive doctorates in a number of scientific fields. The mathematician Sofia Kovalevskaia, for example, became a leader in the field of partial differential equations. She was the first distinguished female mathematician since antiquity.[6]

Many historians of Russian education have underlined the Charter of 1863 as being a turning point in Russian higher learning, but as so often happened in Russian history, the succeeding reactionary period undercut much of the charter's positive potential. In 1866 a new minister of education, Count D. A. Tolstoi, began once again a campaign to eliminate subversive doctrines in the educational system. Tolstoi was also procurator of the Holy Synod, the supreme civil officer in charge of religion. Tsar Alexander announced that education "must be conducted in the spirit of true religion, respect for the right of property, and preservation of the foundations of public order," and that "no school shall tolerate the propaganda, openly or secretly, of destructive notions equally inimical to the advancement of the moral and economic well-being of the people."[7] Tolstoi evidently believed that the best path to these goals was a return to class education. Tolstoi emphasized that the Greek and Latin languages and mathematics, suitably free from contemporary political and social issues, were the best education for the upper classes, while training in the mechanical arts was uniquely designed for the lower orders. A law of 1871 made a classical diploma mandatory for admission to the universities without passing an entrance examination.

Even Tolstoi recognized the need to promote science, however, and increased the budgets of the Academy of Sciences, the research libraries, and the astronomical observatories. He promoted growth of the science faculties at the universities and gave natural scientists greater freedom to form national societies and to hold conferences than he did jurists or social scientists, whom he feared as politically subversive. Although not committed to education and science as reforming influences as his predecessor Uvarov had been, Tolstoi promoted these activities because he hoped that, tied to conservative political interests, they would strengthen the autocracy.

The assassination of Alexander II in 1881 led to a period of deep political reaction. Count Ivan Delianov, minister of education from 1882 to 1897, tried to divert lower class students into trade and technical schools rather than the universities, established quotas in the higher schools for Jewish students, diminished the opportunities for women students, placed the parochial schools under the tight control of the Holy Synod, and pursued a policy of Russification in Poland, the Baltic area, and the Caucasus. By this time, however, education and science in Russia had achieved inertias of their own, and continued to develop under adverse conditions. And even Delianov, in a backhand way, showed his respect for learning when he said he wanted to change the higher schools from "hotbeds of political agitation into hotbeds of science." Between 1884 and 1900 the number of parochial and elementary schools grew impressively, perhaps best indicated by the fact that in 1880 only 22 percent of army recruits were classified literate by their recruiters, but that in 1900 approximately 50 percent fell into this category. At higher educational levels growth also continued, particularly in the technological institutes, which Delianov promoted as a means of supporting Russian industry. Between 1882 and 1895 enrollments in the universities grew from 10,374 to 14,755 and in the technological institutes from 1,777 to 2,826.

LOBACHEVSKII

Nikolai Ivanovich Lobachevskii (1792–1856) was a signal figure in the history of mathematics, a man who has rightly been called "the Copernicus of geometry."[9] The drama and inherent interest of his life are still only poorly appreciated. Here was a person in a remote Asiatic hinterland of the Russian Empire who never studied in or traveled to Western Europe, who even in Russia never received recognition for his work during his life, who began his teaching career in an infant university at that moment undergoing purge and oppression, and yet, in spite of all this, who made a conceptual leap that had eluded geometers since the time of Euclid.

How do we explain the appearance of such a remarkable man as Lobachevskii in a geographic area that modern science had hardly touched? The more we examine his origins the more clearly we will see that Lobachevskii's career and the development of Russian education and intellectual life fit together hand in glove. Indeed, it will become obvious that only a few years earlier a person with Lobachevskii's social origins could not have achieved what he did. The supporting institutions and people simply did not exist.

Lobachevskii was born in Nizhni Novgorod (later Gor'kii) on December 1, 1792, into a family of modest means. His father was a clerk in a land surveyor's office. In 1800, after the death of her husband, Lobachevskii's mother moved her family to the city of Kazan', a trading port on the Volga River where Slavic and Muslim cultures intermingled. Although far from the European influences in St. Petersburg, Kazan' was a provincial capital through which ran important trade routes to Siberia and to Asian lands to the southeast. Several attempts had been made during the reign of Catherine the Great to establish a gymnasium there, but after periods of varying success, the efforts collapsed. Some educational facilities were available in the late eighteenth century to children of the nobility and others for religious training; none was satisfactory for the preparation of scientists nor for students with Lobachevskii's meager financial resources.[10]

Shortly before Lobachevskii's arrival in Kazan' the gymnasium reopened with a broadened curriculum containing more natural science and a system of scholarships available to a few students. Lobachevskii's mother helped her son prepare for the entrance examinations, and, after a year or so, convinced the authorities to give him financial support.[11] Lobachevskii entered the gymnasium in 1802 and studied there for four years.

Lobachevskii grew up academically in step with the city of Kazan'. In 1807 he entered Kazan' University, which had been created only several years earlier in accordance with Alexander I's plan for the expansion of university education. Here he benefited from the unusual philanthropic principles of the Russian educational system in the early years of the nineteenth century. It is true that in Kazan' the support of university education may not have been entirely for altruistic reasons; the Russian authorities were anxious to advance Russian culture in an area where Islamic culture still successfully competed. Thus, the state-supported education of Lobachevskii may not have been quite the glorious application of Condorcet's policy of universal education that it first seems and for which Alexander's liberal advisors declared their support. We have little information on scholarships available to poor Tatar boys (not to speak of girls) in Kazan' and we are probably justified in doubting that many existed.[12] Nonetheless, the public education of Lobachevskii is surely evidence that the policy of fostering educational opportunity can be a success, even if the dimensions of that success are often socially restricted.

At first Kazan' University had a faculty of only six. The mathematics teacher, G. I. Kartashevskii, who also taught in the gymnasium, was a man of unusual teaching ability. Unfortunately, he became involved in one of those political quarrels with the authorities that were so common in Russia, and was forced to leave. During Lobachevskii's second year at the university, a contingent of German professors arrived who were strong in mathematics and physics. The invitation of faculty from Western Europe to take positions for which not enough Russians were qualified was, again, a policy of Alexander's newly instituted Ministry of Education. The Germans were J. M. Bartels in pure mathematics, K. F. Renner in applied mathematics, F. X. Bronner in physics, and J. J. Littrow in astronomy. These men, especially Bartels, brought not only advanced knowledge of European science, but a style of scientific and philosophic thought that profoundly shaped the young Lobachevskii's outlook. Since Bartels was a friend of the greatest European mathematician of the day, Karl Friedrich Gauss, who along with Johann Bolyai, independently developed forms of non-Euclidean geometry, it has often been thought that Bartels may have linked the work of Gauss and Lobachevskii.[13] Recent research shows no direct causal link; the independence of Lobachevskii, who was the first of the three to publish a version of non-Euclidean geometry, is clear.[14] Nonetheless, Bartels imparted to Lobachevskii the basic stance of early nineteenth-century mathematics, which was to examine critically the most fundamental beliefs.

Historians of mathematics have long been intrigued by the fact that a riddle of geometry that had existed for two thousand years was solved almost simultaneously by three different mathematicians, Lobachevskii, Gauss, and Bolyai. One attempt at an explanation has been to attribute the conceptual breakthrough primarily to Gauss, since tentative links can be made between Gauss and both Lobachevskii and Bolyai; as already indicated, recent research has failed to substantiate this explanation, either for Lobachevskii or for Bolyai.[15] An important factor was the spread among European mathematicians in the early nineteenth century of a new critical spirit, a willingness to be suspicious of all unexamined assumptions. This spirit influenced all three men. Yet another influence was the fact that one approach to the problem, namely the attempt to prove the parallel postulate, had exhausted its possibilities.[16] Therefore, the time for alternative approaches was ripe. But the spread of skepticism about the foundations of geometry and the exhaustion of attempts to prove the parallel postulate, important factors though they were, only explain why *someone* might develop a non-Euclidean geometry; they do not explain why these three men were the ones. And here the fact that the first person to make his views known, Lobachevskii, was a complete outsider to the European mathematical community is relevant. Lobachevskii did not fear the outcry of the European mathematical community (as Gauss admitted that he did) because he was not a part of that community.

The young Lobachevskii was independent minded. The author of a history of Kazan' University, N. P. Zagoskin, wrote: "If, however, Lobachevskii's mathematical achievements met with complete approval, his behavior, which was a source of great trouble to the inspectors, appeared in quite a different light to the University administration; the inspectors' reports and the entries in their registers characterize Lobachevskii as a 'stubborn', 'repentless' young man, . . ."[17] His Soviet biographer Kagan tells us that he was disciplined several times by being locked in his room and deprived of his holidays.

Lobachevskii received a master's degree in physics and mathematics from Kazan' University in 1811, and in 1812 he began to lecture there in the same fields. He progressed rapidly in the university ranks, becoming professor ordinarius in 1822 when he was only thirty years old. He also was active in university administration, serving as chairman of the buildings committee (an important task for a new university), dean of the department of physics and mathematics, librarian, and, eventually, from 1827 to 1846, rector, or president, of the university.

It was exactly in the time when Lobachevskii began his academic and administrative duties that the fury of Magnitskii's conservative purge, fostered by Alexander I in the last part of his reign, hit Kazan' University so hard. Magnitskii actually recommended that the university be closed, but Uvarov in St. Petersburg managed to persuade Alexander to leave it open; nonetheless, the university was severely disciplined.[18] Many faculty members were dismissed, and the curriculum was overhauled. It seems a bit odd that Lobachevskii, a person who had so recently been accused of insubordination, would have survived this period and even emerged as head of the university. No doubt the fact that Lobachevskii was working in mathematics rather than history or philosophy helped to protect him. In the early part of his career Lobachevskii devoted himself in his research to a topic the authorities could not understand, the construction of non-Euclidean geometry. It would not be the last time in Russia, or in its successor, the Soviet Union, that mathematics provided an intellectual sanctuary from political oppression. His biographers tell us that he won the support of his faculty colleagues by defending the university as best he could, and, as a result, after Magnitskii's departure from Kazan', Lobachevskii was elected rector by the faculty.[19]

If the Magnitskii terror may have provided Lobachevskii with reason to seek intellectual sustenance in private mathematical reflections, the end of that terror helped him in his effort to get the work published and noticed. The latter effort was not, however, entirely successful, even though Lobachevskii did eventually get his views published. Publication and notice turned out to be two entirely different things.

Euclid developed geometry as a deductive intellectual scheme that logically flowed out of a few initial definitions and axioms. One of these axioms, which states that through a point external to a line one and only one parallel can be drawn, had often been noticed as being less self-evident than the others. Euclid himself probably recognized this feature of the parallel postulate when he listed it last, and only after proving all the other theorems he could without it. Rather than merely accepting it on faith, many geometers in later centuries wanted to prove the parallel postulate, either by deriving it from the simpler axioms, or by assuming that it was false and then showing that such an assumption led to contradictions. Many such attempts were made, and a few, by Girolamo Saccheri in 1733 for instance, uncovered clues that the parallel postulate may not hold. No geometer before Gauss, Lobachevskii, and Bolyai, however, suggested that by making different assumptions about the number of possible parallels to a given line different geometries could be developed, each equally universal and rigorous.[20] Such a suggestion flew in the face of common sense and received philosophic wisdom.

It appears that Lobachevskii first addressed the problem in the process of writing a textbook on geometry. In the draft for the text he made three different attempts to prove the parallel postulate, but noted that none of them was completely successful. Recent scholarship in the former Soviet Union indicates that Lobachevskii's interest in the possibility of a non-Euclidean geometry dates from the year he submitted this text for publication, 1823, and that between 1823 and 1826 he elaborated the basic principles of an alternative geometry. The textbook itself was rejected for publication on the advice of N. Fuss, the last disciple of Euler, who was also the permanent secretary of the Academy of Sciences in St. Petersburg and whose position made him the leading scientific bureaucrat of the day. Lobachevskii did not seem to mind much, since he was now thinking about more creative possibilities in geometry, but this rebuff by the Russian academic establishment was a sign of the future.

In 1826 Lobachevskii presented a paper to his university department of physics and mathematics in which he proposed a new form of geometry. A later historian would comment that he might as well have read the paper standing alone in the middle of the Sahara Desert, for all the response it elicited. Lobachevskii called the new mathematical system "imaginary geometry," by which he meant to point to an analogy in the difference between imaginary and real numbers. To a mathematician, the word "imaginary" here does not mean useless or fanciful, but simply distinct from another system. The word imaginary later haunted Lobachevskii, however, for his university colleagues and students often used the term in its ordinary meaning, indicating that Lobachevskii's professional work was not to be taken seriously. In later years when it was necessary to praise Lobachevskii at ceremonial occasions, reference was made to his administrative work at the university, not to his mathematical research.

The path that Lobachevskii took in constructing his new geometry was to assume that through a point external to a given line an infinite number of parallels can be drawn. On this assumption he constructed an entire system with many differences from Euclidean geometry. For example, in Lobachevskii's geometry the sum of the internal angles of any triangle is less than 180 degrees. Lobachevskii's belief that his geometry might be something other than imaginary in the ordinary sense of the word is revealed by his interest in trying to measure the sum of the angles of the triangle formed by the vertices of two diametrically opposed points on the orbit of the earth and one of the fixed stars. He, like Gauss, was interested in the question of what form of geometry actually occurs in the real world; he never carried out the experiment, however, since he believed that the deviation from 180 degrees would be too small to separate from observational error.[21]

Lobachevskii's election to the position of rector of Kazan' University in 1827 probably helped him to get his ideas published. His first publication on the new geometry was in 1829–30 in a journal issued by the university. It was not easy for the editor of a journal published by a small university to reject a manuscript submitted by the university president. Not many people paid attention to the journal, however, and nobody appreciated what Lobachevskii had done. And one can hardly blame leading European mathematicians for not studying an article on geometry that appeared in a new journal of dubious quality from an obscure university in the provinces of Russia; the fact that an article in the same issue as Lobachevskii's dealt with vampires seemed to say something about the seriousness of the journal.

Lobachevskii then submitted an article to the St. Petersburg Academy of Sciences for publication in one of its scholarly journals with much larger access to the world of mathematics. Once again the leaders of Russian science rejected his work: The reviewer of the article described it as unworthy of publication; someone then wrote an unsigned review for a Russian journal of general circulation ridiculing Lobachevskii's efforts.[22]

Lobachevskii then published several more articles in the accommodating Kazan' journal and also turned to West European publishers in the hope of winning attention. The task was not easy, but in 1837 he managed to publish an account of the new geometry in Crelle's famous journal on pure and applied mathematics and, in 1840, a book in German.[23] No one publicly lauded these works either, but Gauss read the German work, sent to him by Lobachevskii, and became so intrigued that he began studying Russian in order to read the original. Gauss had earlier become convinced that non-Euclidean geometry was valid, but he had confided this belief only to a few friends, and he had sworn them all to secrecy. Gauss privately praised Lobachevskii, but his only overt sign was his nomination of the Russian mathematican for membership in the Göttingen Gesellschaft der Wissenschaften. Genuine recognition of Lobachevskii's achievement did not come until the 1860s, after the deaths of both Gauss and Lobachevskii, and after the publication of Gauss's correspondence, in which the topic of non-Euclidean geometry was so significant.

At about the same time that Lobachevskii was working on non-Euclidean geometry a similar form was developed by Bolyai, a Hungarian who published in 1832. Just as with Lobachevskii, Bolyai was ignored. The only one of the three who would have commanded attention if he chose to take a stand was Gauss, who refused to do so. Gauss was at the pinnacle of European mathematics, the two others were beyond the foothills. Lobachevskii's unenviable position has already been described. Bolyai's father was an able mathematician, but the son had been refused by Gauss as a student and had studied instead at a military academy in Vienna. As a young officer he was sent to a remote area of Transylvania. His mathematical education was not adequate for him to work on the cutting edge of mathematics of the day; the study of the foundations of geometry seemed an appropriate topic.[24]

We thus see some common characteristics in Bolyai and Lobachevskii. Both of them lived in isolated areas and were not in step with current mathematics research. Since both were outside the mathematics establishment, they had little to lose by declaring that the classic works of Euclid were inadequate. Gauss feared the attacks of the "wasps" that would land on the person who tried to overthrow goemetry.[25] Bolyai's father was close enough to the establishment to warn his son fervently to stay clear of the problem of the fifth postulate. "You should shy away from it as if from lewd intercourse," he urged, "it can deprive you of all your leisure, your health, your peace of mind and your entire happiness."[26]

Lobachevskii differed from Bolyai in not having the disadvantage of the senior Bolyai's advice. He was even farther from the ruling orthodoxy in the European centers of mathematics, and hence more willing to defy its rules. His critics in St. Petersburg seemed as distant from him as his reluctant admirer Gauss in Göttingen, who died a year before him. Lobachevskii persevered in his lonely intellectual fortress on the Volga, publishing a new account of his "pangeometry" in the loyal Kazan' University press the year before his death, thirty years after he had made the essential conceptual breakthrough.

MENDELEEV

Dmitrii Ivanovich Mendeleev, who lived from 1834 to 1907, is one of the greatest names in the history of chemistry. In view of his fame it is ironic that today we know so little about him. Despite the existence of a large literature about Mendeleev, there is no complete, certainly no definitive, biography of Mendeleev in any language, not even in that of his native land. The main reason for this lacuna is not the complexity of his scientific work – Soviet historians of science produced some excellent detailed analyses of his work on the periodic table[27] – but the richness of his life. Mendeleev was very active in many different fields, including economics and politics, and it is in the latter areas that our knowledge is skimpiest. Much of the information is still in the archives. An example of Mendeleev's important administrative work is his directorship of the Central Board of Weights and Measures from 1893 to 1907, a position that gradually acquired a much greater significance than the title might seem to indicate. He was, in fact, the government's "science advisor," a role in which he developed a full-scale platform for the economic development of the empire, which was an important part of Count Witte's industrialization program of the end of the nineteenth century. No scholar has yet fully explored this aspect of Mendeleev's life, and it is only one of several such activities in this talented man's career still relatively unknown. The task of the future biographer is daunting: The incomplete edition of Mendeleev's works runs 25 volumes, and the Mendeleev archives occupy several rooms of St. Petersburg University.

Mendeleev was born far from any leading scientific center, in the Siberian town of Tobol'sk, where his father was a teacher of Russian literature and philosophy and later director of the local gymnasium. Mendeleev's mother came from an old merchant family and inherited a small glass factory near Tobol'sk. One of her ancestors was said to have been a Tatar, and her children were proud of their semi-Asiatic and Siberian background.

During Dmitrii Ivanovich's childhood and adolescence the Mendeleev family fortunes steadily declined. His father was physically disabled soon after Dmitrii Ivanovich's birth, and died in 1847. His mother was financially dependent on the glass factory, which, like small estate factories everywhere, was being driven out of the market by large commerical enterprises. The factory was saved from bankruptcy by a fire that totally destroyed it, reducing the family to near poverty. Despite the fact that the decline and destruction of the glass factory occurred while Mendeleev was very young, the family memory of those events may be one of the reasons that, throughout his life, Mendeleev was interested in the relationship of technology to economic development.

One other experience in Mendeleev's youth may also have influenced his later life, his growing familiarity with liberal politics. The Siberian area near his home was a place of exile for political prisoners from central Russia, who were allowed to roam rather freely and who organized local discussion groups. Several of the famous "Decembrists," noblemen who had unsuccessfully revolted against the government in 1825, were living in or near Tobol'sk. Mendeleev and his family became good friends of a number of liberal thinkers, and one of his sisters married a Decembrist. Several of Mendeleev's Soviet biographers probably exaggerated the significance of these political associations, but the early knowledge of political dissent must have had some effect on Dmitrii Ivanovich, the youngest member of the large family. The future chemist was certainly never a radical, but he was always sympathetic with students who got in trouble with tsarist authorities. Many years later he would be forced out of his university teaching position because of his assistance to student demonstrators who had asked him to submit a petition in their name to a government minister.

Mendeleev's early education did not proceed smoothly. In the local gymnasium he was not an outstanding student. He despised the classical languages that all the students had to study in accordance with Count Uvarov's policies. He did well, however, at mathematics and science, which Uvarov also emphasized. Mendeleev's mother noticed her son's excellence in science and decided that he should study at a university. Since he was the last of her large brood and there was nothing left in Tobol'sk to hold her, she took him to Moscow and St. Petersburg herself on a remarkable winter trip by sledge of several thousand miles. Dmitrii Ivanovich's efforts to enroll in Moscow University or St. Petersburg University were unsuccessful because of his provincial background and uncertain academic record. The authorities suggested that he try Kazan' University, which served the Siberian region, but Mendeleev's mother evidently wished a better education for him than she thought he could acquire there. (Kazan' University had actually become rather strong in science under Lobachevskii's leadership, and had just added the promising young chemist Aleksandr Butlerov to its faculty.) After overcoming many obstacles, Mendeleev was admitted to the Main Pedagogical Institute, the same place where his father had studied.

Mendeleev enrolled in the Institute after his classmates had already completed their elementary work, and he had difficulty in catching up.

He also soon gained a reputation for having an intemperate tongue and being impolite to authorities. His mother's death and his own poor health further complicated his education. At the end of two years he ranked 25th out of 28 in his class, and was not promoted to the next class.[28] Eventually he made up his deficiencies, did brilliantly in chemistry, and upon graduation was awarded a gold medal for excellence. His chemistry professor, A. A. Voskresenskii, was enthusiastic about his abilities.

Mendeleev's path to professional eminence after graduation was complicated. As a recipient of a state scholarship during his university studies, Mendeleev was required to accept a government assignment. He was given an undesirable teaching position in a gymnasium in Simferopol, an appointment that his biographers have attributed to his irregular record and reputation as a troublemaker. The gymnasium in Simferopol was closed because of the Crimean War, so Mendeleev instead found work in Odessa, where he continued research he had earlier begun on the relationships between crystal forms and chemical composition. He then won admission to graduate work at St. Petersburg University (where his old teacher Voskresenskii also taught), and in 1856 defended a master's degree there with a thesis in which he pointed to relationships between chemical properties and specific volumes. Shortly thereafter he defended an additional thesis on the structure of silicon compounds. In these works Mendeleev showed that he was a defender of Gerhardt's views on type theory. He regarded chemistry as a physical science dealing with mass and weight, and he resisted Berzelius's electrolytic theory of the formation of compounds.

The pace of Mendeleev's advancement now began to quicken. His teachers at St. Petersburg University were very impressed with his postgraduate work, and in 1857 they managed to persuade the ministry of education to transfer him from Odessa to St. Petersburg, where he began teaching chemistry in the university with the rank of docent. Two years later they recommended that Mendeleev be sent to Germany to improve his knowledge of chemistry. This foreign study was a great break in his professional development, for it allowed him to pursue independent research, to become friends with some of Europe's most noted chemists, and to witness the rapid growth of the science of chemistry, particularly in Germany. Only a few years earlier Mendeleev would not have had this opportunity, since the tsarist government curtailed foreign travel after the revolutions of 1848 in Europe, a policy that changed with the death of Nicholas I in 1855. Just as with Lobachevskii, Mendeleev's career hinged on fortunate changes in government policy at crucial moments. In Lobachevskii's case, a critical development was Alexander I's decision to build public educational institutions in Kazan'. In Mendeleev's it was Alexander II's conviction that reform of his empire required sending young scholars to study European science and technology. Mendeleev went to Germany on a state scholarship at a time when his native land was entering a period of renewal after defeat in the Crimean War.

In Germany in 1859 and 1860 Mendeleev worked at first in the laboratory of Bunsen in Heidelberg, but, finding that he needed independence, later established his own laboratory. In Germany he quickly caught up with the controversies surrounding one of the most interesting times in the history of chemistry. In 1860 he attended the first International Congress of Chemistry in Karlsruhe. This famous meeting was called for the purpose of standardizing the concepts that lay at the base of chemistry, those of atomic, molecular, and equivalent weights. Cannizaro advanced a rational method for calculating atomic weights, making use of Avogadro's principle. Not until this congress had been held was a detailed periodic table possible.

In 1861 Mendeleev returned to St. Petersburg and began teaching at several institutions, including the university. He remained at the university as a professor of chemistry until 1890, when he was forced to resign in the incident already mentioned. In the 1860s and early 1870s he was very active. The development of the periodic table, to be described in more detail below, came in 1869. He also began to study petroleum, and was interested both in its origin (for which he presented an inorganic hypothesis) and its exploitation. He traveled often, and visited the United States in 1876 in order to inspect its petroleum industry. His book titled *Petroleum Production in the North American State of Pennsylvania and in the Caucasus* was critical of the tsarist government for not providing proper incentives for industrial development. In dozens of articles he promoted such diverse activities as polar exploration, exploitation of the minerals of Siberia, the use of balloons for communication, the development of dairy industries, and the implementation of tariffs to protect native industries. He purchased an estate in 1865 to attempt to demonstrate the value of scientific agriculture, and lectured on such subjects as cheese making and fertilizer.

In the 1870s and 1880s in what was probably a midlife crisis, Mendeleev's intellectual and personal interests underwent a transformation. As he remarked, "Much in me was changing; at that time I read on religion, on sects, and philosophy, economic articles." The changes included a new marriage in 1882 and a turning of resurgent energy to new enterprises. In 1887 he made a solo balloon ascension for the purpose of observing a solar eclipse.

Despite the honors that came to Mendeleev after his periodic table was vindicated, he never lost his reputation for being unorthodox and even improper. His flaunting of conventions included the abandonment of his first wife at the age of fifty and marriage to a seventeen-year-old Tatar girl, thus reaffirming his family's Asiatic connections. Mendeleev was criticized by staid society because of his personal habits and by conservative academic factions because of his liberal political convictions. The Imperial Academy of Sciences refused to elect him to membership.

After Mendeleev's loss of his university position in 1890, it appeared that his institutional career was over. However, the strong-minded Count Witte recognized his merits and was willing to ignore what he regarded as idiosyncrasies. Witte, the leading figure of the tsarist government in the period of its most rapid economic development, made Mendeleev his advisor on technical matters. This part of Mendeleev's career is not well known, but he served in this position for well over a decade. Ruth Roosa, in her study of Russian industrialists, described Mendeleev as the major intellectual spokesman and ideologist for the Association of Industry and Trade, a powerful group of businessmen often critical of the government.[29]

The periodic table. One of the most controversial aspects of the periodic table is the question of priority. A few historians of science do not speak of Mendeleev's periodic table, but of Lothar Meyer's, or of Odling's, or even of Chancourtois's or Newlands's. The table has had several aspiring fathers; it is clear that a number of chemists were close to the development of the table by the late 1860s. In what is probably the most exhaustive study yet of the various periodic systems the equitable J. W. van Spronsen concluded that part of the credit should go to no less than six investigators: Chancourtois, Newlands, Odling, Hinrichs, Meyer, and Mendeleev.[30] But even van Spronsen, who made little use of the extensive Russian-language secondary and archival literature on the subject, noted that Mendeleev's table was more complete than the others and the most impressive in terms of the prediction of unknown elements. A Soviet conference on the centenary of the table (1969) compared the various claims and presented persuasive evidence that Mendeleev's was the best.[31]

Western literature on Mendeleev's work on the periodic table is often misleading. It is frequently said, for example, that his method of work was to determine more exactly by laboratory experiment the atomic weights of the elements, and then to arrange them in ascending weights, seeking periodicity of properties. Many writers then repeat the story that Mendeleev, because of illness, asked N. A. Menshutkin to take his place in reading his report of the table to the Russian Chemical Society.[32] These statements are incorrect. There is no evidence that Mendeleev did important laboratory work on atomic weights in the months preceding his elaboration of the table. Furthermore, he did not start out with the assumption that ascending weights were the clue to the system. And, finally, he was unable to read his important paper because he was away on a consulting job on cheese production. All this has been clearly shown by Soviet scholars, especially by B. M. Kedrov, who has given nearly an hour-by-hour reconstruction of Mendeleev's elaboration of the table.[33]

The main task preoccupying Mendeleev in the crucial months of February and March 1869 was the writing of a textbook of inorganic chemistry; this was the work that became his famous *Principles of Chemistry*, later to appear in thirteen editions, the last in 1947. Mendeleev had earlier decided that there existed no satisfactory inorganic chemistry text and so he would write one himself. For heuristic reasons, Mendeleev organized part of his text in terms of families of known elements displaying similar properties. His approach was based on Gerhardt's theory of types, with elements grouped by valence, starting with hydrogen. For weeks he had already been thinking in terms of what we would call the "groups" (following the organization of Mendeleev's later table) of the elements. Placing such a strong emphasis on the individual properties of the chemical elements also fitted with Mendeleev's philosophical commitments; as Bernadette Bensaude-Vincent has emphasized, Mendeleev differed from many of his predecessors in that he did not believe in the existence of a "primary matter," but instead saw the chemical elements as individual, never to be divided or converted into other elements.[34]

In December 1868 he finished the last chapters of the first part of the textbook, which were devoted to the halogens. It had long been known that the elements chlorine, iodine, and bromine displayed similar properties. Methods of producing all three had been devised in roughly a fifty-year period stretching from 1774 to 1826, and in the middle years of the nineteenth century the similarities of these three elements were well known: All three form colored vapors, all three react vigorously with metals, and all three form strong acids when combined with hydrogen.

After finishing the halogens, Mendeleev wondered what group of elements he should discuss next, at the beginning of part II of his book. He decided to begin the discussion of the metals, and he proceeded directly to the alkali metals, such as potassium, sodium, and lithium. Then the question arose, what family of metals should be discussed after the alkali metals? The logical candidates seemed to be alkaline earth metals, such as calcium, magnesium, strontium, and barium. The typical alkali metals had combining powers of one, while the typical alkaline earth metals had combining powers of two. Thus, Mendeleev would follow the principle of ascending combining power in his scheme. However, a serious problem arose at this moment, since there were other metals that had an unclear combining power, or valence, such as copper and mercury. In some compounds these elements displayed a valence of one and in others two. Should they be discussed before or after the alkaline earth metals? Mendeleev decided to discuss the transitional metals such as copper and mercury before the alkaline earth metals, and his first drafts of the new chapters reflected this approach.

Notice that so far the governing principles on which Mendeleev was operating were combining power and similarity of chemical properties, not ascending atomic weight. These are the sorts of principles that would naturally appeal to a writer of a textbook who is seeking teaching principles allowing a great deal of chemical information to be contained in a rather simple scheme. They are not the sort of principles that a person who was attempting directly to prove that periodicity is a function of atomic weight would be likely to adopt in his first approaches. This approach gave Mendeleev an advantage over some of his competitors.

It was only when, on this same day, Mendeleev ran into classificatory difficulties that he began to emphasize different terms. Now he began to think in terms of atomic weights, not in direct order of ascending weights as we would do now, but in order of relationship of weights between the two groups he had already described in his textbook – the halogens and alkali metals – and a hypothetical third group. He began to doodle, listing the halogens on one line and the alkali metals on another:

$$F = 19 \quad Cl = 35.5 \quad Br = 80 \quad I = 127$$

$$Li = 7 \quad Na = 23 \quad K = 39 \quad Rb = 85 \quad Cs = 133$$

Now he could see intervals of ascending weights of approximately equal values *between* certain members of these two lines of elements:

$$F = 19 \quad Cl = 35.5 \quad Br = 80 \quad I = 127$$

$$Li = 7 \quad Na = 23 \quad K = 39 \quad Rb = 85 \quad Cs = 133$$

The intervals varied from 3.5 units to 6. Now, Mendeleev made the crucial step of writing a few elements in a third line, those that fit this rough rule of interval:

$$F = 19 \quad Cl = 35.5 \quad Br = 80 \quad I = 127$$

$$Li = 7 \quad Na = 23 \quad K = 39 \quad Rb = 85 \quad Cs = 133$$

$$Ca = 40 \quad Sr = 87.6 \quad Ba = 137$$

This was the crucial moment; years later, when reviewing his development of the periodic table, Mendeleev wrote that the relationships of these three groups contained the "essence of the whole matter." The transitions $Cl{\rightarrow}K{\rightarrow}Ca$; $Br{\rightarrow}Rb{\rightarrow}Sr$; and $I{\rightarrow}Cs{\rightarrow}Ba$ are still in the periodic table today, although the noble gases were later interspersed after they were discovered.

Mendeleev then decided to try this principle of comparing families of elements on a broader scale. Oddly enough, in several drafts he wrote the elements within families not in ascending order of atomic weight, but in descending order, for reasons that are not clear. It may be a vestige of the fact that Mendeleev had not started out this fateful day with the intention of finding a periodic law. (Indeed, he did not use the term "periodic law" until 1871, two years after his paper announcing his system of elements to the Russian Chemical Society.) Late in the day of March 1, 1869, already deep into the problem that led to the development of the periodic law, he was still more concerned with *families* of elements and sequences of weights between them than he was in finding a periodicity that emerges when one arranges the atomic weights in ascending order.

Eventually Mendeleev hit upon the scheme of writing each element on a separate card and arranging the cards in ascending weights on a table in a pattern similar to the one used in a favorite card game, a form of solitaire called patience. The cards gave a convenient and rapid method of trying to hit upon the most suitable arrangement of the elements. Each card evidently (they are not preserved, though the beginning and final forms of the tables are) contained the element's symbol, its atomic weight, and its most characteristic chemical and physical properties.

Late in the day Mendeleev made a clean copy of his provisional table. He recognized the need to leave blanks for unknown elements, a practice he continued later and which was one of the greatest strengths of his table. He was able to predict the properties of several of these elements with remarkable accuracy, and this strong predictive feature of his table was a reassuring sign of its scientific and not arbitrary character.

Mendeleev worked rapidly that day because he was under great pressure from the Free Economic Society to leave on a consulting trip concerning cheese production. This pressure may have been one of the reasons that he was willing to leave blanks and to place some of the most difficult elements outside the table entirely. This hasty measure was in retrospect brilliant, since some of these elements were in the "long periods" known as transition metals, and to have forced them into a "short form" would have done violence to the system. After finishing his clean draft, he sent off a copy of his table to a publisher. As soon as he was assured that all was well at the printer, he left on his long-delayed consulting trip, and consequently was forced to ask a colleague to present his report to the Russian Chemical Society. Historians later could not imagine that Mendeleev would have purposely been absent from such an occasion, the highlight of his professional career, so they concocted the story that Mendeleev was sick and could not attend the meeting, a myth that apparently will not die.

Looking back on this creative process, we can gain several insights. The usual interpretation of Mendeleev's approach assumes that by studying the external progression of increasing atomic weights, Mendeleev came to see the internal organization of groups of elements. We see, however, that Mendeleev actually began with the internal organization of groups of elements and only later linked them to the principle of ascending atomic weights. By starting with the groups, and then noticing the relationships of weights from one group to another, Mendeleev may have had a methodological superiority over some of his competitors in fitting the difficult elements into groups. All the elements with which Mendeleev began were already in groups. Lothar Meyer was also writing a textbook at the time of his work on a periodic table, but he seems to have been concerned with atomic weights as an ordering principle at an earlier stage than Mendeleev. In Mendeleev's approach the core groups – the halogens, the alkali metals, and the alkaline earth metals – were fairly securely established at an early point. When he ran into subsequent problems he could maneuver freely, and even leave some questions unanswered without losing his basic system. He was even willing to assume that some previously accepted atomic weights were in error, and that some elements needed to fill in gaps were undiscovered, since only on the basis of such an assumption could his groups be retained. The vindication of most of these assumptions in later years was one of the most exciting aspects of Mendeleev's table.

CONCLUSIONS

The evolution of education and science in nineteenth-century Russia was a dramatic and painful story that combined, at different moments, grand ideals and great achievements with political repression and obscurantism. Despite the setbacks and continuing difficulties, the record in science was impressive. By 1900 Russia had produced a number of scientists known throughout the international science community. Science had at last won a place in Russia's rich cultural tradition, alongside religious art, architecture, literature, music, and poetry.

The social acceptance of science was not complete, however. It is not a coincidence that both Lobachevskii and Mendeleev, like their eighteenth-century predecessor Lomonosov, came from modest families in remote geographic areas of the empire. The advantaged noble families in the capital cities where the educational possibilities were much greater made relatively little contribution to science.

The ambitions of the elite sons of Russia were to be close to the court in St. Petersburg, and to enjoy the social and career advantages available there. In such an environment conformity was rewarded, not the sort of youthful rebelliousness that characterized Lobachevskii and Mendeleev.

Independence of spirit was also necessary to break with conventional ideas in science.

Both Lobachevskii and Mendeleev had undistinguished origins, both were educated on state stipends, both experienced trouble with the political authorities, and both made their greatest scientific break-throughs while writing textbooks. The first three characteristics made them unlikely candidates for diplomatic or military careers; the last characteristic is probably a result of the fact that they were pioneering the development of Russian science and therefore felt an obligation to start with the most elementary step: the writing of Russian-language textbooks in their fields. Both non-Euclidean geometry and the periodic table of elements are textbook topics in the sense that they arise in the organization of the basic concepts of geometry and chemistry.

Lobachevskii's and Mendeleev's distance from the West European centers of scholarship also probably increased their boldness in posing radical hypotheses. This advantage of isolation is most clear in the case of Lobachevskii. It is less clear for Mendeleev, who was known among foreign chemists, but Mendeleev moved with a boldness and even haste on that greatest day in his life, March 1, 1869, that suggest that he was as concerned with local issues as he was with his place in the history of chemistry. He brilliantly set aside the elements that would not fit into his system at that time, rather nonchalantly left the reading of his report to a colleague, and hurried on to his other obligations as an agricultural and industrial consultant. The German chemist Lothar Meyer confessed his awe for Mendeleev's willingness to take risks. Mendeleev was a man who was not daunted by social convention, whether it had to do with his personal life or with the rules and practices of the community of chemists.

Both Lobachevskii and Mendeleev benefited from the educational reforms of the tsarist government and both were lucky in having the crucial moments in their educations coincide with progressive periods in tsarist rule. Both attended schools that would have been unavailable just a few years earlier; in Lobachevskii's case, his gymnasium and his university simply did not exist earlier; in Mendeleev's case his German education in chemistry was made possible by a new governmental policy toward foreign study.

The lives of Lobachevskii and Mendeleev are revealing illustrations of the influence of social and political circumstances on scientific creativity. The importance of external factors does not detract, of course, from their own talents. Indeed, the drama of the way in which these two men overcame their remote origins to revolutionize their fields has inspired generations of young Russian science students.

In the next chapter, we will see that the influence of the social environment on science in prerevolutionary Russia extended from the fields of mathematics and chemistry, where Lobachevskii and Mendeleev did their great works, to biology. Darwinian evolution was born in capitalist England. How would it be seen in nineteenth-century Russia, with its feudal institutions, its lingering agrarian socialism, and its new radical political currents?

3

Russian intellectuals and Darwinism

SOME of the first scholars who investigated the history of Darwinism in Russia were struck by the enthusiasm with which it was received by intellectuals there. One historian wrote, "Unlike its reception in the West, Darwinism met almost no opposition in Russia from either the scientists or the social thinkers."[1] A superficial survey of Russian publications in the years immediately after Darwin's views became known seems to confirm this impression. The great majority of writers in Russia of all political orientations accepted the concept of evolution. Even among religious authors there was no attempt to give a systematic refutation of evolution, although a few complaints and reservations about the doctrine appeared in theological journals.[2] No great public debate over evolution took place in Russia in the 1860s and 1870s of the sort that occurred in England and that sporadically flares up in the United States until the present day.

Recent scholarship, however, has gone much deeper and has revealed ironies and paradoxes behind the facade of Russian enthusiasm for Darwinism. What looked earlier as rather simple support for Darwin is gradually turning into a much more complicated picture, with subterranean revisionist or even negative attitudes toward Darwin emerging more and more clearly. Paradoxically, many Russian writers who declared their allegiance to Darwinism were at the same time advancing interpretations with which Darwin himself would have disagreed. These interpretations often failed to acknowledge, or purposely ignored, those features of Darwin's form of evolution that Darwin and many scientists elsewhere found most novel and important. Russian attitudes toward Darwinism thus often contained both overt enthusiasm and implicit criticism.

No progress in understanding this problem can be made unless one distinguishes "Darwinism" from "evolutionism" or "transformism." Darwin was far from the first biologist to argue that existing species of animals and plants have evolved from earlier ones. Among the best known of the evolutionists before Darwin were Geoffroy Saint-Hilaire, Buffon, Lamarck, Maupertuis, and Erasmus Darwin (Charles Darwin's grandfather). Charles Darwin's brilliance and originality were shown by his thesis that natural selection was the primary mechanism by which evolution occurred and by his accumulation of a mountain of evidence in favor of that thesis. In promoting his argument, Darwin took great pains to distance himself as far as possible from previous attempted explanations of evolution, such as those based on teleology or the inheritance of acquired characteristics. And although he did allow room for the inheritance of acquired characteristics and increased attention to it in later editions of his *On the Origin of Species*, he always considered it secondary to natural selection and an intellectual argument of last resort.

Russian commentators on biology in the latter half of the nineteenth century often considered "Darwinism" to be synonymous with "evolutionism." Their enthusiasm for Darwinism was not usually based on the explanatory mechanism – natural selection – for evolution that Darwin proposed, but on his assertion of evolution itself. Many of them were, in fact, uncomfortable with the concept of natural selection as the sole, or even major, mechanism for evolution, and almost all of them rejected Darwin's phrase "struggle for existence," as the American historian Daniel Todes has shown.

The advent of Darwinism to Russia coincided with the era of great reforms promoted by Alexander II. It was a time when Russian intellectuals were ready for unorthodox ideas. As A. O. Kovalevskii, a leading embryologist who as a young man in the 1860s participated in the discussions of Darwinism, wrote, "Darwin's theory was received in Russia with profound sympathy. While in Western Europe it met firmly established old traditions which it had first to overcome, in Russia its appearance coincided with the awakening of our society after the Crimean War and here it immediately received the status of full citizenship and ever since has enjoyed wide popularity."[3]

Reformist Russian intellectuals were critical of the traditional ideology of Russia based on autocracy and religious orthodoxy and looked to science for new guidance. Darwinism supplied both opposition to religious creationism and, at least implicitly, support for social and political change.

The Orthodox Church was a logical source of resistance to Darwinism, but it possessed few priests with the modicum of scientific education necessary for advancing a reasoned opposition. Instead of trying to win debating points in clashes with scientists, as Bishop Wilberforce in England attempted to do against Thomas Huxley, priests in Russia relied on the church's traditional emotional appeal to the faithful. The religious issue was thus never fully joined in Russia even though many liberal and radical intellectuals believed that evolution was implicitly contradictory to religious teachings.

When some Russian intellectuals lost hope in the reforms of the tasrist

government after the sixties and turned to more radical visions for society, including socialism, their enthusiasm for Darwinism remained high.
It was not difficult to turn Darwinism into a message of anthropological
materialism, even atheism, and the more radical members of the Russian
intelligentsia exploited this possibility. Political commentators and jour-
nalists used Darwinism to advance their own ideological visions, without much worry whether Darwin would have agreed with the glosses
they placed on his views. Natural scientists, especially biologists, were
more attentive to the details of Darwin's theory, but even they were
influenced by the political currents that ran so strongly through Russian
society in the latter half of the nineteenth century. As young students
many of them had either been involved in, or sympathized with, political protests, and some of them chose science as a means to help reform
society. Science was regarded by many Russian intellectuals as a natural
ally of political change and as a natural enemy of tyranny and religious
orthodoxy. Darwinism was the latest and most exciting weapon in this
struggle.

The most influential early Russian reaction to Darwin's *On the Origin of
Species* was a review by the radical literary critic Dmitrii Pisarev in the
well-known journal *The Russian Word*. The review, as long as a short
book, appeared in 1864, the same year as the first Russian translation of
the *Origin*. Many educated Russians first learned of Darwin's views
through Pisarev's summary and interpretation of them. Drama was
added to the episode by the fact that Pisarev wrote the review in prison,
where he had been confined by the tsarist authorities for his oppositional political opinions.

Pisarev saw Darwinism as a vindication of rationalism and materialism, and he gloried in its denial of creationism and catastrophism. Furthermore, he took Darwin himself as a model for a new type of critical
thinker, one who studies facts as they really are, unburdened by metaphysical or religious prejudices. In the cooperation between Darwin and
other scientists in England, such as Hooker, Lyell, and Wallace, Pisarev
saw a form of "free association" that for him symbolized the form that a
society of the future should take.

Not for a moment did Pisarev doubt that Darwin's account of the
evolution of plants and animals was correct. He went to enormous
lengths to give his readers, the great majority of whom knew little of
botany or zoology, a summary of Darwin's evidence and arguments. He
discussed domestication of animals, pigeon breeding, intentional and
unintentional effects of man upon animal variation, the struggle for
existence, the complex interrelations between species and their environments, natural selection, sexual selection, speciation, the effects of selection on animal morphology and behavior, the role of instinct, the formation of insect societies such as those of the bees and ants, geological

evidence for evolution, embryology and comparative anatomy, and many other aspects of Darwin's work. Throughout, Pisarev put his emphasis on Darwin's arguments as they appeared in the *Origin*, and, consequently, said very little about the origin of man, although it is quite clear that he accepted the full implications of evolution for humans. In this sense, he was loyal to Darwin's views as expressed in the *Origin*.

For all these reasons a number of historians of the reception of Darwinism in Russia, both Western and Soviet, have described Pisarev as a dedicated Darwinist, referring to him as an illustration of their thesis that Darwinism was accepted by intellectuals in Russia more enthusiastically than in other countries. And yet a careful reading of Pisarev will reveal a great irony: For while considering himself an ardent Darwinist, and exerting considerable talents to the task of converting his readers to the new doctrine, Pisarev misunderstood and misinterpreted Darwin on precisely those points and emphases that Darwin considered most crucial, namely those aspects of Darwin's doctrine that distinguished it from the host of pre-Darwinian evolutionary views.

The two aspects of evolution on which the differences between Darwin and Pisarev show up most clearly are the effects of use and disuse on inheritance and the role of volition, or goal seeking, in evolution. On both of these subjects Darwin went to considerable lengths to define his position as exactly as possible, but Pisarev missed the subtleties of Darwin's arguments. For example, on the question of the effects of use and disuse on inheritance, Pisarev was correct in seeing that Darwin believed that some phenomena, such as blindness in cave fish, can be explained only in that fashion. But Darwin stated quite clearly that, while he believed the effects of use and disuse could be inherited, he was convinced that selection "is by far the predominant Power."[4]

Yet Pisarev relied very heavily on use/disuse doctrines in explaining evolution, citing this factor in many instances where Darwin obviously preferred natural selection alone. The drooping ears of domestic animals, the thick legs of chickens, the wing shape and beaks of woodpeckers, the claws of sparrows, even the size of the human brain – all these were explained by Pisarev, first of all, by exercise of the relevant organs and only secondarily by natural selection. To Darwin, the inherited effects of exercise of an organ was an explanation to be used only when pure natural selection among variations seemed inadequate for the phenomenon to be explained; to Pisarev exercise seemed the preferable explanation, and there is good reason to doubt that Pisarev ever got straight in his mind what natural selection really meant to Darwin. This detail, at first sight perhaps insignificant, is indispensable to an understanding of Darwinian evolution.

Pisarev departed even more from Darwin on the question of the role of volition in evolution. The issue arose in Pisarev's discussion of insect societies. Darwin wrote in the *Origin* that the production of neuters or sterile females of several distinctly different physical structures in insect communities "at first appeared to me insuperable, and actually fatal to my whole theory."[5] The problem, of course, was that since these infertile insects did not have progeny, it was impossible for natural selection to work through them alone; and even if one believed, as Darwin did, that natural selection worked on the family, rather than the individual, it would still be difficult to explain the discontinuous castes of workers, rather than a continuous gradation. Darwin managed, finally, to explain this phenomenon on the basis of natural selection alone by hypothesizing that in this case selection worked through the fertile parents, rather than the infertile progeny, and that, furthermore, the intermediate structures had become extinct, leaving only the distinct castes.

Since Pisarev did not understand how important it was to Darwin to try to explain all this through natural selection alone, he did not believe that he was violating the spirit of Darwinism when he proposed a wildly speculative clarification of Darwin's explanation. According to Pisarev, the different castes could be explained by assuming that social insects, "like humans, completely consciously strove at each given moment to achieve what seemed useful or convenient to them."[6] Therefore, according to Pisarev, the worker insects fed and trained the larvae in such a way that, through the influence of the environment and education (*vospitanie*), different castes were produced.

In putting so much emphasis on the volition of conscious ants, Pisarev realized that he was going somewhat beyond Darwin's text, but he failed entirely to understand why Darwin had not gone down the same path as he. At one point Pisarev noticed that Darwin always "ignored completely the conscious activities of ants." He asked himself, "Why does Darwin do this? I don't know. Perhaps he does not want to enter into details not having direct relationship to his theory. . . ."[7] Yet the avoiding of goal seeking as a causal factor in evolution was fundamental to Darwin's whole approach, and his spurning of such explanations was not a detail, but the heart of the matter.

The main point to the Russian reader, however, was not the mechanics of evolution, but the fact of it. If we define Darwinism as merely the doctrine that plants and animals evolve, then Pisarev was a defender of Darwinism. But it would be much more accurate to say that what Pisarev took to be Darwinism was actually transformism of a sort that had been around Europe for decades, and that Pisarev missed the essential novelty of the Darwinian view.

A fellow radical on the writing staff of the journal *Russian Word*, V. A. Zaitsev, combined Pisarev's enthusiasm for Darwinism with eagerness to extend the doctrine into the social realm. Although Zaitsev was only twenty-two years old at the time he entered into the discussion of Darwinism in 1864 he was already well known to members of the Russian intelligentsia through his trenchant analyses of literature and politics. He prided himself on his willingness to follow all arguments to their logical ends, no matter how shocking his conclusions might be to social sensibilities. He was an ardent follower of mechanistic materialism in the spirit of Vogt and Moleschott, and saw Darwin as standing sqaurely within the same tradition.[8]

Basing his arguments on his understanding of Darwinism, Zaitsev published a racist, pro-slavery article in the *Russian Word* at a time when the editorial board of the journal had already taken an abolitionist position and openly sympathized with the North in the ongoing American Civil War. The result was intense embarrassment for the journal and among its pro-emancipation readers. The controversy that ensued boiled over onto the pages of several publications and lasted more than a year.

Zaitsev believed that Darwin's analysis should be extended to man, and that an inevitable conclusion followed that "human races are just as much separate species as horses and donkeys."[9] Without bothering to show exactly where Darwin had drawn this conclusion, Zaitsev plunged on to observe that these separate races were obviously not all of the same worth, and he maintained that there was not a single scientist in Europe who considered "colored tribes" biologically equal to the white ones. When he discussed evolution, Zaitsev abandoned his usual radical skepticism and totally embraced the most racist expressions of nineteenth-century European biologists and anthropologists, a rich field for exploitation in this fashion. Darwinism itself only provided him with a platform for his views, not a source of concrete evidence. Zaitsev believed that within the "colored races" some were more advanced than others; lowest of all were American Indians and Polynesians, whom he believed to be incapable of social relations, and who lived "not in societies, but in herds."[10]

To Zaitsev, the "lower races" of man could never be given the rights and privileges of the white race. When a colored race comes in contact with the white race, he wrote, "the best outcome which the colored man can hope for is slavery."[11] He continued: "The sentimental enemies of slavery can only cite texts and sing psalms, but they cannot point to one fact that would indicate that education and freedom can make a negro equal to a white mentally."[12]

Zaitsev made a grotesque effort to combine these views with a radical, even socialist critique of the capitalist West. Critics of slavery such as Harriet Beecher Stowe, he said, should stop worrying so much about the position of the colored race, and instead concern themselves with the exploitation of their fellow whites who barely survived at the bottom of the economic ladder. England so exploited the Irish, he wrote, that in a few more centuries the desperate environmental conditions in which they lived could make of the Irish a "new race, already having lost forever the higher capabilities which distinguish Caucasians."[13] All these ramblings were considered by Zaitsev to be directly derivable from biological science. As he made his observations about the effects of the environment on producing inferior races he evidently never noticed Darwin's observation in the *Origin* that he placed "very little weight on the direct action of the conditions of life"[14] as an influence on heredity. Some of Darwin's writings indicated that he was not free of the common Victorian prejudices about races, but nothing that he said justified Zaitsev's wild extrapolations. Indeed, on the *Beagle* Darwin had quarreled with the ship's captain Fitzroy about Brazilian slavery, which Darwin eloquently opposed.

Zaitsev's outburst caused a crisis among Russian radicals, revealing a contradiction in their beliefs. They prided themselves on their devotion to science and to materialism; Zaitsev's contemptuous attitude toward sentimental philanthropy and idealistic morals was shared by most of them. They thought that the new and superior social order for which they yearned would have to be based on science and cold reason, not on religious principles or vague altruism. Yet Zaitsev's compulsive extension of Darwinism into the social realm and his consequent defense of slavery made many of them intensely uncomfortable on moral grounds. In the 1860s Russia had abolished the servitude of estate serfs as the United States had abolished the servitude of plantation slaves, and discussions of the evils of human bondage had occupied intellectuals throughout Europe and North America. In these discussions, the Russian radicals had earlier taken the side of freedom; if they were often critical of the emancipatory acts it was not because they disagreed with the purpose underlying the legislation but because they believed the emancipations had not gone far enough in providing for the security and freedom of the oppressed. Now, however, one of their number had proclaimed that modern science justified slavery based on race.

The member of the Russian radical intelligentsia who first tried to answer Zaitsev was Nikolai Nozhin, a young student of biology and zoology. Nozhin had been to Europe, where he associated with Bakunin and the followers of Proudhon.

A reading of Nozhin's critique of Zaitsev clearly shows that his main objection was a moral one; is it possible, he asked, that Darwin's theory could cause "new tears and sorrows for mankind?"[15] And he queried, "Is Mr. Zaitsev trying to fool the public, or in the innocence of his soul does he not understand what slavery and bondage really are?" Whatever his motive, Nozhin castigated Zaitsev for writing an inhumane article issuing from an incredible lack of feeling for the sufferings of others, an insensitivity that would have been reprehensible, he said, even if it had been displayed only toward animals, not to speak of human beings.

But Nozhin shared with Zaitsev the assumption that science had a significant role in deciding social and moral questions. Furthermore, Nozhin, like Zaitsev, revered science as objective truth, and could not dream that some of the giants of biological science had allowed social and economic prejudices to influence their writings. Therefore, Nozhin constructed a critique of Zaitsev's views on slavery that accepted almost all of Zaitsev's biological arguments and rejected only his conclusion. And he thought that even the flaw of that conclusion was not, in the final analysis, based on its immorality, but on its alleged misunderstanding of biology.

The following quotation reveals Nozhin's tortured effort to ally himself with a science that he believed differentiated the human races by intrinsic worth yet at the same time disassociate himself from Zaitsev:

> The essential difference between white and black tribes is recognized at the present time by all, and, together with Huxley and Vogt, we recognize that the negro is, in terms of structure, lower than the white man and constitutes a transitional step between the latter and other mammals. Having declared this fact, Mr. Zaitsev considers himself justified in fastening on to this declaration a very improbable *consequently*, saying that *consequently* slavery is unquestionably justified. But the lower development of women with respect to men and of the lower classes of society with respect to the higher are facts of completely the same sort as those cited by Mr. Zaitsev. Is not every thinking person obliged to protest and to struggle to the limits of his strength against this *consequently* which real life attaches to itself à la Zaitsev?[16]

What legitimate reason, according to Nozhin, did the thinking person have for this protest against Zaitsev if the reason were not to be moral? Zaitsev's mistake, said Nozhin, was that he failed to understand that Darwin's very emphasis on progress in nature and the mutability of species meant that every organism was capable of progress, however low its position. Inferior to whites though blacks may be, believed Nozhin, blacks should be allowed to improve themselves in accordance with Darwin's evolutionary laws, and therefore slavery was impermissible. Thus Nozhin tried to cover his moral revulsion against slavery with a scientific defense that was loyal to the common materialistic assumptions of his fellow Russian radical intellectuals. In the process he had provided a notable example of extreme ratiocination.

Nozhin struggled to remain loyal to biological Darwinism in his criticism of Zaitsev's social Darwinism. He castigated the members of the Academy of Sciences in St. Petersburg like von Baer who resisted Darwin's interpretation of evolution. But Nozhin, in the final analysis, differed dramatically with Darwin himself. The difference arose on the question of competition within a given species. Nozhin maintained that competition occurs only between organisms that are structured differently, such as hosts and parasites, and not between individuals of the same species. He preferred not to notice how clearly Darwin had stated in the *Origin* that "the struggle almost invariably will be most severe between the individuals of the same species. . . ."[17] Nozhin's political vision of cooperation among humans in accordance with Proudhon's doctrine of *mutualité* overwhelmed his attachment to Darwin. In the end, he called Darwin a "bourgeois-naturalist" whose theory rested on mistaken Malthusian assumptions.[18]

Nikolai Chernyshevskii was a radical social critic who had great influence on the ideas that powered the Russian revolutionary movement of the late nineteenth and early twentieth centuries. Arrested in 1862 for his activities as editor of a journal critical of the government, Chernyshevskii spent two decades in Siberian exile. His main comments on Darwinism were written in 1888, only a year before his death.

Like many of the earlier Russian radicals, Chernyshevskii was also a materialist, and he firmly believed in the scientific approach to reality. But above all he was a political and social critic, and science was therefore a secondary subject in his intellectual hierarchy. He once remarked, "Everybody who has reached intellectual independence has political convictions and judges everything from the standpoint of those convictions."[19] Darwinism, then, was something to be judged politically before one drew conclusions about its general validity. Unlike Zaitsev and even Nozhin, Chernyshevskii would not take Darwinian evolution as a given, the latest fruit of objective science, and then speculate about what its sociopolitical implications might be. On the contrary, Chernyshevskii evaluated Darwinism within his political worldview and concluded that it was sorely wanting. On social and political grounds Chernyshevskii became an implacable foe of Darwinian evolution. He defended instead a form of Lamarckian transformism that continued to have great influence among Russian radicals for decades.

In a long essay on evolution published in 1888, Chernyshevskii said that Darwinian natural selection was based on the false assumption that good results can arise from the evils of hunger and suffering. Chernyshevskii agreed that evolution has occurred in nature, and he believed that the changes have been progressive. The fact that careful Darwinists usually defined progress only in terms of reproductive success, not in terms of social morality, did not concern Chernyshevskii. He turned his attention to the fact, troubling to many intellectuals of the nineteenth century, that Darwinism seemed, by implication, to justify violence by postulating that the evolution of organisms was based on competition and struggle. As a socialist, Chernyshevskii found this hypothesis unacceptable. He turned, therefore, to an attack both on Darwin as a person and on his evolutionary theory.

According to Chernyshevskii, Darwin was a poor scientist, a person who could not distinguish fundamental issues from unimportant ones. Indeed, Chernyshevskii called Darwin a "simpleton" of "childlike naïveté" who had wasted thirty-eight years of his life working on "petty details" like the habits of earthworms and the coloration and configuration of orchids. Chernyshevskii's attack on Darwin's method of work was an expression of impatience with scientific research itself, belying Chernyshevskii's stated reverence for science:

> Darwin, in his predilection for the monographic exhaustion of issues, constantly forgot that trivial details are no more than trivial details, that a significant question is decided on the basis of a few essentially important facts or broad ideas; a thousand details cannot in any way have an appreciable weight in the evaluation of arguments about significant questions.[20]

According to Chernyshevskii, not only did Darwin work by poor methods, but he drew the wrong conclusions. His theory of natural selection was simply incorrect. If evolution worked the way Darwin said that it did, the result, Chernyshevskii believed, would not be progress, but degradation. The dire struggle for existence would ruin the health of all organisms engaged in that struggle, and they would give birth to progency with similarly ruined health. "And if the course of life goes along this line through several generations," maintained Chernyshevskii, "then with each new generation the dimensions of the result increases, because it is the sum of the faults of the earlier generations."[21]

The Lamarckian assumption that underlay this criticism also was at the base of Chernyshevskii's attack on Darwin's belief that the selective breeding of domesticated animals gives an accelerated example of the effects of natural selection. Chernyshevskii maintained that Darwin made an "enormous scientific mistake" in comparing selective breeding with natural selection, since animal breeders do not submit all the members of a herd to adverse conditions in order to eliminate the inferior ones; nature, on the contrary, *does* submit all members to the conditions of hunger and privation that Darwin maintained resulted in the survival of the most fit. Thus, in order for Darwin's comparison to have validity, said Chernyshevskii, it would be necessary for the animal breeder not only to strike the inferior animals with the killing blow of an ax, but also to beat all the other animals at the same time. This critique, which possesses a certain superficial cleverness, is inconsistent with Darwinians' belief that the important influence on animal heredity is not the conditions of the environment, but the genetic constitutions of the parents. In other words, it was based on a fundamental misunderstanding of Darwinism. Yet this misunderstanding was shared by many educated Russians, including the great novelist Leo Tolstoy, who called Chernyshevskii's criticism of Darwin "beautiful" and "powerful."[22]

RECEPTION IN THE SCIENTIFIC COMMUNITY

The concept of evolution was widely accepted among Russian scientists. Indeed, even before the publication of *On the Origin of Species* Russia had possessed a number of scientists who believed in one or another form of organic evolution, including K. F. Rul'e, A. N. Beketov, K. F. von Baer, and L. S. Tsenkovskii. Some of them, like Rul'e and Beketov, favored views very similar to Lamarck's, and placed heavy emphasis on the inheritance of acquired characteristics; others, like von Baer, advanced a teleological interpretation of evolution, believing that the natural purposiveness of life resulted in favorable variations. Even though their views were different from Darwin's, the existence of these evolutionists in Russia before the advent of Darwin's theory meant that the concept of evolution itself would not be a surprise to Russian biologists.

The best-known evolutionists in Russia after the publication of the *Origin* were K. A. Timiriazev (1843–1920), I. I. Mechnikov (1845–1916), P. A. Kropotkin (1842–1921), V. O. Kovalevskii (1842–1883), and A. O. Kovalevskii (1840–1901). Born within five years of one another, all were very young men when Darwin's works first became known in Russia. All praised Darwin, all were excited by the intellectual vistas opened up by his work, all deeply believed in organic evolution, and all devoted their lives to research and teaching in which evolution occupied the central place. An examination of their works, however, shows that several of them were sharply critical of some aspects of Darwin's theory; not even the most energetic and passionate Darwinist among them, Timiriazev, fully accepted Darwin's terminology. The presence of these elements of resistance to Darwinism in the viewpoints of these five leading Russian evolutionists should not automatically be seen as flaws in their views of a sort that did not afflict Darwin; after all, several of them maintained that Darwin's analysis had been one-sidedly affected by *his* socioeconomic environment, namely the capitalistic society of nineteenth-century England. These Russian evolutionists lived in a different sort of society, one in which capitalism was not yet dominant, and it would not be surprising if they saw evolution a bit differently from Darwin. The Malthusian metaphor, in particular, did not go down easily with intellectuals who were interested in socialism, as some of these men were. Two different models of socialism could be found in Russia, the vestigial peasant socialism in the ancient agricultural communes or the new Marxist socialism that began to win adherents in the last decades of the century.

The topic of the comparison of external influences in different cultures on concepts of evolution is still an open one. The evaluation of the variations in interpretations among biologists calling themselves Darwinists but who differed on specific details has not been completed by historians of science, either by those working on Darwin or by those working on the history of Russian biology. In the last few years important scholarship on this topic has begun to appear.[23]

The Russian biologist with the securest claim to being an orthodox Darwinist was K. A. Timiriazev. If any Russian deserved the title of "Darwin's Bulldog" to match that of Huxley in England, it was Timiriazev. A plant physiologist, Timiriazev combined scientific knowledge with radical politics. Throughout his career he was engaged in a running battle with established authorities, even though, as a teacher in St. Petersburg University, and, later, a professor at Moscow University, he became something of an authority himself, particularly among radical intellectuals. His political scrapes included expulsion as a student from St. Petersburg University and, many years later, dismissal from the faculty of Moscow University as a result of his continuing radical sympathies.

Timiriazev was the most popular defender of Darwinism in all of Russia. His books, *A Short Sketch of the Theory of Darwin* and *Charles Darwin and His Theory*, were published in fifteen editions between 1883 and 1941. His influence was so great that it could still be felt well into the middle of the twentieth century. In an interview in Moscow in 1970 academician A. I. Oparin, a well-known authority on origin of life, described the lectures on Darwinism Timiriazev gave at the Polytechnical Museum in Moscow when Oparin was a boy as the most important influence on his professional development. According to Oparin, Timiriazev described Darwinian evolution and revolutionary political thought as being so intimately connected that they amounted to the same thing. In this view, Darwinism was materialistic, it called for change in all spheres, it was atheistic, it was politically radical, and it was causing a transformation of thought and politics.

No matter how much Timiriazev may have exaggerated the sociopolitical significance of Darwinism, he was loyal to the spirit of Darwin in downplaying the inheritance of acquired characteristics and stressing the central importance of natural selection. He strenuously fought against all deviations from Darwin's teachings, a position that eventually led him to oppose Mendelism and, indeed, all theories of heredity, seeing these theories as unnecessary speculations going beyond the positive facts of science.

But at least one aspect of Darwin's explanations of evolution would not fit with Timiriazev's political preferences. Like many other radical intellectuals living in a society only beginning to experience capitalism, he disliked Darwin's term "struggle for existence." Darwin entitled Chapter 3 of the *Origin* the "Struggle for Existence" and built much of his theory around this concept. Yet, as Todes has pointed out, Timiriazev avoided the term whenever possible, and in his famous article "Factors of Evolution" shunned it entirely.[24] Timiriazev's hesitation about using "struggle for existence" grew with time, leading one to believe that he may have learned that it was not received well by Russian audiences. Timiriazev came to prefer the term "harmony" to "struggle," seeing in the elimination of nonadaptive variations the achievement of a sort of natural balance. "Natural selection," a term that Timiriazev celebrated and defended, was in his mind a description of nature's way of achieving harmony, not a mechanism whose primary result was cruel competition. Indeed, in 1910 Timiriazev wrote that he had defended Darwinism for twenty years without "uttering that unhappy expression 'the struggle for existence'."[25]

Timiriazev's molding of Darwin's teachings to fit his own political viewpoint did not do much damage to their scientific core. However, his social and political preferences are evident in his writings, and not only in acts of omission, such as the avoiding of references to the struggle for existence, but also in acts of commission. An act of commission was his much stronger linkage of evolution to "progress" than in the writings of Darwin. To Darwin, evolution was a story of the survival of variations that represented "progress" primarily in the sense of reproductive success. To Timiriazev, however, the whole story of evolution was one of overall progress, measured by increasing complexity of organization and function. One senses that Timiriazev came to see this story of success as extending beyond the animal world into human history, with capitalism progressing into socialism. Timiriazev tightly linked science and politics in many of his writings; not surprisingly, he welcomed the Revolutions of 1917.

I. I. Mechnikov was a very different sort of scientist. One of the internationally best known of all Russian evolutionists, he moved almost as easily in Western Europe as in his native Russia, and spent the last twenty-eight years of his life in Paris, where he was given a laboratory by Pasteur to continue his work on immunology, for which he shared a Nobel Prize with Paul Ehrlich in 1908.

Mechnikov was not greatly interested in political causes, and sought refuge from the sort of turmoil that Timiriazev seemed to relish. But while he was far less politically radical than Timiriazev, he held much more radical views toward Darwin than Timiriazev. Mechnikov praised Darwin frequently as a person who had substantiated biological evolution, but he differed strongly with Darwin's description of the causes of evolution. Like Timiriazev he objected to the concept of the struggle for existence and also believed that Darwin was mistaken in placing so much emphasis on Malthusian overpopulation. Darwin, furthermore, exaggerated the significances of intraspecific competition, according to Mechnikov. Mechnikov wished to replace natural selection as the major factor of evolution with several other factors, including the inheritance of acquired characteristics, and an internal "special tendency to perfection" in the organism. Mechnikov did not see this last factor as teleological or idealistic, since he believed it could be explained on materialistic grounds. Only in that way, he thought, could the apparent progress of organisms toward ever more complex forms be explained.[26]

Mechnikov's lack of interest in politics means that any association of his biology with his politics in the way that seems likely in the case of Timiriazev would be mistaken. Yet Mechnikov shared with almost all other Russian commentators on Darwin an aversion to the notion of the struggle for existence, as Todes has shown, a characteristic that can be connected to conditions of Russia such as its relative lack, compared to Western Europe, of economic competition.

The importance of political factors becomes manifest in the case of P. A. Kropotkin, yet another well-known Russian evolutionist. Kropotkin's analysis of evolution makes an insightful comparison to that of Darwin himself. Darwin cast his brilliant theory of evolution in a language studded with violent terms: competition, survival, struggle. No doubt such terms were necessary in some significant degree in order for the core evolutionary theory and its mechanism of natural selection to be explained adequately. Biological evolution turns, after all, on survival ability. Recent scholarship has indicated, however, that Darwin used these terms rather freely, and in some instances when they were not necessary.[27] It seems likely that the mores and economic practices of industrialized Europe and Victorian England influenced Darwin unconsciously as he elaborated his theory. His reference to Malthus's version of political economy as an aid in his construction of the theory has been well documented by historians of biology.

Kropotkin was firmly committed to Darwin's theory of evolution, but he interpreted the theory in a different way. To him, important terms for interpreting evolution included "sociability," and "mutual aid." Indeed, where Darwin saw intraspecific *competition* as an important feature of evolution, Kropotkin saw intraspecific *cooperation* as the guiding force of evolution. In his 1902 book *Mutual Aid* Kropotkin cited a host of examples of cooperation in nature among birds, wolves, lions, rodents, and monkeys. He then extended this argument to human history, emphasizing cooperation among members of primitive tribes, medieval guilds, and modern labor unions. The modern state system of Western Europe was an aberration temporarily exaggerating the competitive side of evolution. The longer view revealed, he maintained, that "sociability is as much a law of nature as mutual struggle."[28]

Kropotkin, a member of an old aristocratic family, was a typical repentant nobleman who was swept up in the currents of radical, populist thought of Russia in the 1860s. He joined the subversive "Chaikovsky Circle" in 1872 and was subsequently arrested and imprisoned for political activity. The populists with whom Kropotkin associated opposed the development of capitalist industry in Russia, which they saw as immoral, and pinned their hopes for the salvation of Russia in the still surviving peasant communes, where an ancient agrarian form of socialism was practiced. Land in these communes was regarded as common property and was often tilled collectively. Kropotkin rhapsodized about these communes, affirming: "the sight of a Russian commune mowing a meadow . . . is one of the most inspiring sights; it shows what human work might be and ought to be."[29]

The brothers Aleksandr and Vladimir Kovalevskii were also prominent evolutionists. Vladimir Kovalevskii was, of all Russian evolutionists, the one who had the closest personal contact with Darwin, translating his work, meeting with him in England, and using his theory as a guide for his research in paleontology. Darwin described Vladimir's research on the phylogeny of ungulates as brilliant illustrations of his theory. The mechanism by which Vladimir described evolution was quite similar to Darwin's, combining natural selection with elements of the inheritance of acquired characteristics, particularly through use and disuse of individual organs. He placed great emphasis on adaptive radiation. A number of subsequent researchers have described Kovalevskii as a "forerunner of Neo-Lamarckism in paleontology," although a prominent Soviet historian of biology, L. J. Blacher, rejected this view as "groundless."[30] Work still needs to be done on both Kovalevskii brothers to determine what were their similarities and differences with Darwin. Both were influenced by radical politics, and it would not be surprising if they shared some of the general antipathy toward the Malthusian "struggle for existence" that was common among Russian socialists. Vladimir sympathized with the radicals of the 1870 Paris Commune, among whom he lived for a month while doing scientific work.

POLITICAL AND RELIGIOUS REACTIONS TO DARWINISM

The most substantial criticism of Darwinism in nineteenth-century Russia came from N. Ia. Danilevskii, which appeared too late (1885–7) to have much effect. Nonetheless, it created a temporary stir in some circles and was used by writers opposed to Darwinism who emerged in the increasingly conservative last decades of nineteenth-century tsarism.[31]

Danilevskii came to his crusade against Darwinism after taking sides in another great ideological battle in Russian culture; his attitude toward Darwinism is best understood as an outgrowth of that earlier struggle: the venerable dispute between Slavophiles and Westernizers over Russia's fate. Danilevskii agreed with the Slavophiles, and his attack on Darwinism was not so much a criticism of biological theory as it was an assault on Western science as a whole.

Danilevskii as a young man had been a member of the unorthodox Petrashevskii circle, a group of intellectuals exploring the ideas of European radicals like Proudhon and Fourier. Along with the other members of the group, Danilevskii was arrested by the police and exiled to distant locations. He managed to get back into the establishment, and at the same time acquire knowledge of ichthyology, by joining in the 1850s an expedition of Karl von Baer investigating the fishing resources of the Caspian Sea area. Danilevskii's political views began to move to the right. He gradually became convinced that Russian culture was a unique entity threatened by extinction by encroachment from the West.

In his first major work, *Russia and Europe* (1869), Danilevskii emphasized what he regarded as the ancient struggle between the Slavs and Western Europe. He saw this battle not as one over material resources but as a conflict of ideologies. The Western peoples, especially the English, were, in his view, violent and individualistic, while the Eastern peoples preferred harmony and cooperation. He saw this difference even in sports, where the Englishman "boxes one-on-one, not in a group as our Russians like to spar," and he extended this comparison to many other areas of Western and Slavic culture.[32] Fearing the triumph of the Western ethos because of its industrial and military foundations, he called for a federation of Eastern nations, with the Russians as the leaders; this alliance would include not only all the Slavs of Eastern Europe but also other Orthodox Christians such as the Greeks and Romanians, and would have Constantinople as its capital. He hoped that this coalition would have the cultural, political, and military power to resist Western domination.

In his second major effort, a two-volume work requiring over a decade of preparation and appearing only after his death, Danilevskii focused his anti-Western views on science, and especially on Darwinism. Western science, he maintained, was materialistic, atheistic, and intellectually superficial. Nonetheless, Danilevskii agreed that Darwin was one of the most talented scientists to appear within the tradition of Western science, limited though that tradition was. And Danilevskii did not deny the possibility of evolution, only Darwin's version of it. Because of his talent and his prodigious capacity for accumulating research evidence Darwin had, according to Danilevskii, persuaded many intellectuals, including Russians, of the truth of his particular variation of evolution. Only with time had the flaws in Darwin's reasoning become evident. Danilevskii cataloged what he considered to be those errors, drawing heavily on Western critics of Darwin such as A. J. Wigand, Georges Cuvier, Louis Agassiz, A. Kölliker, A. de Quatrefages, and many others.

Danilevskii believed that in emphasizing the randomness of variation Darwin erroneously denied purposiveness or teleology in organic change. Danilevskii preferred to assign a role to supernatural or divine regulation of the process of organic change. He also criticized Darwin for not allowing sufficient room for "leaps" in organic transformation. And Danilevskii found unacceptable Darwin's assignment of a primary role to the "struggle for existence" as a mechanism of natural selection.[33]

Danilevskii's opposition to the struggle for existence drew on his experience during expeditionary work in the vast areas of the Russian Empire. Shortage of food and overpopulation were only local problems, he maintained; if a certain species faced such problems in one locality, it would not elsewhere. He contrasted Darwin's studies of island populations to his own continental ones and indicated that his approach was more valid. In this form of argumentation we see, as Todes has pointed out, that some Russian opposition to Darwinism was not merely political or ideological in nature (although that resistance was a major source) but also geographical, reflecting the characteristics of the Russian Empire, the largest land mass possessed by any nation on earth.

With the possible exception of this strong geographical emphasis, Danilevskii's arguments were entirely unoriginal. His work on Darwin was a grab-bag collection of earlier criticisms of Darwin. It served as a rich source for later Russian critics of Darwinism, some of whom did not entirely agree with Danilevskii's overall interpretation. A few religious writers, for example, used Danilevskii to oppose evolution outright, failing to notice that Danilevskii preferred a teleological form of evolution to a static view of the organic world.

After Danilevskii's death the role of defending his views fell to the conservative publicist N. N. Strakhov. His principal opponent was K. A. Timiriazev. This battle of bulldogs in the late 1880s probably came closer than any other to paralleling the controversies over Darwinism in England and America. It also displayed characteristics unique to Russia. As Alexander Vucinich has observed:

> The Strakhov–Timiriazev debate went far beyond the limits of Darwinian controversy. The question whether or not Darwin's theory had solved the riddle of organic evolution became part of the much larger question of the place of science in modern culture. This was a time of rising government oppression, which encouraged attacks on "natural science materialism" as a pernicious ideology. Populists, anarchists, Marxists, and a wing of academic liberals were unyielding in their determination to present scientific knowledge as the only sound path to the salvation of Russia. At the same time, the government received unlimited aid from idealistic philosophers, led by V. S. Solov'ev and Boris Chicherin, who concentrated on the intellectual narrowness and materialistic underpinnings of scientific thought.[34]

Debates like these do not usually have clear-cut victors or losers. Indeed, some of these issues would surface later in Russian and Soviet history. Echoes of them can even be seen in the differing dissident views almost a century later of Alexander Solzhenitsyn and Andrei Sakharov. Nonetheless, Danilevskii's opinions failed to sway the majority of Russian intellectuals in late nineteenth-century Russia. Science continued to have great appeal to all but the most conservative thinkers. As intellectuals became more radical in the last years of tsarism many embraced Darwinism fervently as a symbol of progress and rationality.

V. V. Rozanov was a religious philosopher and literary critic who analyzed Darwinism. A person of idiosyncratic views, Rozanov was not a part of the Russian religious establishment. In fact, he deeply criticized Christian attitudes toward sex and family life, maintaining that Christianity suppressed emotional enjoyment. Despite his reservations, Rozanov strove to find a hedonistic version of Christianity that he considered worthy of his support, one that would celebrate the emotions.

Not surprisingly, Rozanov in his book *Nature and History* found Darwin and Darwinism emotionally unsatisfying.[35] Rozanov simply did not like Darwin as a person, and was convinced that the unattractive features of Darwin's personality had distorted his science. Darwin, according to Rozanov, was emotionally inert, incapable by his own admission of enjoying music, poetry, or art. Each of these requires not merely external observation, but internal understanding and emotional participation. Darwin approached nature, according to Rozanov, passively and superficially, and when he wrote about the biological world he simply expressed his own personality. Darwin's theory of evolution was a mere system of external classification that did not penetrate below the surface of phenomena, did not explain the origin of species, and did not reach a deep level of understanding. Rozanov thought great science should be similar to great poetry. Darwin's science, said Rozanov, could be compared to "poetry" presented to the public by a writer who simply pasted together verses borrowed from others.[36] Individual verses might rhyme and even be beautiful, but the overall work had no meaning.

Such criticism obviously had no effect on the Russian scientific world. Because of Rozanov's religious heterodoxy and ethical hedonism, it also had little effect among theological writers. Rozanov's views were extreme expressions, however, of an attitude not infrequently found among mystical Russian conservatives, the belief that Darwinism reflected an arid approach to reality that fitted poorly with the subjective and religious sensitivities inherent in some of Russia's oldest traditions.

The church establishment in Russia slowly awoke to the challenge of Darwinism but had great difficulty presenting a sophisticated response. At first the theological journals merely reproduced criticisms of Darwinism that had appeared in Western journals. Not until near the end of the nineteenth century did the church manage to produce independent and knowledgeable critiques of Darwinism. Perhaps the most sophisticated (although far from original) of these publications came from S. S. Glagolev, who wrote a series of works on biology between 1894 and 1913.

Glagolev was a convinced creationist, but he was careful to present his arguments in a restrained manner. He observed:

> I do not claim that my arguments against the evolutionary theory are incontrovertible. But my survey shows that the origin of species has continued to be a question without an answer. There is not a single species for which the question of origin has been answered. Evolution is merely a hypothesis: only the future will tell how, and to what degree, it is true.[37]

Glagolev was able to take advantage of new work in science, such as that of Gregor Mendel and Hugo de Vries, work that even professional biologists had not yet managed to integrate into a new Darwinian synthesis. No doubt enjoying the fact that Mendel was a monk, Glagolev maintained that Mendelian genetics discredited Darwinism and pointed to a new union between science and religion. In addition, the work of de Vries on mutation showed, he thought, that Darwin's emphasis on gradual change through variation and natural selection was faulty.

By citing Mendel and de Vries, Glagolev was referring to problems for Darwinism acknowledged throughout most of the scientific world. Glagolev went on beyond his citation of new scientific work, however, to a defense of Lamarckian views tied to "transcendental teleology." It became clear that his citation of science was merely a means to reintroduce the concept of a divinely regulated universe. Few scientists or even secular laypeople were won over by Glagolev's writings, especially since most of them appeared in theological journals not widely read outside religious circles. But their existence does show that the Orthodox Church was capable of producing a religious critique of Darwinism similar to those appearing in many other countries.

The reception of Darwinism in Russia makes an interesting comparison with other countries. Among the characteristics of Russia that shaped the reception that Darwinism received there was its relative lack of capitalism and economic competition, the growing politicization of many of its intellectuals, the absence of strong interest among leaders of the Orthodox Church in scientific questions, and the little intellectual influence exerted by fundamentalist Protestant sects (which provided strong resistance to evolution in the United States).

Russian attitudes toward Darwinism and evolution were not, of course, unique; all of them could be found elsewhere. However, the particular mixture of attitudes in Russia and the relative strengths of each of the elements of the mixture differed from those in other countries where the reception of Darwinism has been studied. And this constellation of ideas about Darwinism would continue to have influence in Russia well into the twentieth century.

Almost no writers in Russia of prominence objected to evolution itself, only to the version that Darwin presented. Radical Russian intellectuals usually supported Darwinism because it was antitraditional, gave a materialist evolutionary account of organisms and man, and opposed, at least implicitly, a priori principles derived from religion and idealistic philosophy. Once they made these points they often differed with Darwin (sometimes without realizing it) by giving natural selection less weight than he did and the inheritance of acquired characteristics more. Most of them believed strongly in the concept of progress, easily seeing it in the biological world and hoping for it in the human one.

Natural scientists and biologists were usually closer to Darwin on the details of his theory of evolution, but even they often objected to the concept of "the struggle for existence." In contrast to the nonscientists who commented on Darwin, most of the scientists knew when they were disagreeing with him. They wished to make several revisions and additions to Darwin's theory, while retaining natural selection as one way among several in which evolution could proceed.

PART II

Russian science and a Marxist revolution

4

The Russian Revolution and the scientific community

MANY years ago Crane Brinton compared different revolutions in modern states and concluded that in the Russian case "events were telescoped together in a shorter period than in any of our revolutions. . . ."[1] According to Brinton, who described the Russian Revolution in terms taken from the French Revolution, the Russian Thermidor, or counterreaction, succeeded the events of the political overturn itself fairly rapidly. Brinton concluded that the Russian Thermidor, the end of the Revolution, came in 1921 with the institution of the New Economic Policy (NEP). Thermidor itself, he said, "comes as naturally to societies in revolution as an ebbing tide, as calm after a storm, as convalescence after fever, as the snapping-back of a stretched elastic band."[2]

Brinton's observation leads us to attempt to establish some sort of time boundary for the Russian Revolution as we turn to science in the Russian Revolution. If we follow Brinton and define the Revolution as the period 1917–21, then we must study the impact of political and social turmoil on Soviet scientific institutions in this same period. If we accept this periodization, however, we will omit most of the important events for science in early Soviet history, which predominantly occur after 1921.

In this chapter, then, I will define the Russian Revolution much more broadly than the years 1917 to 1921, and will consider events in the period 1917 to 1932, for it is within these years that most of the important changes for Soviet science either occurred or were prefigured.[3] I am not suggesting 1932 as a terminal date, as a time for Thermidor; I believe that Brinton's model for revolution does not fit the Russian case. The metaphor of the "snapping-back of a stretched elastic band" will not serve here. Rather, I would suggest as a heuristic alternative Andrei Amalrik's comment that the Russian Revolution was like the expending of the tension of a tightly wound mainspring.[4] It was a process that sometimes occurred rapidly, sometimes slowed, then speeded up again. Although the unwinding was largely controlled by political leaders in its later stages, it still possessed an inherent energy of its own. No date will serve well as the moment the tension was entirely released, but we can agree with Amalrik's observation in 1975 that by then the spring had ceased to unwind.[5]

In this chapter I will try to give something approaching a synthetic overview of the transformation of Soviet science from 1917 to 1932. In the concluding section I will return to the question of a comparison of the impact of the French Revolution on science with that of the Russian Revolution, particularly with respect to the striking fact that while France abolished its most influential prerevolutionary center of science, Russia moved its comparable center to an unprecedented level of influence.

Russian science before 1917 lagged behind the leading countries of the West, but it was already developing in an impressive and promising fashion. In certain areas – mathematics, soil science, physiology, astronomy, and some aspects of physics, biology, and chemistry – Russian scholars had proven their international standing by the turn of the twentieth century. Names such as Mendeleev, Dokuchaev, Sechenov, Lobachevskii, Chebyshev, Mechnikov, Kovalevskii, Pavlov, and Butlerov occupied solid positions in the history of the sciences. The founding dates of professional societies such as the Russian Physical-Chemical Society (1869) were comparable to the birth dates of similar societies in the United States, another nation that, like Russia, turned to Western Europe for scientific sustenance.[6] The Russian Empire possessed ten universities at the beginning of World War I, the oldest (Moscow) dating to 1755. Its Imperial Academy of Sciences (1725) continued to sponsor valuable research throughout the nineteenth and early twentieth centuries. If nineteenth-century Russia was often thought of in the West as a country outside the scientific tradition, a nation where forms of Slavic mysticism and Orthodox Christianity not conducive to science were the principal intellectual trends, it is quite clear, to the contrary, that by the end of that century Russia possessed a developing and capable scientific community already rooted in an institutional base.[7]

Prerevolutionary Russian science and technology also suffered from a number of weaknesses stemming from the recent history of the empire. Since Russian industry borrowed its techniques from abroad, where much of the capital also originated, industrial research was weakly developed. Even large chemical and machine industries often left research and development to foreign sources. University science, although stronger than industrial research, was also still immature. It was necessary for Russian graduate students to go abroad, often to Germany, in order to obtain a first-class scientific education. Furthermore, the political difficulties of the last decades of the Empire often obstructed the flowering of scientific talent, since the best students frequently became embroiled with the political authorities in the course of their studies, resulting in the growth of political opposition among intellectuals and the weakening of scientific professionalism and institutional support. Russian universities in the early twentieth century were several times paralyzed by strikes and political demonstrations for lengthy periods, and in 1911, the tsarist minister of education, L. A. Kasso, either fired or, by his repressive policies, forced the resignations of over one hundred university professors, some of them the best in their fields.

Organizationally, Russian science was still in flux in the first decades of the century. The tsarist government had neither the funds nor the commitment to match the pace at which other leading nations were beginning to support advanced science and education. Furthermore, the underdevelopment of native Russian capitalism meant that philanthropic and private support of science, of increasing importance in Western Europe and the United States, was only beginning.[8]

In later Soviet years, the most influential and interesting scientific institution would be the Academy of Sciences, which became far more influential than the national academies of Western nations. The story of how that growth toward preeminence occurred is one of the most important aspects of the history of Soviet science, particularly since there were moments when abolition seemed a possible, even likely, fate for the Academy. Already by 1917, however, the Imperial Academy of Sciences possessed several characteristics that distinguished it from its foreign counterparts. In contrast to West European scientific academies, the Imperial Academy of Sciences still sought to be the leading scientific institution in the country. The tsarist government distrusted the politicized university professors more than it did the members of the Academy, and the latter institution benefited from this distrust. The permanent secretary of the Academy in the years after 1904, Prince S. F. Ol'denburg, dreamed of a renaissance of Russia, a blooming of its scientific and cultural potential, with the Academy of Sciences playing the leading role. When the Communist leaders inherited this extraordinary institution they faced a decision – abolish it, as in the French Revolution; support it at the existing level while expanding research in other institutions such as the universities; or build a structure of scientific research in which it would be the central and critical element.[9] They decided to adopt the last choice.

THE REVOLUTION AND EARLY ATTITUDES TOWARD SCIENCE

Of the two revolutions in Russia in 1917 the first, in February, brought to power a government in which Westernized intellectuals played a large role. The short-lived provisional government, pressed by more urgent tasks than the development of a policy toward science, would undoubtedly in less strained moments have promoted a science establishment modeled on the more advanced West European states and based upon similar assumptions about the place of science in the economy and the intellectual life of the nation. In the few months of 1917 when the liberals and the democratic socialists were in power, reforms were enacted that had influence on the future of Soviet science. The universities adopted new structures of faculty governance, the professional societies asserted their independence from state control, and the Academy of Sciences, for the first time in its history, elected its own president, the geologist A. P. Karpinskii. The permanent secretary of the Academy, S. F. Ol'denburg, also served as minister of education in the provisional government. Both of these men retained their positions at the Academy for many years after the second revolution, and they were instrumental in the development of the new Soviet scientific establishment even though their own positions were never secure.

While most of the scientists, engineers, and physicians in the Russian Empire greeted the February Revolution as an event that held much promise for both political freedom and scientific research, their reaction to the advent of the Bolsheviks in October was one of suspicion and hostility. The overwhelming majority of Russia's scientific community considered the Bolsheviks to be extremists likely to damage the political and intellectual future of Russia. There were exceptions, of course – that small group of scientists and engineers who either were members of the Bolshevik Party or shared its goals – but they were an insignificant portion of the technical intelligentsia.[10] No member of the Academy of Sciences was also a member of the Communist Party, and none would be for many years.

It is one of the paradoxes of the history of science in early Soviet Russia that the scientific institution that was generally acknowledged as being the most conservative, the Academy of Sciences, met the Bolshevik Revolution with less overt resistance than did the universities and other scientific institutions. Indeed, the Academy not only refrained from the hostile declarations and acts that were characteristic of many learned organizations and professional societies immediately after the October Revolution, but cooperated with the Soviet government from a fairly early date. Yet it is clear that the members of the Academy were, as individuals, no more sympathetic to the young Bolshevik government than were their colleagues in the universities and professional societies. Their greater willingness to tolerate the government not only reflected the prevailing wish of leading researchers to keep politics separate from science, but also of the special place that the Academy already held in Russian society.

Beginning in late 1917 and continuing sporadically throughout the first few years of Soviet power, groups of intellectuals and specialists demonstrated their disapproval of the new regime by passing anti-Soviet resolutions, declaring strikes, waging informal boycotts, and ignoring

Soviet orders. Transport often stopped, electricity frequently did not flow, schools were shut down, even hospitals closed. The Pirogov Society, an association of physicians, on November 22, 1917, censured the seizure of power by the Bolsheviks and called for strikes by medical workers.[11] The resulting walkouts were most effective in Moscow and Petrograd.

The Bolsheviks viewed such strikes as outright sabotage, and answered with repressive measures. Intellectuals suspected of organized resistance were sometimes summarily executed. The official policy higher up was a combination of persuasion for the rank-and-file intellectuals the Bolsheviks hoped to win over, and imprisonment only for overt "treasonous acts." Yet the interpretation of what was treason was often a subtle affair.[12]

In the first few weeks after the October Revolution the most vigorous intellectual opposition to the new government came from the ranks of those whose professions put them in close contact with the social turmoil: some of the military specialists, the teachers in the lower and middle schools, the physicians in the medical societies and hospitals, the engineers in industry, and the government bureaucracy.[13] These intellectuals met almost daily with political activists, radical workers, and students, and they were soon involved in polemical exchanges with reformers who proposed to "democratize" local institutions; great disputes erupted over efforts to reelect professors in the universities, and to submit engineers to workers' control in the plants. The professors, engineers, and doctors realized that their careers and interests were at stake, and many of them offered at least passive resistance throughout the winter of 1917 and spring of 1918.

The situation in the Academy of Sciences and in some of the more theoretical research institutions was considerably different. The work of scholars there did not force them to meet the hostile and militant elements directly, and many of them asked for no more than to be left alone. The principle of keeping science out of politics, strong in the Central European universities where many Russian scholars had studied, was often expressed by senior researchers as their "political neutrality." If the Soviet regime would not meddle with science, the scientists would not meddle with politics. Thus, in the early days after the Bolshevik Revolution the Academy of Sciences defeated a proposal by several of its members that it, like the Pirogov Society, should deny the legitimacy of the new regime.[14]

Yet it seemed unlikely that the senior researchers could remain aloof. Within a few months Soviet Russia entered its most fervent ideological phase, that of War Communism (1918–21), when all enterprises were nationalized and an early abortive effort was made to create a command economy. As all institutions inherited from the tsarist regime came under heavy criticism, the Imperial Academy of Sciences seemed a probable target for reconstruction or elimination. The analogy with revolutionary France stirred the air, with much talk of the "Russian Jacobins" and institutional leveling.

As a historian Ol'denburg was familiar with the fate of the Académie Royale des Sciences in Paris in 1793. He knew that as a member of a princely family and a former minister of the provisional government he should be prepared for both personal and institutional attacks. Deciding that the best defense was a demonstration of the usefulness of science and learning to the new regime, he cast aside his initial pessimism about the development of science in Soviet Russia and, together with other leaders of the Academy, sought an accommodation with the Soviet government.[15] They were criticized by some of their more anti-Soviet fellow scholars for negotiating with the Bolsheviks; nonetheless, Ol'denburg and his friends saw these complaints as the price to pay for preserving culture.[16] In early 1918 President Karpinskii somewhat stiffly replied to an inquiry by A. V. Lunacharskii, the new Soviet commissar of education, that the Academy, in keeping with its tradition of service to the state, would help develop the productive forces for national needs.[17] In response the government began to release funds for the Academy's operations.

While the leaders of the Academy were taking defensive measures, revolutionary spokesmen began to formulate plans for a reorganization of Russian science that would eliminate or diminish the role of the Academy. Several different radical plans emerged, coming from different quarters and evidently having little or no connection with one other. Probably the most ambitious plan to reorganize Russian science came from the commissariat of education of the union of communes of the northern area (Sevpros), which included Petrograd. The goal of this project was no less than to "win science for the proletariat," to destroy the "fetishization of pure science" allegedly endemic in traditional scientific institutions. The northern radicals wanted to abolish the Academy of Sciences and similar "old forms of the social organization of science," and called for their replacement by a "homogeneous set of scholarly-pedagogical institutes" in which teaching would be a primary function. Even those institutions that had traditionally been involved only in theoretical research would be required to assume major pedagogical responsibilities. The guiding motif of the new organization would be the "unity of scholarly and teaching work." Direction of the new network of institutes would be centralized on the all-governmental level in a Department of Proletarian Labor Science.[18]

Undoubtedly, the success of the Northern Communes Project would have fulfilled the worst fears of the leaders of Russian science. The scarcely hidden agenda behind the calls for a democratization of science and the union of teaching and research – which as rational reforms, carefully executed, might have been progressive steps – was the destruction of the finest scientific institutions of Russia and the creation of new organizations in which the preservation of scientific standards would have been nearly impossible. The promoters of the Northern Communes Project called the traditional scientific laboratories "completely unneeded vestiges of the pseudoclassical epoch of class society."[19]

Ol'denburg probably had such projects in mind when he later described the early revolutionary period in the following way:

> Especially great was the danger of the loss of culture, which was deeply rooted in the old, previous way of life and therefore appeared to be unacceptable for the new order. And it was therefore exceedingly difficult to find a path along which we could proceed. There were moments when it appeared that culture and science would perish, when it seemed that no one needed them during the great overturn that was being executed so rapidly.[20]

Another project, more carefully planned, but still radical in its implications, was one worked out in Moscow in 1918–19 by the Scientific Department of the Commissariat of Education. Although not as far-reaching as the project of the Northern communes, it would nonetheless have transformed Russian science through its creation of a Russian Association of Science, an organization patterned on the Soviet system of government, with levels of elected governing councils, or *soviets*, proceeding upward from the local level to a centralized Association Conference. The word "association" had been popular even before the Revolution among some Russian scientists as a proposed framework for the reform of Russian science, and they often cited the models of the British and American Associations for the Advancement of Science, as well as the German Gesellschaft deutscher Naturforscher und Arzte. However, the resemblance to foreign models would have been more in name than in fact, for the proposed Russian Association was to be carefully centralized, with close supervision by the Commissariat of Education. And since the primary units of the Association would have been organized on disciplinary lines (there were to be vertical associations of institutes in each of the fields of biology, physics, etc.) the old organizations such as the Academy of Sciences would have been either disbanded or greatly reduced in significance.[21]

The Academy of Sciences decided to meet these new challenges by submitting a reorganization plan of its own. In June 1918 the geologist A. E. Fersman, in the name of a group of scholars in the Academy, submitted a plan calling for a Union of Scholars, an organization designed to bring under one umbrella all the scholars and scholarly organizations in Soviet Russia. The Academy's proposal differed from the others in two important respects: (1) the new organization would be organized on functional and not disciplinary lines (and therefore would preserve an important place for the Academy of Sciences); (2) the primary function of the Union of Scholars would be to provide for a means of governmental financial support for scholarship, not for internal control of scientific research. The Academy's proposal, in contrast to all other plans, did not subordinate science to a central planning agency. Clearly, the Academy was making a plea for the reform of science on the basis of existing institutions, not by starting de novo in the manner of the Northern Communes plan. While the Moscow plan for an association appears to have straddled the midpoint of these two alternatives, it, too, probably would have led to the demise of the Academy as an independent agency.

None of these plans was implemented. (In fact, they have only recently been unearthed from the archives.) The reform of science turned out to be an immensely complex task. As Soviet Russia was engulfed by civil war, economic crisis, and famine, the task of finding a revolutionary framework for science moved lower and lower on the table of priorities.[22] The country needed technical advice on armaments, expert assistance on transportation, help in running electrical power systems, and new sources of food and fuel. On all these matters the scientists and engineers in the established institutions were the best people available. At this moment of great need, a radical overturn of these institutions seemed ill advised. Ideas about recasting the organization of science and technology to fit the new proletarian order were postponed.

One new project posing a strong implicit challenge to the Academy of Sciences did succeed in the early years, however. Several prominent Bolshevik scholars working on the draft of a new constitution of the Russian Republic conceived the idea of a new Academy designed to cultivate the social sciences from a Marxist point of view. Since the universities and other academic institutions were under the control of non-Marxist scholars, the new Academy would serve as an alternative scholarly center for the development of Marxist interpretations of society and the training of Marxist scholars. Founded in June 1918, the Socialist (later Communist) Academy ultimately developed a small section in the natural sciences, and more than one commentator saw it as a rival to the "bourgeois" Academy of Sciences. It was never able, however, to compete successfully with the older academy in the natural sciences. In the social sciences it enjoyed a period of flowering in the twenties, and produced some of the best Marxist scholarship of Soviet history. In a sense, it succeeded too well in this area, since Stalin did not like independent-minded Marxists offering views on social issues that might challenge his own. He abolished the Communist Academy in 1936 at the beginning of his mass purges.[23] The library of the Communist Academy survived, first as the Fundamental Library of the Social Sciences, and later as the kernel of the collection of the still-existing Institute of Scientific Information of the Social Sciences. (In the 1960s, 1970s, and 1980s that library provided more free access to sensitive materials than any other in the Soviet Union, and was, in fact, an important source for this book.)

THE EFFECTS OF MILITARY AND ECONOMIC CRISIS

During World War I the Academy of Sciences had acquired considerable experience advising the tsarist government on military needs, an expertise needed by the new Soviet regime.[24] Through the Commission for the Study of Natural Resources (KEPS), geologists such as V. I. Vernadskii and A. E. Fersman, engineers such as P. Pal'chinskii, and chemists such as V. N. Ipat'ev had recommended means of development of mineral resources and fuel supplies, and had consulted on chemical warfare.[25] The new Soviet government needed similar help. There often seemed to be a frustrating inverse ratio between commitment to Bolshevism and ability to provide concrete and useful answers to technical problems. Pressed by the military situation, the Soviet leaders turned to experts who were often not politically compatible with the new order but who were nonetheless extremely competent. When the assistance came from these experts, the government began to respond with increased financial support to the institutions in which they were located, despite the fact that these organizations were integral parts of the old, despised tsarist government. A result was that organizations politically committed to the Soviet regime such as the Communist Academy received less financial support than bourgeois institutions such as the Academy of Sciences, a fact much resented by Marxist scholars.

In the memoirs of prominent scientists and engineers who worked in Russia before and after the Revolution, the similarities in the work they did in both periods can be seen clearly.[26] The chemist Ipat'ev, for example, had been an industrial consultant to tsarist industries, and during World War I he headed the Chemical Committee on Defense. After the Revolution, Ipat'ev became the chairman of the Technical Section of the War Council, and, later, chemical advisor to the Supreme Council of National Economy. During NEP (1921–6) he continued to advise the chemical industries and served as director of the Scientific Technical Administration (NTU). It is obvious that Ipat'ev, an outstanding chemist, member of the Academy of Sciences, and thoroughly middle class in his outlook,[27] was much more valuable to the Bolshevik regime in its time of dire need than any number of orthodox revolutionaries calling for the proletarianization of science. Thus the utopian schemes of reform were shelved as traditional personnel and institutions played larger and larger roles. Yet the revolutionary impulse did not wane. The more militant elements among the Bolsheviks remained convinced of the need for radical reforms of the scientific establishment. In fact, the temporary restraints on the militants intensified their underlying revolutionary resentment.

THE GREAT DEBATE OVER TECHNICAL SPECIALISTS

The increasing reliance of the new Soviet government on technical specialists and economic managers indistinguishable from their prerevolutionary counterparts was viewed by many zealous communists and workers' groups as a betrayal of the Revolution.[28] Was not Soviet Russia supposed to be a workers' state? Just who should run Soviet industry, the proletariat or white-collar technical and managerial specialists? Who should decide which expensive scientific research projects would be financed by the new government, the workers in their assemblies (the *soviets*), or the scientists themselves?[29] To one colorful leader of the radical segment of the Russian proletariat the answer was clear: The technical specialists were "remnants of the past, by all their nature closely, unalterably bound to the bourgeois system that we aim to destroy."[30]

In the fall of 1922, a running debate appeared in the pages of the main Bolshevik newspaper *Pravda* over the legitimacy of the proletarian culture (Proletkul't) movement.[31] In September the chairman of the central committee of the organization, V. Pletnev, published an article titled "On the Ideological Front," in which he not only defended strongly the idea of a new proletarian culture that would differ markedly from the culture of the prerevolutionary period, but also made it clear that he included science and technology among those cultural artifacts that were destined for major overhaul.[32] Pletnev denied that his view was a destructive one – we will preserve the *material* creations of bourgeois culture, he reassured his readers – but he called for the destruction of the ideology on which these monuments stood, replacing the individualist principles of bourgeois culture with the collectivist ethos of proletarian culture. It was hopeless and wrong, Pletnev wrote, to call for the proletariat to work closely with bourgeois scientists and engineers because the members of the proletariat were "alien to the *intelligent*, the doctor, the jurist, the engineer, those people who are educated on the principles of capitalist competition." An attack on bourgeois science by the proletariat was inevitable, he wrote, brought about by "the very process of the Revolution itself."

Pletnev was much less clear about the sorts of science and technology that would take the place of the traditional forms. But he was certain that science and technology would become much more practical and public spirited in their orientations. The proletariat was not interested, he said, in "science for science's sake," but instead in making science a direct servant of the proletariat's needs. Pletnev believed that new proletarian engineers who transcended the narrow frameworks of capitalist industries were needed, people who would devote themselves to electrification and great public works projects. For such tasks the new experts would need to know as much about economic, legal, and social planning as about technology. The proletarian specialists of the future would be "social engineers" capable of "uniting all fields on a grand scale." The old division of labor involved in capitalist industry would disappear. Pletnev called for a socialized science that would reveal "the interconnection of things," and would create a unique scientific methodology that would simultaneously be easier to master than the "arid, specialized" disciplines taught in the traditional universities. The new science would differ from the old in its "essence, method, form and scale." Practical experience, not diplomas, would be the qualification for the new scientists and engineers.

Even more important to Pletnev than the content of the new science and technology was the primacy of proletarian control in its development and utilization. The toleration of the bourgeois specialists in socialist industries even in subordinate positions was already an irritation to Pletnev, and he found totally unacceptable, even incomprehensible, their being given supervisory functions and high salaries.[33] Any bourgeois specialist who has a change of heart and who "comes over to the proletarian cause" will be an "isolated instance," a person without general significance, Pletnev maintained, since the old specialists were constitutionally incapable of understanding the new era. The solution to the problem of mastering technology must come, he thought, from the ranks of the proletariat itself. The workers, not diplomaed engineers and business managers, must be the true masters of the workers' state.

Following the appearance of Pletnev's article, Pravda ran replies and criticism from other authors, including N. K. Krupskaia (Lenin's wife), I. I. Skvortsov-Stepanov, and Ia. A. Iakovlev.[34] We know that Lenin himself became agitated about Pletnev's views, since a copy of the article was later found with his sarcastic remarks in the margins.[35] One of Lenin's comments inquired what percentage of Pletnev's loyal proletarians knew how to build locomotives; another ridiculed his belief that scientists would always think in terms of immediate, practical needs, and still another disputed Pletnev's contention that the proletariat was capable of producing through its own efforts the new engineers with whom Pletnev was so enamored. Lenin had earlier defended bourgeois specialists from attacks from the left, and he had also warned the critics of the Academy of Sciences not to harm it.[36]

The major responsibility of answering Pletnev fell to Krupskaia and Iakovlev. Krupskaia was the more moderate critic, observing primarily that the exact sciences were based on the accumulation of many centuries' experience, and that "To throw these achievements overboard would be laughable and barbaric."[37]

Iakovlev zeroed in on Pletnev's underestimation of the difficulty and sophistication of science and technology by pointing to his statement that the new, proletarian science would be much simpler to master than university science. Pletnev seemed to believe, Iakovlev implied, that novel forms of valid science could be obtained in a way similar to the production of art and sculpture in the experimental Proletkul't studios, by disregarding rigidified academic conventions. On the contrary, Iakovlev argued, modern science was an immensely difficult and valuable achievement of civilization that must be studied with traditional rigor. Not only was this no time for irresponsible and romantic calls for proletarian controls over science, Iakovlev maintained, but, indeed, "The very existence of Soviet power is a question of studying at the feet of the professor, the engineer, and the public school teacher who were inherited by us from capitalism."[38]

THE EFFECTS OF THE NEW ECONOMIC POLICY

With the introduction of NEP, which brought tightly controlled elements of capitalism back to Soviet Russia, the policy of keeping prerevolutionary technical specialists on the job gained greater strength. And with the dramatic shift in economic policy, some of the specialists themselves began to reconsider their basic attitudes toward Soviet power. If the previous years of military strife and command economy, surrounded by communist rhetoric, had appealed to romantic revolutionaries of apocalyptic bent, the twenties appealed to the practical managers, economic entrepreneurs, and technical specialists. There were factories to restore and reequip, communications with Western Europe to reestablish after years of disruption, diplomatic and trade agreements to sign.[39] Soviet scientists began to appear for the first time at congresses of scholars in Western Europe.

The prerevolutionary scientists and engineers, many still quite unsympathetic to the ideology of the Bolshevik leaders, rationalized their growing cooperation with the regime in various ways. A few experienced genuine political conversions and joined the Communist Party, as did others for reasons of political opportunism. The predominant attitude, however, was based on allegedly apolitical arguments, and was often called the Development of Productive Forces school of thought. According to this interpretation, whatever political future Russia was to have, it would need a strong economy to sustain its population and to resist the encroachments of foreign powers. As a Soviet historian wrote, supporters of this view often said to themselves:

We will not pay attention to any party programs whatsoever. Without respect of class considerations, we will consider everything good that fosters the development of the productive forces of the country and everything bad which hinders the natural recuperation of the national economy. In the light of this theory capitalist tendencies by themselves are not considered harmful nor socialist ones healthy. Thus, the development of productive forces can equally help socialism or capitalism.[40]

Outside the sphere of the scientific-technical intelligentsia the attitudes of some educated Russians began to shift in favor of the Soviet regime. Among the large community of emigrés living abroad who had fled the Revolution one of the most important indicators of this intellectual change was the "Change of Landmarks" movement, named after a book published by a group of Soviet emigrés in Prague in 1921, and soon picked up by other intellectuals in various cities in Soviet Russia and abroad.[41] The leaders of the Change of Landmarks movement argued that continued resistance to Soviet power was pointless, that the new economic program of Soviet Russia illustrated that the extreme revolutionary platform had been abandoned, and that the opportunity existed for Soviet Russia to evolve more and more toward a democratic state, whether socialist or capitalist. Cooperation with such a state, they argued, was not only possible, but desirable.

Though temporarily muted, the same voices that had criticized the use of bourgeois specialists in the early years remained strong throughout the 1920s. Indeed, among militant youth groups and some workers' organizations, "specialist-baiting" (*spetseedstvo*) was a barely controllable phenomenon. These currents ran deeply in Soviet society, and as the twenties wore on, grew in intensity as the Marxist student body in the universities increased in size. The militant communists still considered the use of bourgeois specialists a temporary necessity, and they were awaiting the time when this unpleasant expedient could be cast aside.

The life-style and mental outlook of the prerevolutionary generation of intellectuals constantly reminded the radicals of the gulf that divided them. In dress and language the young radicals set themselves apart from their teachers and professors, creating a contrast between the proletarian fashion of Komsomol activists and their middle-class teachers, who preferred the style of the West European academics. In Soviet slang, the word "bourgeois" was firmly implanted as a descriptive masculine noun; a *burzhua* was a person you could recognize as he walked down the street: He wore a necktie and a hat, and he lived in an apartment indistinguishable in its furnishings from his West European models. It was further assumed that such differences in appearance were only a superficial indicator of much deeper ideological differences.[42]

Although these differences in style and demeanor were demagogically exploited by the political bosses whenever it seemed advantageous to have another wave of criticism of the old intellectuals, this hostility was not an artificial creation. The animus toward specialists educated under the old regime was deeply rooted among radical youth groups, workers' organizations, and left-wing intellectuals.

The older intellectuals tried to maintain communication with Western Europe, following general intellectual trends and keeping up with scientific and technological developments, and therefore were easily susceptible to the charge of spreading ideologically dangerous doctrines of Western origin among Soviet youth. When a group of Soviet writers became interested in Spengler's *Decline of the West*, which was a cause célèbre in Western Europe in the twenties, the Soviet radicals attacked them for falling prey to bourgeois pessimism and fatalism about the future of civilization at exactly the moment when the Russian Revolution had opened the doors of a bright, new era.[43] When another group became admirers of Bergson's philosophy of Creative Evolution, the radicals accused them of defending idealistic and anti-Darwinist descriptions of nature.[44] When some Soviet intellectuals tried to discuss Freud and the development of psychoanalysis, the critics answered by connecting Freud's views to the bourgeois culture of middle Europe, with its guilt-ridden neuroses.[45] When a group of Russian academicians and lay-persons became interested in the eugenics programs so popular at that time in many countries, it was easy enough for the radicals to point to the conservative social ideology often hidden in such movements.[46]

There are many, many examples of such conflicts in Soviet intellectual life in the twenties, and they are crucial for an understanding both of that period and of the immediately following one. When some economists raised doubts about the ability of the Soviet Union to achieve unheard-of industrial growth rates in face of limited resources, radical economists accused them of trying to slow the development of the socialist state.[47] When geologists preferred theoretical studies of the age of the earth's crust to practical searches for minerals and petroleum, the political observers thought they were trying to avoid participation in socialist construction.[48] When certain Soviet authors tried to interpret the meaning of the new developments in relativity and quantum physics, showing that the old concepts of materialism and causality were no longer adequate, the Soviet critics often replied that bourgeois intellectuals were engaged in a reactionary effort to discredit scientific materialism, the ideology of the victorious revolution.[49] When Soviet public health authorities studying mortality and morbidity rates adopted the statistical methods of their American and West European colleagues, the Soviet radicals pointed out that these methods concealed the degree to which disease and death are linked to social class, and called for alternative methods of analysis.[50] When engineers involved in electrification projects for the Soviet Union turned to the logical Western sources for expertise, the German and American capitalist companies developing the electrical industry, the suspicion was always present that the engineers would borrow not only technique, but the economic worldview of their foreign mentors.[51] And in fields such as anthropology, where concepts of racial difference were at that time widespread in scholarship throughout the world, it was easy for the Russian radicals to develop a biting criticism, just as socialists in Western Europe were doing at this time.[52]

The assumption in each case cited above, and in the many more that could be given, was that the old intellectuals were importing, along with some necessary knowledge, a large amount of dangerous ideological baggage. And not until old intellectuals had been removed from positions of influence would this danger pass.

The fact that these radical Soviet attacks on Western intellectual trends contained some elements of truth helps to explain the curious amalgam of political control and popular ferment that characterized the twenties and energized the coming Cultural Revolution. Western European concepts of the twenties on anthropology, for example, were often based on the assumption of the superiority of the Caucasian race. A number of Western writers on relativity physics and quantum mechanics, such as James Jeans and Arthur Eddington, strained to favor religion or combat materialism on the basis of the new science. Western public health and medical authorities often underplayed issues of class, gender, and race, as the radicals noted. Indeed, in almost all the controversies there lurked genuine issues of power and politics, topics appropriate for social analysis. Stalin's ability to capitalize on this ferment and ruthlessly turn it to his own benefit distinguished him from his competitors. With the intrusion of political coercion, controversies that in less heated circumstances would have remained intellectual discussions became, in many cases, mortal combats. As the result of such confrontations people were fired from their positions, many went to prison, and a tragic number were executed or died in Siberian exile. And in the end the intellectual content of the original issues was squeezed out. Only the necessity for political obedience remained.

SCIENCE IN THE CULTURAL REVOLUTION

The greatest change in the scientific establishment of the Soviet Union occurred in the years of the Cultural Revolution, 1928–31. The changes in the Academy that occurred in these years were only a small part of the "Great Break" (*velikii perelom*) that engulfed Soviet society, a vast campaign of cultural and economic transformation. It swept through Soviet industry, where private enterprise was eliminated and the first five-year plan was initiated; Soviet agriculture, where the collectivization program was launched with staggering intensity and violence; and Soviet educational, artistic, and scientific institutions, where reelections of scholars, purges of staffs, and campaigns for political militancy became routine.[53]

Coming after Stalin's successful struggle against his rivals (the "left" and the "right" oppositions), the Cultural Revolution has appeared to most historians in the West as being entirely the creation of Stalin, designed by him to mold the society in accordance with his own narrow ideological vision, and ultimately aimed toward the acquisition of total personal control. Stalin's power by this time was immense, and he certainly was the crucial element in many of these events. Yet it is also clear that in the initial phase of the Cultural Revolution Stalin was not running an exhibition with the control of a puppeteer, but was unleashing powerful semiautonomous forces, those revolutionary groups who believed that the reconstruction of Soviet society had been delayed during the twenties, and who now seized the opportunity to act in accordance with their conviction that bourgeois institutions must be renovated. The campaign was not one carried out in accordance with a rational vision, however radical, but instead was a primitive assault that vented deep and violent class hatreds, and was executed on the local level in a variety of forms. As Sheila Fitzpatrick pointed out, it was a struggle "waged by young against old, and junior against senior" in accordance with "aggressive, disorderly, anti-authoritarian and iconoclastic instincts," a "militant confrontation of the class enemy."[54]

For the Academy of Sciences the Cultural Revolution was the time for the long-feared assault. A newspaper campaign outside the walls of the Academy raised many old charges and a few new ones: The Academy was a "citadel of pure science" aloof from the needs of socialist construction; it was under the control of an "academic caste" that refused to teach communist youth; some of its members were allegedly plotting the overthrow of the Soviet government and had even prepared a list of ministers for the new government; the library was being used as a repository for "anti-Soviet documents"; the scholarship of the Academy was linked to the prerevolutionary period, as revealed by the fact that it contained groups studying the history of religion; engineers and scholars interested in applying their knowledge were excluded from membership; and the system of electing new members was designed to prevent communists from becoming members.[55]

The more specific charges were false, and the general ones were exaggerated or distorted.[56] Yet behind the false particulars lay not only genuine class and political differences but a truly alternative vision of the place of science in society. To the critics, tying scientists closely to political and social demands was not a loss for science or society, but a gain. Since the critics saw the goals of the Soviet state as being correct, the enlistment of scientists to serve these goals could only be beneficial.

Most of the scientists in fundamental research, however, believed that insistence on social relevance within the framework of a dogmatic ideological prescription for the future would do great damage to theoretical science. They resisted the encroachments of the political activists in a battle that dragged on for months; although eventually they capitulated to the election of Communists to Academy membership, in the process they managed to preserve both the traditional institutional base of Russian science and much of its intellectual framework.

The Academy that emerged from the Cultural Revolution was a greatly altered institution, but it survived. Indeed, it continued most of the lines of work in the natural sciences that it had previously promoted, and for some researchers life afterward was different only in tone, not in substance, from the previous period. For other members, it was a time of personal and professional tragedy. Several hundred research assistants and workers were dismissed, many imprisoned. Four full members of the Academy were arrested and sent to labor camps, only one to return.[57] Ol'denburg lost his position as permanent secretary. A new charter brought the Academy's research much closer to governmental supervision than before, and, for the first time, engineers joined fundamental researchers as members. The first Communist members formed an internal Party structure that tried to bring the other scholars in line with the decisions of the Party leadership, though not always with complete success.[58] The Academy's publications were henceforth submitted to a standard system of censorship controls. The political neutrality of the Academy was ended.

The new Academy of Sciences of the USSR that emerged from this traumatic experience became the centerpiece of fundamental science in the Soviet Union. It grew into a vast institution that, together with its various branches, comprised hundreds of research institutes spread throughout the Soviet Union. That this venerable institution, closely linked in its origins to monarchical government, should win the central place in the organization of science in a socialist state dedicated to proletarian rule must strike many observers as paradoxical (as it did many of the radicals we have discussed). But Stalin, ruling from the Kremlin of the tsars, shared with them the desire to have institutions that could be centrally controlled. The old Academy, placed within a new Soviet framework but still subject to hierarchical authority, was more attractive to him than a decentralized system controlled by workers or scientists themselves.

Impressive a scientific institution as the Academy became, the fact that its main organizational features were formed under Stalinism would provide fuel for a new wave of criticism emerging many years later. Reformers emerging in the late 1980s would question whether such a centralized system of science was the best way of fostering creativity. Their criticism was not based on the desire to make Soviet science distinct from that of the rest of the world, as had been the hope of the radicals of the 1920s, but to make it more like that of Western Europe and America. This story of new hopes for reform will be discussed in Chapter 9.

The greatest change in the scientific establishment of the Soviet Union occurred in the years of the Cultural Revolution, 1928–31. If one is to compare the impacts of the Russian and French revolutions on scientific institutions, these years, starting over a decade after the advent of the Bolsheviks to power, are the ones best suited for discussion. And here we see one of the greatest differences of the two revolutions in terms of their impacts. The French Revolution peaked rather quickly, disbanding the Parisian Academy in August 1793, only four years after the beginning of the crucial political events; the major attack on the Academy in Russia, however, came many years after the political overturn itself, and, indeed, after a long period of relatively calm economic recovery. Rather than proceeding more rapidly, as Crane Brinton believed, it appears that the Russian Revolution proceeded more slowly than others, and even accelerated its pace after a period of quiescence.

Probably the most important reason for the survival of the Academy of Sciences in the Soviet Union was the necessity to postpone its radical reform in the initial phase of the Russian Revolution because of the military and economic emergency. Radical critics had been ready with plans for its demolition or renovation, as we now know from the archival records. If any of the early plans for the reorganization of Soviet science had been adopted the resulting reforms would undoubtedly have been much more far-reaching in institutional terms than those that finally ensued. In this most militant phase of the Revolution a defense of the Imperial Academy of Sciences would have been difficult if the debate had ever reached full volume.

The need of governments for science, including revolutionary ones, had increased since the time of the French Revolution, and, consequently, the risks involved in tampering with science were correspondingly greater; Lenin, Lunacharskii, and other leaders of the Communist Party who were concerned with science and education were aware of these risks, and they frequently expressed their fears.[59] Lenin, in particular, was highly skeptical of the proletarian culture movement's belief that science in the new era would be radically different from traditional science, and he repeatedly called for the preservation of established centers of scientific and technical expertise.

Although the Academy of Sciences in Petrograd had many radical critics, no one opponent held sufficient stature among the revolutionary leaders to be able to play a decisive role in determining Communist Party policies. In fact, most of the enemies of the Academy were minor figures. And here another difference with revolutionary France emerges:

The most effective critic of the Parisian Academy was Jean Paul Marat, a radical journalist and prominent politician who also harbored scientific aspirations.[60] Before the Revolution, Marat had advanced a series of scientific discoveries that he hoped would place him in the top rank of French scientists. When he was rebuffed by the Academy, he turned venomously against the institution; after the Revolution he launched a series of attacks in his paper *L'Ami du Peuple* that were instrumental in bringing revolutionary vengeance upon the Academy.

No equivalent influential critic of the Russian Academy of Sciences existed in the years after 1917. Maxim Gorky, who had been refused membership in 1902 as a result of the intervention of the tsar, became one of the most energetic protectors of science and scientific institutions after the Revolution, appealing for special food rations for scientists and clemency when they were charged with counterrevolutionary activities.[61] M. N. Pokrovskii, the Marxist historian who headed the Communist Academy, was a caustic critic of bourgeois science, but he reserved his most scathing attacks for university professors. The fact that he headed an academy of his own further illustrated that the existing institutional form suited his own ideas about the organization of learning.[62]

In his study of the ill-fated Parisian Academy of Sciences during the French Revolution Roger Hahn emphasized that one of the major ideological criticisms of the Academy was based on "anti-corporate bias" that issued from two different sources: (1) belief in economic free trade and an analogous cultural liberalism that rejected self-propagating and guildlike academies, and (2) admiration of the English pattern of an "institutionally open society organized around common-interest groups rather than hierarchic corporate structures." Thus, the model for revolutionary French science became the *société libre*, not the Academy with its centralized structure inherited from the monarchy.[63]

The Bolshevik leaders of revolutionary Russia had very different biases and models; to them, the idea of a centralized Academy was not itself ideologically repugnant, only the particular political views held by the members of the Russian Academy. Soviet Russia was to be a centralized state in which the economy would be directed according to a rational economic plan. The Soviet leaders valued science for the help that it could give in the process of economic advancement, and they wished to have a scientific organization that could be easily linked to the economic planning organs. They saw a "free association" of scientific institutions, as proposed by several critics of the Academy, as a decentralized model of science appropriate perhaps for states with a capitalist economy based on laissez-faire economics, but hardly for the first socialist economy.[64] The question, then, was not whether a central organ should be created for Soviet science, but *which* of the available candidates should be the central organ.[65]

By the time the Cultural Revolution came in the late twenties, this question was being answered on the basis of actual performance. The Communist Academy had proved its inability to perform research at a high level in the natural sciences, and the various early reform schemes to replace the traditional organization of science were now seen more and more as utopian dreams. When the assault on the Academy of Sciences came in 1929, the demands voiced by the attackers were not so much for a totally new institution as they were for access to an old institution by Marxist scholars and an integration of that institution with the new economic plans. Lenin's and Stalin's policy of preserving the old forms of intellectual and cultural institutions inherited from tsarism had by now survived the criticisms from the left and was firmly ensconced. The militants, still nursing grievances against these institutions, correctly saw that the conversion of the older scientists to the service of Soviet socialism was at best partial, but they were not able to enact a fundamental reform of science in the Soviet Union. Science that was traditional in its assumptions continued to prosper, interrupted only occasionally by ideological intrusions.

Paradoxically, it is possible that had the radical efforts to renovate Soviet science succeeded, the later political intrusions would have been less frequent. An early utopian effort to construct a totally new Soviet science, launched in the first years of the Soviet regime, would have released the tension behind the radical call for a proletarian science, would likely have gone to costly extremes that fully revealed the flaws in the initial romantic vision, and then would have led to a retreat to different, more realistic forms of scientific organization. Such was the case in France, where a true Thermidorean snapping-back of the elastic band occurred. In the Soviet Union by the time the Cultural Revolution came in the late twenties, the old scientific institutions were well established in the Soviet context. The ideologists knew, however, just how incomplete the transformation of Soviet science had been, and they continued to suspect the leading scientists of bourgeois deviations. This continuing suspicion provided a reservoir of potentialities for political intrusions in later years, such as the Lysenko affair in the thirties and forties, when academic biologists were accused of hampering the advancement of Soviet agriculture.[66]

Out of the first fifteen years of the existence of Soviet Russia an organization and ethos of science thus emerged that was an amalgam of prerevolutionary institutions and revolutionary ideology. The revolution that the most radical reformers hoped to attain was never completed, but the old forms were irretrievably altered. The result was a unique scientific establishment, one that remained in force until the reforms of the late 1980s and 1990s (see Chapter 9).

5

The role of dialectical materialism: The authentic phase

O NE of the remarkable aspects of the Russian Revolution was that it presented not only a prescription for a different political and economic order but also an alternative form of knowledge of the natural world; it called for a Marxist interpretation of nature consciously opposed to existing "bourgeois" descriptions. No other revolution in history contained a radical epistemological and cognitive system to the same degree. It is true that in both the English and the French revolutions one can see a few similar tendencies; several divines in the wake of the English Revolution in the seventeenth and early eighteenth centuries tried to construct new models of both the universe and the social order using Newtonian metaphors[1] and the encyclopedists and several of the later radical Jacobins of eighteenth-century France tried to develop distinct philosophies of nature.[2] In none of these previous instances, however, did the movements toward an alternative knowledge last as long nor go as deep as in the Russian case.

Of course, not all Soviet intellectuals who were sympathetic to Marxism after the Revolution supported dialectical materialism, the Bolshevik philosophy of science. And even among those who did support dialectical materialism there was considerable disagreement over just what it meant. Nonetheless, most of the major leaders of the Bolshevik Party in the early years of the Soviet state – including Lenin, Trotsky, Bukharin, and Stalin – espoused a distinctly Soviet Marxist philosophy of nature. Almost all of the Marxist philosophers did so as well. In these writings and speeches there was a common core of ideas, despite different emphases – ideas that have continued to have a life down to the present.

As a historical phenomenon the significance of dialectical materialism is still underestimated outside the former Soviet Union. One reason is that observers have often assumed that Marxism in the Soviet Union was similar to Marxism in the West – that is, a theory of economics and society. They often failed to take sufficient notice of the Soviet claim that Marxism has two halves: historical materialism, which is a theory of social development, and dialectical materialism, which is a philosophy of science. Another reason for the frequent lack of understanding in the West of Soviet dialectical materialism is that by the time the Western world learned much about Soviet philosophy of science it had already become a calcified and dogmatic system that hobbled science. The best-known case of such damage was the disaster of Lysenkoism in biology, but there were other ideological intrusions in science as well. As a result of these violations of the freedom of science, Western observers have almost entirely overlooked the more intellectually interesting side of dialectical materialism; this aspect of Soviet Marxism was illustrated by the existence of a number of talented Soviet scientists and philosophers who were dedicated dialectical materialists but who saw this system of thought as an innovative option, not as a scholastic dogma. Some of them were influenced by dialectical materialism as they wrote publications that won recognition by the scientific community. Many of them were horrified when this system of thought was used as a cudgel to enforce orthodoxy. Instead, they believed it should be treated as a strong philosophy of science that, when utilized in the way that scientists have used other philosophies of science, may help scientists to develop new and creative theories about human and physical nature.

In this chapter I will examine the works of several outstanding Marxist scientists of the former Soviet Union whose work was in one way or another influenced by the Marxist philosophy of nature. Their work was internationally recognized, although few people in the West believed that Marxist philosophy of science had anything to do with their work. On the contrary, without the proper recognition of the role of Marxism, the viewpoints of these scientists cannot be properly understood.

DIALECTICAL MATERIALISM

Dialectical materialism is a form of traditional philosophical materialism, one that postulates that all nature can be explained in terms of matter and energy. Objective reality exists, external to the human mind, and this objective reality obeys natural laws. Knowledge derives from the influence of the material world on the knowing subject, who is also, in the final analysis, a material being. Dialectical materialism denies divine influence on nature or the existence of a deity; it also opposes the view that there exist any forces or phenomena that are inaccessible, in principle, to scientific explanation.

In these characteristics dialectical materialism is quite similar to traditional materialism as espoused, for example, by the classical Greek atomists or nineteenth-century scientific materialists. However, dialectical materialists differ from most of these earlier materialists by their sharp criticism of reductionism, a belief that all phenomena in nature, including human behavior, can ultimately be explained in terms of the simplest interactions of matter. Greek atomists like Democritus and Leucippus were severe reductionists, as were nineteenth-century materialists such as Buchner and Moleschott. In contrast to the views of these scholars, dialectical materialists have differentiated various "levels of being" in nature, such as the physical, the biological, and the social. Phenomena on a higher level of being, such as the social level, cannot be exhaustively explained in terms of laws pertaining to a lower level, such as the physical level.

While dialectical materialists deny the validity of reductionism, they also oppose vitalism, the belief that the "higher" phenomena of nature, such as living beings and especially humans, are based on nonmaterial principles outside the explanatory schemes of science. Dialectical materialism is thus an effort to steer between the extremes of reductionism and vitalism, an effort that has often been described by its critics as erring to one side or the other. Like the reductionists, dialectical materialists believe that only matter and energy exist in the universe, but unlike the reductionists, they do not see physics as the most basic of all sciences, the eventual key to all scientific explanations. In the biological and social sciences in particular they have usually been quite skeptical that such phenomena as life and consciousness can be explained in elementary terms taken from physics, chemistry, or even biology. Instead, they look to "social principles," usually derived from Marxist social and economic theory.

Dialectical materialists are heavily committed to evolutionary viewpoints, not only the Darwinian form of biological evolution, but also to the evolution of nonliving matter, both before and after the origin of life in the history of the universe. The long-term development of matter during the history of the earth from the simplest chemical and geological stages on up through the origin of life and eventually to human beings and their social organizations is seen by dialectical materialists as a series of quantitative transitions involving correlative qualitative changes. These qualitative changes result in the operation of different natural laws on different "dialectical levels." The British Marxist scientist J. D. Bernal once described this aspect of dialectical materialism as "the truth of different laws for different levels, an essentially Marxist idea."[3] Within the framework of this conception, social laws cannot be reduced to biological laws, and biological laws cannot be reduced to physicochemical laws.

As a nonreductionist materialism, dialectical materialism attempts to preserve a place for uniquely human values while at the same time basing itself on the principle that nothing exists in nature but matter and energy. It tries to be compatible with science without surrendering to the reductionist sciences of physics and chemistry. Many of the controversies about dialectical materialism have emanated from the difficulties of performing this task.

The principle of nonreductionism in dialectical materialism provided valuable safeguards in the Soviet Union against scientistic exaggerations based on biological or behavioral explanations of social values (such as eugenics, sociobiology, behaviorism). None of these forms of reductionist scientific explanations of social life ever became as widespread or as popular in the Soviet Union as in some Western countries. Ironically, the same principle strengthened the authoritarian rulers of Soviet intellectual and political life by providing a justification for the existence of a special group of Marxist social theorists. Since dialectical materialism postulates the existence of different laws on each level of nature (physical, biological, social), it therefore seems to follow that each level of nature should be authoritatively interpreted by an individual group of specialists. Physicists were to stick to physics, while particularly qualified experts (read: Party theorists) evaluated and directed the social organism. Thus, the Marxist philosophy of science became useful as a state doctrine for rationalizing an authoritarian government in which a political party representing a minority of the population bore special responsibility for discerning and applying socioeconomic regularities.

The form of dialectical materialism developed in the Soviet Union can be summarized in the following tenets. Most of these principles are similar to the assumptions of many working scientists in countries all over the world, although most of these scientists do not consider themselves dialectical materialists:[4]

1. The world is material, and is made up of what current science would describe as matter-energy.
2. The material world forms an interconnected whole.
3. Human knowledge is derived from objectively existing reality, both natural and social; being determines consciousness.
4. The world is constantly changing, and, indeed, there are no truly static entities in the world.
5. The changes in matter occur in accordance with certain overall regularities or laws.
6. The laws of the development of matter exist on different levels corresponding to the different subject matters of the sciences, and therefore one should not expect in every case to be able to explain complex biological and psychological phenomena in terms of the most elementary physicochemical laws.
7. Matter is infinite in its properties, and therefore human knowledge will never be complete.
8. The motion present in the world is explained by internal factors, and therefore no external mover is needed.
9. The knowledge of human beings grows with time, as is illustrated by their increasing success in applying it to practice, but this growth occurs through the accumulation of relative – not absolute – truths.

Dialectical materialism evolved slowly during the Soviet period, and the complete system summarized here does not represent accurately the views of all persons who, at one time or another, called themselves dialectical materialists. The writings that were most important in the early years of the Soviet Union were Lenin's *Materialism and Empirio-Criticism* and Engels's *Anti-Dühring*. As David Joravsky has shown, many Marxists in the 1920s were rather simple materialists, or even mechanists, who did not put much emphasis on the nonreductive principles of dialectical materialism.[5] After Stalin consolidated his power in the late twenties, dialectical materialism increasingly was interpreted in terms of "dialectical laws": the notorious "Law of the Transformation of Quantity into Quality," the "Law of the Unity and Struggle of Opposites," and the "Law of the Negation of the Negation." Often these laws were interpreted in dogmatic fashion. A more subtle interpretation of the dialectical laws was held by sophisticated Marxists according to which the first of these laws was seen merely as a statement of nonreductionism; the second, an attempted explanation for the presence of energy in nature; and the third, an assertion of constant change in the universe.

LEV SEMENOVICH VYGOTSKY (1896–1934)

Perhaps no Soviet scholar better represents the generation of Marxists who tried to revolutionize knowledge in the 1920s and 1930s than the psychologist Lev Semenovich Vygotsky. Vygotsky is of interest today because his views are exercising great influence in education and the social sciences. The American philosopher of science Stephen Toulmin has called Vygotsky "the Mozart of psychology," comparing him to Freud and Piaget.[6] Jerome Bruner observed that "Vygotsky was plainly a genius."[7] James Wertsch asked in 1988, "Why is it that this Soviet scholar is having so much influence on Western thought more than a half century after his death?"[8]

Those few scholars who have carefully examined Vygotsky's original writings and his life agree that Marxism was an important stimulus to his thought. Yet many psychologists in English-reading countries have missed this element in Vygotsky's work because they knew only the first American translation of his classic *Thought and Language*, an abridged edition in which references to Marxism had been systematically deleted. The translators evidently believed these references were extraneous to Vygotsky's core ideas.[9] All references to Lenin, for example, were eliminated. Only in 1986, over a half-century after it was written, was the complete text of *Thought and Language* published in the English language.[10] In it the role of Marxism clearly emerges. Wertsch, an American psychologist who has studied Vygotsky's work carefully, stressed "Vygotsky's sincere dedication to creating a Marxist analysis of mind."[11]

In the original Russian it is clear that Vygotsky's effort to show the importance of sociocultural context for a theory of mind was based on the Marxist concept that "being determines consciousness." Vygotsky believed that the development of children's thought could be understood on the basis of Lenin's epistemology, which emphasized the influence of objective reality on the knowing mind. In particular, Lenin's emphasis provided a basis for criticism of the views of the leading French psychologist Jean Piaget, who spoke of children's "autistic" use of language without reference to the influence, from early in the child's development, of the environment. Vygotsky believed that Piaget's views were based on a form of epistemological dualism and idealism incompatible with Marxism. A Marxist approach, Vygotsky wrote, emphasizes the "external" or "social" origins of language, rather than the isolated and independent activity of the mind. Furthermore, Vygotsky noted Friedrich Engels's analysis of the importance of tools in human evolution as a similar Marxist emphasis on how higher mental functioning is mediated by the environment – in this case, socioculturally evolved tools. Vygotsky even cited Engels's dialectical laws by speaking of the "unity and struggle of the opposites of thought and fantasy" in cognition.[12]

Vygotsky came to maturity in Soviet Russia in the 1920s, a time of great enthusiasm among many young students and teachers for the Marxist reconstruction of knowledge. Russia had just passed through a war involving staggering losses, a traumatic revolution, and a divisive civil war. The problems of illiteracy and economic backwardness facing the country were enormous. Psychologists like Vygotsky who were committed to the goals of the new regime hoped to find a way of creating "a New Soviet Man" on the basis of a theory of mind that assigned primary importance to the role of society. If the development of individuals is based primarily on inherent personal characteristics, the prospect for rapid improvement seemed small; but if a reorganized society could exert strong influence on the development of the personalities of its citizens, the chances for a rapid transformation seemed much greater. This issue explains both Vygotsky's opposition to Piaget's concepts of "autistic behavior" and "inner speech" (both resistant to social influence) and his support for the view that mental development is heavily conditioned by its sociocultural context.

The problem that lay at the center of much of Vygotsky's work was the interrelation of thought and speech. He objected to Piaget's description of speech in a child as proceeding along an evolution from an early individualistic stage to a later social one. He believed that Piaget's approach was, at bottom, based on a Cartesian distinction between mind and body that Marxists reject. As Vygotsky described his differences with Piaget:

> The development of thought is, to Piaget, a story of the gradual socializa-
> tion of deeply intimate, personal, autistic mental states. Even social speech
> is represented as following, not preceding, egocentric speech.
> The hypothesis we propose reversed this course. . . . We consider that
> the total development runs as follows: The primary function of speech, in
> both children and adults, is communication, social contact. The earliest
> speech of the child is therefore essentially social. At first it is global and
> multifunctional; later its functions become differentiated. At a certain age
> the social speech of the child is quite sharply divided into egocentric and
> communicative speech. (We prefer to use the term *communicative* for the
> form of speech that Piaget calls *socialized* as though it had been something
> else before becoming social. From our point of view, the two forms, commu-
> nicative and egocentric, are both social, though their functions differ.) Ego-
> centric speech emerges when the child transfers social collaborative forms of
> behavior to the sphere of inner-personal psychic functions. . . . In our con-
> ception, the true direction of the development of thinking is not from the
> individual to the socialized, but from the social to the individual.[13]

This transition of language from the social environment to the mind of the child, where it has an impact on the very modes of thought of the child, Vygotsky called "the internalization of speech," the concept for which he is best known.

> Piaget believes that egocentric speech stems from the insufficient socializa-
> tion of speech and that its only development is decrease and eventual
> death. Its culmination lies in the past. Inner speech is something new
> brought in from the outside along with socialization. We believe that ego-
> centric speech stems from the insufficient individualization of primary
> social speech. Its culmination lies in the future. It develops into inner
> speech.[14]

Vygotsky's concept of the internalization of social speech presented him with a problem unlike that faced by Piaget. Does a child in its earliest months, before it has internalized any speech, have the ability to think? Vygotsky believed that such a child can indeed think, even though it possesses no language medium with which to think; it was necessary, therefore, for Vygotsky to find quite different roots for thought and language. Thought would originally come from the "in-side," while language would originally come from the "outside," al-though later thought and language would affect each other so intimately that the fact that they had separate roots would become almost invisible.

If prelinguistic thought comes from the "inside," how was Vygotsky's psychology any less "dualistic" or Cartesian than Piaget's? Was not this concept of the internal origin of prelinguistic thought at variance with Lenin's emphasis on the reflection in the mind of external reality?

Vygotsky was at pains to point out that prelinguistic thought had material origins in biological evolution and could also be explained in Marxist terms:

> The thesis that the roots of human intellect reach down into the animal realm has long been admitted by Marxism; we find its elaboration in Plekhanov [an early Russian Marxist]. Engels wrote that man and animals have all forms of intellectual activity in common; only the developmental level differs; animals are able to reason on an elementary level, to analyze (cracking a nut is a beginning of analysis), to experiment when confronted with problems or caught in a difficult situation. . . . It goes without saying that Engels does not credit animals with the ability to think and to speak on the human level. . . .[15]

Prelinguistic thought in the child, according to Vygotsky, is therefore somewhat similar to the embryonic thought of some animals. If so, was not Vygotsky erring to the side of vulgar materialism, failing to notice that dialectical materialism warned against reducing the "social" activity of humans to the "biological" activity of animals? Vygotsky turned to Marxism for his explanation, emphasizing that the dialectical interaction of thought and language transformed the child's thought processes into something qualitatively distinct from those of animals.

Vygotsky believed that the crucial moment for the child came when he or she realized that every object has a name. At this moment the curves of development of thought and speech meet for the first time, and from that time forward these curves never develop entirely separately. "Speech begins to serve intellect, and thoughts begin to be spoken."[16] Once the child has seen the link between word and object, thought becomes verbal and speech becomes rational.

Once again Vygotsky presented his views in Marxist terms, placing the development of the child's psychology within the framework of the irreducible "biological and sociohistorical levels of being" of dialectical materialism. He wrote that after the child has discovered the connection between objects and names:

> The nature of the development itself changes from biological to sociohistorical. Verbal thought is not an innate natural form of behavior but is determined by a historical-cultural process and has specific properties and laws that cannot be found in the natural forms of thought and speech. Once we acknowledge the historical character of verbal thought, we must consider it subject to all the premises of historical materialism, which are valid for any historical phenomenon in human society. It is only to be expected that on this level the development of behavior will be governed essentially by the general laws of the historical development of human society.[17]

Thus Vygotsky developed for the explanation of the interrelation of thought and language a scheme that contained a high degree of inner consistency and arrived eventually at Marxist conceptions of social development.

Vygotsky then took the analysis of thought and language to a higher stage, one in which higher mental functioning is heavily influenced by culturally conditioned language. As the child learns to read and becomes more sophisticated, language influences his or her thought in ever more subtle ways. Growing to adulthood, his or her mode of thinking is heavily influenced by all the literary and cultural media of his environment. A corollary of this thesis is that people in distinctly different environments, for example, those in advanced industrial societies and those in primitive tribal ones, think in distinctly different ways. In order to understand better the way in which thought and language interact in this most advanced stage of the adult in modern society, Vygotsky paid much attention to literary analysis, semiotics, and linguistics. His early work was known to the great Soviet linguist Mikhail Bakhtin, who similarly emphasized the influence of society on modes of thought. Bakhtin has also in recent years attracted great attention in the West, becoming almost a cult figure among some intellectuals.[18]

At the time that Vygotsky developed these views, in the late twenties and early thirties, they were a brilliant breakthrough in the analysis of thought and language. They opened up new areas of research in the Soviet Union and abroad. Vygotsky launched a major new school in Soviet psychology. Several of his students, including A. R. Luria and A. N. Leont'ev, would dominate Soviet psychology in the post–World War II period. Luria before his death in 1977 became a world figure in neuropsychology.

Vygotsky's theories about early child psychology were clearly connected with Marxism as he interpreted it. But under Stalin, independent thinkers like Vygotsky frequently suffered. Stalin fancied himself an intellectual and commented without hesitation on many scientific theories. During the time when Stalinist influences were dominant in Soviet scholarship, from 1936 to 1956, Vygotsky's writings were in disfavor. In 1950, in his *Marxism and Linguistics*, Stalin wrote that "bare thoughts, free of the language material . . . do not exist," contradicting Vygotsky's conception of prelinguistic thought.[19]

After Stalin's death, a rebirth of Vygotsky's influence gradually occurred. By the 1960s Vygotsky was the major force in Soviet psychology even though he had been dead for thirty years. A. R. Luria remarked, "All that is good in Russian psychology today comes from Vygotsky,"[20] and dedicated his important monograph *Higher Cortical Functions in Man*, published in Moscow in 1962, to Vygotsky's memory. Luria observed that his own work could in many ways "be looked upon as a continuation of Vygotsky's ideas."[21]

Western interest in Vygotsky began in 1962 with the publication in

English of the abridged version of his *Thought and Language,* gained strength from the attention drawn to his views by Jerome Bruner in the sixties, and grew further as a result of the work of Michael Cole and his coauthors (Cole is an American who studied in Moscow in the 1960s with Vygotsky's student Luria).[22] By the late 1980s Vygotsky was being discussed so widely in the United States that Wertsch objected that "distortions are already becoming manifest as American psychologists try to use Vygotsky's approach piecemeal or to unwittingly 'individualize' a theory that is grounded in inherently collectivist assumptions."[23]

Some of Vygotsky's views are now being questioned by psychologists placing far more emphasis than he did on the genetic foundations of psychology. The central point for the historian about Vygotsky, however, is that he advanced a brilliant theory of psychology in which social influences, and, particularly, the theories of Marxism, played central roles. This fact is a valuable antidote to the common Western view that the influence of Marxism on Soviet science has been uniformly destructive, as illustrated by the Lysenko affair.

ALEKSANDR IVANOVICH OPARIN (1894–1980)

The two scientists responsible for reawakening interest in the twentieth century in the topic of the origin of life were Aleksandr Ivanovich Oparin, a Russian, and J. B. S. Haldane, a Scot. Both presented materialistic hypotheses and both subsequently declared that Marxism had been an important influence on their scientific thought. Although his approach possessed several interesting features of its own, Haldane graciously admitted that Oparin had precedence over him. Oparin's work, like Vygotsky's, is an example of early Soviet research in which dialectical materialism was an intellectual component.

The possibility that Marxism may have played a role in the early work on origin of life has tantalized a few scientists and historians of science. The distinguished British geneticist C. H. Waddington wrote in 1968:

> In the late Twenties and early Thirties the basic thinking was done which led to the view that saw life as a natural and perhaps inevitable development from the non-living physical world. Future students of the history of ideas are likely to take note that this new view, which amounts to nothing less than a great revolution in man's philosophical outlook on his own position in the natural world, was first developed by Communists. Oparin of Moscow, in 1924, and J. B. S. Haldane, of Cambridge, England, in 1929, independently argued that recent advances in geochemistry suggested that the conditions on the surface of the primitive earth were very different from those of today, and were of a kind which made it possible to imagine the origin of systems that might be called "living". . . .[24]

Oparin was a prominent biochemist in the Soviet Union for decades. He was a member of that generation of young Russian intellectuals fundamentally affected by the growth of political radicalism in the last years of tsarism and during the Russian Revolution. Oparin believed that radicalism and science fit together hand in glove. As mentioned in Chapter 3, in an interview in Moscow in 1971 Oparin related how, as a boy before the Revolution, he was thrilled by the lectures on evolution given by "Darwin's Russian Bulldog," K. A. Timiriazev.[25] Oparin then went on to study under the political revolutionary A. N. Bakh, who published on Marxism as early as the 1880s.[26] At the time of the Russian Revolution Oparin was in his early twenties, and by then was quite willing, even eager, to apply radical ideology to scientific research. In numerous books and articles over the next half-century he wrote about the relevance of Marxism to biology. He eventually became one of the best-known biologists in the Soviet Union, a professor at Moscow University, a full member of the Academy of Sciences, and a leading administrator of biological research. At the height of Stalinism he supported Lysenko and discredited himself in the eyes of Soviet geneticists. However, he never incorporated Lysenko's views into his own research. After Stalin's death, Oparin resisted several of Lysenko's followers.[27]

Although Oparin was a political radical from his early youth, at the time of his first publication on the origin of life in 1924 he had no deep knowledge of Marxism or of dialectical materialism. In these early years he equated radicalism with nineteenth-century materialism, what the more sophisticated Marxist philosophers called "vulgar materialism." In other words, he was a thorough reductionist. Many people in Russia immediately after the Revolution believed that Marxism was based on materialism and determinism precisely in this nineteenth-century sense, and it is likely that the currency of such views influenced Oparin's thought. As he learned more and more about Marxism in the late 1920s, Oparin incorporated the nonreductive principles of dialectical materialism into his work.

Oparin's approach was materialistic, but it was not based on the simple hope for spontaneous generation that had marked such earlier speculators as Felix Pouchet in his famous debate with Louis Pasteur in the 1860s. Oparin noted that all living microorganisms, including those that Pouchet described, are extremely complex bits of matter. It seems improbable that a highly ordered piece of protoplasm capable of supporting the coordinated processes of metabolism could accidentally have arisen from a relatively disordered and formless mixture of organic compounds, as Pouchet supposed. Such an assumption, believed Oparin, required a metaphysical leap, a violation of the scientific approach of finding the simplest and most plausible explanation of natural phenomena.[28] A much more sensible avenue of investigation, continued Oparin, was to go back to the very simplest forms of matter and to extend the Darwinian principles of evolution to inanimate matter as well as animate matter. Oparin declared his intention of tying the "world of the living" and the "world of the dead" together by examining them both in terms of their historical development. He noted that the origin of any finely structured entity, alive or dead – whether a one-celled organism, a piece of inorganic crystal, or an eagle's eye – seems inexplicable unless it is examined in historical, evolutionary terms. He declared his intention of finding the origins of the simplest living beings not by starting with their immediate environments, as Pouchet had unsuccessfully attempted, but by delineating the extremely long evolutionary prehistories of such organisms, including environments strikingly different from their present ones.

The hypothesis that the earliest prehistory of life occurred in environments quite different from the present ones was a thought that Oparin subsequently developed into the Marxist view that the evolution of life passed through several "levels of being" that were necessary for its origin. For example, he believed that the prior nonexistence of life was one of the necessary conditions for the origin of life, and that consequently, now that life exists on earth it cannot originate again, or, at least not in the same way in which it first did. The reason for the non-repeatability of the phenomenon was that in a world in which life already exists the ubiquitous bacteria or other microorganisms would consume any primordial life substances before they could develop to the stage of the truly living.[29] In subsequent editions of his work, Oparin explained this process within the framework of dialectical concepts of natural law: On each level of being different principles prevail; therefore, the laws of chemistry and physics that operated on the earth in the absence of life were different from, and superseded by, the biological laws that qualitatively emerged with the appearance of life. In the case of man, the biological laws were transcended, in turn, by social ones.

Oparin favored a colloidal suspension as the most likely medium for the origin of life. He noted that such suspensions in which complicated organic molecules are present are unstable and frequently form coagula or gels. In subsequent years Oparin's hypothesis of the origin of life was frequently described as a gel or "coacervate" theory. This aspect of Oparin's hypothesis later incurred sharp criticism from other biologists, and is currently not considered a strong contender among rival views on the origin of life. Oparin is still given credit, however, for awakening interest in the problem and for presenting one of the first plausible twentieth-century hypotheses describing how life might have arisen.

Throughout his long life Oparin's interest in the philosophic aspects of Marxism continued to deepen, and his writings clearly illustrated his melding of Marxism and biology.[30] Dialectical materialism heavily influenced the very structure of his analysis and the organizational schemes of his books.[31] Oparin became not only a nonreductionist, but a vehement one. The point to which Oparin returned again and again is that dialectical materialism is a *via media* between the positions of frank idealists and vitalists on the one hand, and mechanistic materialists, exuberant functionalists, and supporters of spontaneous generation on the other. He asserted that dialectical materialism was indeed a form of materialism and therefore opposed to the idealistic view that the essence of life was "some sort of supramaterial origin which is inaccessible to experiment."[32] But dialectical materialism was equally opposed to the view that all living phenomena could be explained as physical and chemical processes. To take the latter position would mean, said Oparin, "to deny that there is any qualitative difference between organic and inorganic objects. We thus reach a position where we must say either that inorganic objects are alive or that life does not really exist."[33] Dialectical materialism provides a means, Oparin continued, of accepting the principle of the material nature of life without regarding "everything which is not included in physics and chemistry as being vitalistic or supernatural."[34] To dialectical materialists, life is a "special form of the motion of matter," one with its own distinct regularities and principles. Life was a flow, an exchange, a dialectical unity: "An organism can live and maintain itself only so long as it is continually exchanging material and energy with its environment."[35]

These points of view may have helped Oparin as he labored in the thirties and forties over the evolution of coacervates, but they got him into difficulty in the fifties and sixties as the field of molecular biology suddenly blossomed. The structure of DNA as presented by Watson and Crick was definitely mechanistic, and their approach was thoroughly reductionist. They regarded Oparin's approach as unrigorous and speculative. Furthermore, crystallized viruses, bits of DNA, were often considered by molecular biologists to be alive, yet they clearly did not fit Oparin's definition of life as "a flow, an exchange, a dialectical unity." These differences brought Oparin into conflict with some of the molecular biologists. Oparin believed, for example, that there were serious reasons for excluding viruses from the realm of the truly living, since they can exist in crystalline form. To Oparin, metabolic flow was the essence of life.

This period of controversy coincided with the Lysenkoite period in Soviet biology, to be discussed in the next chapter. Politically, Oparin and Lysenko were linked, however far apart they may have been in intellectual sophistication. Both had won favor from the Stalinist regime, both had built their careers within it, and around both there had arisen schools of biology that were officially described as "Marxist-Leninist." They benefited from the government, and they repaid the government in political praise and cooperation. As a high administrator in the biological sciences in the Soviet Union during the late Stalinist and early post-Stalinist years, Oparin was involved in the perpetuation of the Lysenko school.[36] The emigré Soviet biologist Zhores Medvedev wrote in his history of Lysenkoism that in 1955 a petition directed against the administrative abuses of both Lysenko and Oparin was circulated among Soviet scientists.[37] Medvedev reported that in the final struggle with Lysenko, Oparin took a neutral position.[38]

Oparin spanned the period of Soviet science from the early time when serious intellectuals were still interested in Marxism and believed that it could help them in their work to the later time when the calcification of Marxism in the Soviet Union and simultaneous political repression caused most independent thinkers to turn away from Marxism in anger and disgust. Oparin's most creative work coincided with the early period.

V. A. FOCK (1898–1974)[39]

Vygotsky and Oparin were Soviet scientists whose initial interest in Marxism developed in the twenties and early thirties. They were attracted to the relationship of Marxism and science in the early, idealistic period of Soviet history before Stalinist ideological controls squeezed out much of the intellectual content in dialectical materialism. The noted Soviet physicist Fock came to the topic later, and for a somewhat different reason. In his early years Fock, like many young physicists, was interested primarily in physics, not in philosophy. He was a member of that generation of physicists who wrestled with the development and refinement of relativity physics and quantum mechanics. He had excellent connections with West European physicists like Niels Bohr and Werner Heisenberg, spent time at Bohr's institute in Copenhagen, and was thoroughly in step with the development of modern physics. He defended relativity physics and quantum mechanics strongly. Along with such other leading Soviet physicists as M. P. Bronshtein, L. D. Landau, and I. E. Tamm, he was a member of what was often called the "Russian branch" of the Copenhagen School of physics.

In the mid-1930s Fock's interest in philosophy and Marxism began to grow. In precisely these years militant Marxist ideologues in the Soviet Union, of whom Fock was certainly not one, began a series of attacks on modern physics, maintaining that the prevailing interpretations of quantum mechanics and relativity physics revealed the "bourgeois, idealistic" presuppositions of their authors. The Marxist critics said that Western physicists such as James Jeans, Arthur Eddington, and even Bohr himself were trying to use physics to destroy Marxism; these Western authors asserted that relativity meant the disappearance of matter and the end of philosophic materialism, and that quantum mechanics discredited causality and therefore contradicted philosophical determinism. Marxists could not remain neutral in this struggle, the ideologues maintained, since dialectical materialism is based on materialism and determinism.

There was an element of truth in the Soviet ideologues' charges, even though they often grossly and unfairly exaggerated this element. Arthur Eddington, for example, was a devout Quaker whose hostility to philosophical materialism was no secret and who used the new developments in physics to make his philosophical and religious points. In the Gifford Lectures in 1927, in which he discussed relativity physics, he stated:

> I do not claim spiritual power only for the religious mystic; but the closing of the eyes to the overwhelming supremacy of that power over material and even intellectual power is the gross error of the widespread materialism of this and other ages – a materialism far more deadly than the philosophic doctrine of that name which I have been combatting in earlier parts of these lectures.[40]

And Bohr in a heady moment in 1927 asserted that quantum mechanics showed "a complete rupture" with causal descriptions,[41] a position from which he later retreated. Many Western physicists enjoyed the popular resonance that came from proclaiming to the general public that the new physics showed that the old reputation of physical scientists as heartless materialists who allowed no room for free will or the human spirit was no longer accurate. Paul Forman has shown how this effort to please the public was particularly strong among physicists in Weimar Germany.[42] In Revolutionary Russia the popular resonance was quite different.

So long as these issues were mere subjects of intellectual discussion physicists like Fock were not much concerned. However, in the Cultural Revolution of the late twenties and early thirties, the tone of the Soviet criticisms of Western physics became much nastier and threatening. For the first time the possibility arose that an alliance of Marxist ideologues and Stalinist bureaucrats would result in restrictions on physics itself, perhaps even to the proscription of relativity physics and quantum mechanics.

Fock was one of the most prominent of a group of talented Soviet physicists who believed that the alleged opposition between Marxism and physics was false. Fock was a materialist and he was sympathetic to the goals and ideology of the Soviet regime. At the same time, he considered quantum mechanics and relativity physics the greatest achievements of twentieth-century physics. He was convinced that a careful examination of the new physics would show that it did not contradict the principles of dialectical materialism. The problem, he thought, was that Marxist philosophers with little knowledge of physics were writing the influential Soviet commentaries on the disputes. He decided that he would enter the fray and remove the misunderstandings based on ignorance of physics.

This effort turned out to be quite complex. Fock could not simply defend all the prominent Western proponents of the new physics, for many of them had, in fact, clothed their expositions of physics in their personal philosophical viewpoints, many of which were contrary to Marxism. Fock, then, had to engage in a long scholarly effort, one that lasted for decades, to separate the physics from its philosophical clothing – or, more accurately, to replace the old philosophical clothing of the physics with the new one of dialectical materialism. As Fock observed about quantum mechanics in 1963, "Having begun, like many physicists, with a formal application of the mathematics of quantum mechanics, I later began to think about philosophical questions. . . . I finally came to the conclusion that Bohr's views could be completely separated from the positivistic coating that at first glance seemed to be intrinsic to it."[43]

Many Western scholars who came upon Fock's writings in which he attempted this philosophical reformulation discounted them as mere ideological ornamentation. A more accurate assessment would be to grant Fock the same right to his philosophical commitment to Marxism that one grants to, for example, Eddington's Quakerism, even if one does not subscribe to either view.

In quantum mechanics Fock decided that he would redefine the Copenhagen Interpretation in terms of its minimum rather than maximum claims. (This "core meaning" of the Copenhagen Interpretation was once described by the American philosopher N. R. Hanson as "a much smaller and more elusive target to shoot at than the *ex cathedra* utterances of the melancholy Dane."[44]) Fock believed that the way to do this was to show that quantum mechanics left plenty of room for a realist or materialist interpretation of nature, and for a continued assertion of causality, suitably reformulated.[45]

In order to pursue this goal, Fock went to Copenhagen in February and March 1957 and engaged Bohr in a series of long discussions on the philosophical meaning of quantum mechanics. The discussions occurred in Bohr's home and at the Institute of Theoretical Physics, which he headed. Fock later wrote about the conversations:

> From the very beginning Bohr said that he was not a positivist and that he attempted simply to consider nature exactly as it is. I pointed out that several of his expressions gave ground for an interpretation of his views in a postivistic sense that he, apparently, did not wish to support. . . . Our views constantly came closer together; in particular it became clear that Bohr completely recognized the objectivity of atoms and their properties . . . ; he further said that the term "uncontrollable mutual influence" was unsuccessful and that actually all physical processes are controllable.[46]

Soviet writers declared that from this time onward, Bohr's views became acceptable to them, and they claimed at least partial credit for the alleged change. Fock, for example, said that "After Bohr's correction of his formulations, I believe that I am in agreement with him on all basic items."[47] The Soviet philosopher Omel'ianovskii maintained that after publication of Fock's criticisms Bohr "made a definite advance toward a materialist approach to quantum mechanics."[48]

This version of events is misleading, for it portrays Soviet dialectical materialists as having been correct all along about quantum mechanics, and it describes Bohr as having "come over to their side." In fact, by far the most significant changes in position on quantum mechanics during the last fifty years were among the dialectical materialists rather than among the supporters of the Copenhagen School. Not only did the dialectical materialists repudiate the concept of complementarity for many years, they even on occasion denied the legitimacy of quantum mechanics as a science. There were moments in the 1930s and 1940s when it appeared that the dialectical materialist criticism of quantum mechanics might result in the suppression of the theory, just as actually happened in the case of genetics. Fortunately, the intellectual tragedy of genetics was not repeated in physics.

On the specific point of Bohr's attitude toward the concept of "uncontrollable influence," however, the Soviet commentators made claims that deserve to be examined more carefully than they so far have been. It *does* appear that Bohr changed his mind on this subject. Writing in 1935, Bohr spoke of "the impossibility, in the field of quantum theory, of accurately controlling the reaction of the object on the measuring instruments, i.e., the transfer of momentum in case of position measurements, and the displacement in case of momentum of measurements."[49] Elsewhere in the same essay he wrote that "the impossibility of controlling the reaction of the object on the measuring instruments" entails "a final renunciation of the classical ideal of causality and a radical revision of our attitude towards the problems of physical reality."[50]

Many years later, in Bohr's 1958 essay titled "Physics and Philosophy" (written after the discussions with Fock) we find a different tone. Here Bohr did not use the concept "uncontrollable interaction"; furthermore, he said that he was against using expressions like "disturbance of phenomena by observation" or "creation of physical attributes by measurements." In addition, he added that the description of atomic phenomena "has in these respects a perfectly objective character, in the sense that no explicit reference is made to any individual observer."[51] And, on the subject of causality, he concluded "far from involving any arbitrary renunciation of the ideal of causality, the wider frame of complementarity directly expresses our position as regards the account of fundamental properties of matter presupposed in classical physical description, but outside its scope."[52]

Soviet philosophers described this essay as a fundamental change of position by Bohr, but Bohr's son Aage has explained it in somewhat different terms.[53] Writing shortly after his father's death, Aage Bohr said that "My father felt that, in this little article, he had succeeded in formulating some of the essential points more clearly and concisely than on earlier occasions."[54] The question therefore arises: Does the 1958 article represent a change in philosophical position when compared to the 1935 one, as the leading Soviet interpreters of quantum mechanics maintained, or is it merely a clarification, as indicated by Aage Bohr?

More work on this issue is in process, but it already seems rather clear that some transition did occur in Bohr's thought, a transition away from emphasis on the interaction of the measuring instrument and the micro-object as the key to quantum mechanics, away from a renunciation of causality, and toward a greater recognition of the physical reality of the microbodies of quantum mechanics. And it is entirely possible that the conversations with Fock were an important cause for Bohr's shifts, although proving this causation would be quite difficult.

Fock attempted a similar reformulation of Einstein's relativity physics.[55] Fock noted that far from diminishing the importance of materialism, relativity physics heightens it, since according to general relativity the density of matter in the universe determines the configuration of space-time. Matter therefore acquires a significance of which eighteenth- and nineteenth-century materialists could not have dreamed. Fock indicated that he believed that Einstein made a mistake in placing the term "relativity" in the name of his great theory, since this led some people to conclude falsely that the theory implied that "everything is relative," when actually the theory is based on an absolute standard, that of space-time. Fock preferred the term "theory of gravitation" to the term "general theory of relativity," and used the former term in his own publications, a number of which were translated into other languages and attracted attention in the West. In many of these publications he advanced a rather idiosyncratic mathematical method, that of harmonic coordinates, designed to show that there is a preferred system of coordinates for use in relativistic descriptions of the universe. Here again Fock was attempting to show that physical relativity does not result in philosophical relativity.

Prominent world physicists disagree about Fock's criticisms of the term "general relativity." His interpretation has been challenged both in the former Soviet Union and abroad. One should notice that it was mainly a philosophical and methodological critique of relativity, not a fundamental revision of the theory. Fock sought a way of maintaining all the potency of relativity theory for working scientists while changing some of its terms and philosophical assumptions. Fock's variation of the theory commanded respect and attention as a defensible and interesting point of view. At international conferences Fock's approach was regarded as a serious one, and won at least partial praise from such prominent Western scientists as Hermann Bondi of England, André Lichnerowicz of France, and Stanley Deser and John Wheeler of the United States.[56] Usually, of course, Fock's effort at these conferences was judged entirely on the basis of its physical validity. Few people in the West understood or cared that one of Fock's major goals was to show the compatibility of modern physics with dialectical materialism.

In this chapter I have concentrated on the work of three Soviet scientists of international rank – the psychologist Vygotsky, the biologist Oparin, and the physicist Fock – who exemplified that group of Soviet scholars who considered Marxism important to their work. Scientists of this type were more common in the early years of the Soviet Union than after Stalinist dogma destroyed the creative spirit in Soviet Marxism. However, scientists of this sort existed throughout the entire history of the Soviet Union, and even in the former Soviet Union today there are some who look upon Marxism, shed of its Stalinist trappings, as a helpful philosophic framework. They included the psychologists A. R. Luria, S. L. Rubinshtein, and A. N. Leont'ev; the physiologist P. K. Anokhin; the biologists A. S. Serebrovskii, I. I. Agol, and the young N. P. Dubinin; the mathematicians A. D. Aleksandrov and A. N. Kolmogorov; the astronomer-mathematician O. Iu. Shmidt; the physicists S. Iu. Semkovskii, D. I. Blokhintsev, and G. I. Naan; the astrophysicists V. M. Ambartsumian and A. L. Zel'manov; and many others.[57]

Two of the scientists with interests in Marxism, O. Iu. Shmidt (1891–1956) and A. N. Kolmogorov (1903–1987), deserve a bit more discussion. Like Vygotsky, Oparin, and Fock, both of these men were internationally famous. Shmidt was better known as a polar explorer than as a mathematician or astronomer, although he was active in all three fields. Kolmogorov was one of the century's great mathematicians, known throughout the world of mathematics.

In addition to being an active researcher, Shmidt served as editor of the first edition of the *Large Soviet Encyclopedia*. This edition of the major Soviet reference work will always be regarded as the classic one of the revolutionary period, retaining interest even yet as a novel effort to present an alternative form of knowledge on all cultural and intellectual issues. The volumes that were published before 1930 are the most interesting, since the later ones fell under the Stalinist controls that spread throughout Soviet life, extinguishing the spirit of innovation and experimentation that had been present in the Soviet cultural and intellectual worlds in the twenties.

Shmidt's intellectual agenda contained both the daunting task of editing a revision of universal knowledge in the major Soviet encyclopedia and also the more narrow one of applying a Marxist critique to his own specialty, the origin of the solar system.[58] Shmidt believed that the most popular works on astronomy of his day – those written by the British astronomers James Jeans and Arthur Eddington – pandered to the bourgeois prejudices of Western society by surrounding their scenarios of the creation of the planetary system with a miraculous aura. According to Shmidt, these Western writers relished the rarity of events contained in their descriptions of cosmogony, emphasizing such improbable occurrences as the grazing of two stars, and implying that perhaps only the "finger of God" could account for such happenings. Shmidt, in his own writings on the origin of the planets, eschewed all elements of mystery and gave a highly naturalistic and plausible account. While his writings seem dated in the light of the wealth of astronomical data that has been collected since his time, his critique of Jeans and Eddington contained some valuable elements, even if Shmidt often mistook their moments of poetic license for serious scientific thought.

A. N. Kolmogorov, one of Shmidt's authors in the first edition of the *Large Soviet Encyclopedia*, wrote the entry "Mathematics." In fact, Kolmogorov was the author of the article on mathematics in all three editions of that encyclopedia, his last version appearing in 1974. I will briefly compare Kolmogorov's article with those in the *Encyclopedia Britannica* written by Frank Plumpton Ramsey and Alfred North Whitehead at approximately the same time as the first edition of Kolmogorov's article.

The points of difference arise on the most essential questions of mathematics: What are the origins of mathematics? and What is the relationship between mathematics and the real world? According to Kolmogorov, mathematics is "the science of quantitative relations and spatial forms of the real world."[59] It arose out of "the most elementary needs of economic life," such as counting objects, surveying land, measuring time, and building structures. In later centuries mathematics became so abstract that its origins in the real world were sometimes forgotten by mathematicians, but Kolmogorov reminded them that "the abstractness of mathematics does not mean its divorce from material reality. In direct connection with the demands of technology and science the fund of knowledge of quantitative relations and spatial forms studied by mathematics constantly grows."[60] Kolmogorov then went on to sketch a history of mathematics in which its growth was intimately related to economic and technological demands. His views were consistent with Lenin's insistence on the material world as the source of human knowledge, and Engels's emphasis on technical needs as a motivating force in the development of knowledge. In the articles in all three editions of the encyclopedia, Kolmogorov quoted Engels's views on how mathematics was a reflection of material relationships and answered practical needs in its early history, later grew to be a highly abstract field, but never lost its organic tie to material reality.

Looking at the *Encyclopedia Britannica* entries "Mathematics, Foundations of" and "Mathematics, Nature of" written by Ramsey and Whitehead in roughly the same years, one sees a very different analysis. According to Ramsey, mathematics is not a reflection of material relationships, but a logical system about which truth or falsity is not an important question for the mathematician. Ramsey asserted, "as a branch of mathematics, geometry has no essential reference to physical space";[61] the mathematician "regards geometry as simply tracing the consequences of certain axioms dealing with undefined terms, which are really variables in the ordinary mathematical sense, like x and y. And he demands of his axioms, not that they should be true on some particular physical interpretation of the undefined terms, but merely that should be *consistent* with one another."[62] In a somewhat similar vein, Whitehead defined mathematics as the "science concerned with the logical deduction of consequences from the general premises of all real reasoning," with no reference to the influence of the material world. In fact, Whitehead maintained that the "act of counting" is "irrelevant to the idea of number."[63]

Marxism continued to have some influence in science in the former Soviet Union even quite recently, and the influence could not always be dismissed as harmful or trivial. In his 1988 best-selling book *A Brief History of Time* the British astrophysicist Stephen W. Hawking linked some Soviet opposition to the "big bang" theory of the universe to Marxism. Some of his Soviet colleagues were not willing to accept easily the idea of a beginning to all time, seeing a possible link to religion or mysticism. Hawking was in close contact with a number of these Soviet researchers, including E. Lifshitz, I. Khalatnikov, and A. Linde, and eventually came to oppose the big bang theory himself, supporting instead a version of an "inflationary model" worked out by Linde and others in the 1980s. Meanwhile, historians of science in the West began to examine the possibility that the popularity of the big bang theory in Western Europe and North America in the middle of this century may indeed have had something to do with religious and cultural attitudes.[64]

In this chapter I have emphasized what I have called the "authentic phase" of dialectical materialism, those Soviet writings in which Marxism was a spur to criticism and originality. There is, unfortunately, another side to the story of Soviet Marxism and science, a much darker, catastrophic side. In the repressive atmosphere fostered by Stalin, unprincipled scientists soon learned that Marxism could be used as a cudgel to overwhelm their opponents. Many Marxist philosophers and even quite a few natural scientists criticized their colleagues for expressing "anti-Marxist" views in the hope that they would win the favor of the political authorities. Often they succeeded, and these ideological campaigns resulted in the imprisonment and deaths of many innocent scientists. This story represents the most egregious trampling of the principles of scientific freedom in the history of science. In the next chapter we will consider the case of Trofim Lysenko, the pseudogeneticist who was the primary actor in the most tragic of these cases.

6

Stalinist ideology and the Lysenko affair

UNDER Stalin, dialectical materialism was used to terrorize scientists. If a certain scientific theory was branded by Stalin's ideologists as "idealistic" or "bourgeois," the scientist who continued to defend that theory immediately came under suspicion of political disloyalty. The scientist would be subjected to humiliating criticism at political meetings in his or her institute or laboratory. Demotion or imprisonment (often the same as a death sentence) were then clear possibilities, although the pattern of criticism and subsequent purge was never consistent enough to be entirely predictable. Uncertainty about one's fate was one of the methods used by the security organs to destroy independence of spirit.

I have divided the history of dialectical materialism into two phases, the "authentic" and the "calcified" periods. It was much easier to be an authentic dialectical materialist before the purges of the 1930s. The idealism and sincerity of Soviet Marxism died with the multitude of victims in the cellars of the Lubianka, headquarters of the secret police in Moscow, and with the even larger numbers who succumbed in labor camps (*gulag*) spread throughout the "Gulag Archipelago" of the Soviet Union. Yet the authentic and calcified periods cannot be cleanly divided in chronological terms. Even in the 1920s there were careerists and dogmatists as well as true scholars, and even at the height of Stalinist dogmatism, in the late 1940s, there were genuine scientists who continued to believe that Marxism was a defensible and valuable philosophy of science. When I speak of authentic dialectical materialists, then, I mean those people who tried, at whatever moment of time, sincerely to unite Marxism and science; when I discuss the dogmatists I mean those people, at any moment, who used Marxist philosophy for primarily political goals and had little interest in the intellectual sides of questions.

In this chapter I will discuss the dogmatists who, led by Stalin, brought about the calcification of Soviet Marxism, an unattractive yet fundamentally important aspect of Soviet science. Even today, science in the former Soviet Union has not fully recovered from the damage it suffered in those years, though the situation has vastly improved. Some of the Stalinist habits of administration and even of thought still play roles in Russian science. In recent years much new information on the damaging influence of the dogmatists has appeared.[1]

The episode of this period best known in the West is Lysenkoism in genetics. Lysenkoism was only the most extreme of many manifestations of philosophical dogmatism and political oppression under Stalin.

In the 1920s in the natural sciences ideology was not an intrusive influence. If some scientists, such as Oparin, Shmidt, and Vygotsky, were interested in the Marxist philosophy of science, their interests derived largely from their own intellectual pursuits; Marxism was not forced upon working scientists. This began to change as the political atmosphere of the whole Soviet Union became much more strained. After triumphing over his rivals Trotsky and Bukharin, Stalin adopted the most sweeping and traumatic of the available plans for changing industry and agriculture, which required the total mobilization of society. Collectivization was a violent trauma for the entire country, involving the dislocation of millions of peasants, and the frequent use of troops or police to ensure obedience to the new rules. Such changes echoed throughout Soviet society, even the relatively insulated world of science. In these years the Academy of Sciences and the universities were purged and reorganized.

The intellectual tone of academic discourse changed. The shifts were most dramatic in the social sciences, but they could be seen in the natural sciences as well. The historian who today leafs through Soviet journals of the late twenties can easily perceive a transformation around 1929, the year of "the Great Break." Before that date the contents of the journals are heterogeneous in outlook, and genuine intellectual controversies occur. After 1929 the journals become thinner and a veil of orthodoxy is pulled over all discussions.

By the end of the twenties and the beginning of the thirties ideologists in the Soviet Union often classified science *itself* as "bourgeois" or "idealistic," rather than just some *interpretations* of science. This was a major transition. Even Lenin had usually made a distinction between science proper and the interpretations appended to it by philosophers or philosophically minded scientists. By the midthirties that distinction was no longer so clear. Relativity theory, quantum mechanics, and Mendelian genetics were increasingly labeled by ideologists as inherently linked to the capitalist world in which they originated. The scientists in these fields overwhelmingly rejected these criticisms while they went on with their work, but a feeling of unease about possible intrusions into the intellectual core of science began to spread among Soviet researchers.

Theoretical scientists in the thirties were also worried about the increasing emphasis by Communist Party and government leaders on applied science and engineering rather than fundamental research. To a certain extent, this reorientation in science policy was understandable.

Like many underdeveloped nations with only a thin stratum of people with university educations, tsarist Russia had a stronger tradition in theoretical subjects than in industrial engineering. Had the turn toward engineering been taken at a more moderate pace, with careful attention to the maintenance of quality in both fundamental and applied science, it would have been a valuable reform. However, the new priority was forced through so rapidly that it resulted both in poorly trained engineers and erosion of some of the traditional strengths of Russian science in fundamental areas. Outstanding theoretical scientists were pressured to give much time to industrial consultation, even if they were not best suited for this role.

Many Western commentators on the Soviet Union have noted that despite Marx's praise of internationalism and his disdain for patriotism and chauvinism, national pride was a prominent part of Soviet political culture. Not surprisingly, the nationalistic element became particularly prominent during and after World War II.

By the late thirties and forties the Stalinist system of controls over Soviet intellectual life was complete. The purges had destroyed the will of most people to resist. Administrators on lower levels of power looked to their superiors for signals indicating current policy, and hurried to obey as soon as the signals were discernible. The granting of higher degrees, personnel assignment and promotion, scientific publishing, academic research and instruction – all were subject to the control of Party officials who followed such signals. Censorship of all publications, including scientific journals and books, was institutionalized. The central Party organs closely scrutinized the appointment of the top positions in Soviet academic life, such as ministers of education, health, and agriculture, presidents of the Academy of Sciences of the USSR and the other specialized academies, editors of the leading journals, and directors of the various research institutes. Textbooks for secondary schools were under close control. These characteristics of the Soviet power structure help to explain the way in which Stalin was able in 1948 to give Lysenko's biological theories official and monopolistic status despite the opposition of the established geneticists.

LYSENKOISM

The most common explanation given for the rise of Trofim Lysenko is that his views on the inheritance of acquired characteristics fitted with Soviet ideological desires to "create a new Soviet man." If people can inherit improvements acquired from the social environment, so the argument goes, then revolutionary changes in society can quickly result in the improvement of human beings. Therefore, Lysenko's view of heredity extended hope to the revolutionary leaders of the Soviet Union that backward peasants could be transformed in a few generations into outstanding citizens who were both environmentally and genetically transformed.

This explanation of Lysenkoism is not accurate. Lysenko never claimed that his views on heredity were applicable to human beings. Indeed, he castigated eugenics and all other attempts to alter human heredity as examples of bourgeois influence on science. All the careful studies of the history of Lysenkoism that have been written during the last twenty years agree that Lysenkoism was not based on human genetics, yet the myth lives on.[2]

The roots of Lysenkoism lie not in Marxist ideology, but in the social and political context of Soviet Russia in the 1930s. Lysenko originated his ideas outside the circles of Marxist philosophers and outside the community of established geneticists. He was a simple agronomist who developed ideas about plants not very different from those of many practical selectionists of the late nineteenth and early twentieth centuries, but who was able to promote those ideas to an unheralded prominence because of the political and social situation in which he found himself. An extremely shrewd but basically uneducated man, he learned how to capitalize on the opportunities that the centralized bureaucracy and ideologically charged intellectual atmosphere presented. Seeing that his ideas would fare better if they were dressed in the garb of dialectical materialism, with the help of a young ideologist he recast his arguments in Marxist terms. But to the end of his days he never applied his biological scheme to human beings.

Lysenko began his career in Ukraine and in Azerbaidzhan, southern regions of great agricultural importance to the Soviet Union.[3] Even as far south as Azerbaidzhan, however, winter crops were threatened by occasional freezing temperatures. Lysenko tried to shorten the period of growth of cereals and other plants so that they could be harvested before the lowest temperatures arrived. He developed a process known as "vernalization" in which moisture and cold were applied to seeds before planting in order to accelerate their growth. Such techniques had been used in Germany and the United States in earlier years, but the opinion of researchers outside the Soviet Union was that they usually involved greater losses than gains. The soaking of seeds often spread fungi and plant diseases, and is an extremely labor-intensive process.

Lysenko used the term "vernalization" for almost anything that he did to plants, seeds, or tubers. For example, when he planted potatoes he first allowed the sections of potatoes to sprout before placing them in the ground. This is a practice known the world over and used by gardeners for centuries, but to Lysenko it was vernalization. He often soaked seeds before planting them, calling this vernalization, despite the fact that farmers and gardeners have also long done this to certain seeds.[4] Lysenko also attempted several spectacular feats, such as the "conversion" of winter into spring wheat (to be discussed below), but these efforts and their subsequent failures were only a small part of his agricultural program and therefore were not at first as prominent as they later appeared to be. It took a number of years before even Lysenko's critics realized what a disaster the widespread use of his more radical techniques would cause. The less radical techniques were often trivial from the scientific point of view. Quite a few of the plants from which he produced crops using vernalization techniques might have provided as good or better harvests without vernalization; in the absence of rigorous controls and careful statistics (Lysenko never mastered statistics or scientific methods of agronomy) it is impossible to determine just what effects Lysenko's techniques had on the success or failure of his harvests. In his later years it became clear that his methods did great damage to Soviet agriculture, especially because his monopoly over biology caused the Soviet Union to miss the agricultural revolution based on modern genetics that was sweeping the rest of the world.

Lysenko usually presented vernalization as a method of hastening the growth of a traditional crop so that peasant farmers would have a better chance of harvesting it before frost. The difference between bringing in a successful harvest and losing it to frost could be a matter of a few days. These are the kinds of experiments in which the evidence can be manipulated very easily, or where even careless recordkeeping can conceal results from an honest but unsophisticated researcher. The ideological fervor with which Lysenko surrounded his work, his constant declarations that he was transforming socialist agriculture for the good of the Soviet state, meant that few people were willing to criticize his results; such criticism could easily be interpreted as a lack of enthusiasm for the goal of socialist agriculture. Lysenko frequently cast his critics in that light. This political atmosphere fostered the emergence of a circle of sycophants around Lysenko who enthusiastically claimed victory after victory for his methods. Such enthusiasm went a long way in the circumstances of the time, which included unsystematic data collection, almost no control groups, irregular weather conditions, hastiness in drawing conclusions, readiness to discount contradictory evidence on the grounds of peasant recalcitrance, impure plant varieties, and small samples.

The last factors – impure plant varieties and small samples – were probably the clues to Lysenko's most exorbitant claims, such as his alleged conversion of winter into spring wheat.[5] In 1937 Lysenko announced his success in this effort after a "prolonged" experiment. Examination of the spotty records shows that this experiment had lasted only slightly longer than one year and involved two (!) plants of the *Kooperatorka* variety of winter wheat, which Lysenko claimed to have transformed into a spring wheat. One of these plants perished during the experiment, so the result was actually based on one plant of impure genetic stock. It is a gross understatement to say that this "experiment" proves absolutely nothing. The *Kooperatorka* was probably heterozygous. But Lysenko had no patience with people who suggested that only experiments with thousands of plants of known genetic composition with careful comparison to control groups over a period of many years would allow one to make such an ambitious claim as the conversion of one species into another. Such a person, according to Lysenko, did not realize that Soviet agriculture could not wait for such academic niceties.

It is one thing to be a sloppy researcher who makes unjustified claims. It is quite another to become the dictator of an entire academic field, as Lysenko eventually did. How did he make this stunning transition? Lysenko turned out to be a clever and cruel political maneuverer, but his success is only partially explicable in terms of his own characteristics; even more important was the Stalinist atmosphere in which he operated, which gave an unprincipled careerist like Lysenko unusual possibilities for advancing his own fortunes and destroying those of his rivals.

Lysenko was conducting his field work at a time when Soviet agriculture was in crisis as a result of the recent massive collectivization, and this desperate moment presented him with extraordinary opportunities to win the authorities' attention with his alleged solutions to agricultural problems. The collectivization program had been incredibly violent; involving the deportation and eventual deaths in camps of hundreds of thousands of peasants. A famine that followed in Ukraine, in particular, resulted in the deaths of millions. The resistance of the peasants had been powerful and bitter; as a last resort, many of them burned their crops and slaughtered their animals. The damage done to Soviet agriculture was so great that it was decades before recovery; some critics who emerged in the former Soviet Union over sixty years later wrote that the damage was still visible in the low productivity of the peasants and their guarded attitude toward the regime.

At the time Lysenko began his campaign for a socialist agriculture in the thirties there were few agricultural specialists who were willing to work energetically for the success of the new and troubled collective farms. Many agronomists of the time were educated before the Revolution; even among the younger ones with Soviet educations many disagreed with the collectivization policies, seeing damage that had been done in the countryside. Among the biologists in the leading universities and research institutes the most exciting topic of the time was not agriculture, but the new genetics arising out of research on the fruit fly *Drosophila melanogaster*. Only later would it become obvious that this research had great agricultural value, producing many agricultural innovations like hybrid corn. In the late twenties and early thirties it was easy for radical critics like Lysenko to castigate the theoretical biologists as they bent over trays of fruit flies in their laboratories at a time when famine stalked the countryside. Since many of the professional biologists had bourgeois backgrounds their political loyalties were always suspect to the regime. The unwillingness of many theoretical biologists to work directly on agricultural problems was seen by the radicals as purposeful "wrecking," an effort to disable the Soviet economy and cause it to fail, rather than the result of the common division the world over between theoretical and applied biology.

Lysenko was strikingly different from the majority of biologists and agronomists. He came from a peasant family, he was a vociferous champion of the Soviet regime and its agricultural policies, and he offered his services to agricultural administrators. Whenever the Party announced plans to cultivate a new area or plant a new crop, Lysenko came up with practical suggestions on how to proceed with the plan. He developed his various nostrums so rapidly – from cold treatment of grain, to plucking leaves from cotton plants, to removing the anthers from spikes of wheat, to cluster planting of trees, to unusual fertilizer mixes, to methods of breeding cows – that before the academic biologists could show that one was valueless or harmful, Lysenko was off announcing another technique. The newspapers invariably applauded Lysenko's efforts and questioned the motives and political backgrounds of his critics. In this environment, a peasant agronomist who promised a revolution in agriculture had enormous political advantages over sober academic geneticists who constantly appeared to be restraining progress by crying "not so fast!" or "inadequate verification!"

The man who is remembered as the antipode of Lysenko is Nikolai Vavilov, a famous biologist who in the 1920s organized expeditions throughout the world for the collection of varieties of agricultural plants. Vavilov was an indefatigable promoter of research and an organizer of research institutes. He came from a wealthy merchant family, was urbane and well educated, spoke many languages, and dressed in the tie and starched collar of the old Russian professoriate, something always noticed in these ideologically passionate years by lower class critics in peasant or worker dress. Nonetheless, Vavilov was a moderate supporter of the Soviet regime who cared deeply about agriculture, unlike many of his biologist colleagues.

Vavilov in his most important work, *The Centers of Origin of Cultivated Plants*, published in 1926, argued that the greatest genetic divergence in cultivated plant species could be found near the places of origins of these species, a conclusion that led him to expeditions to remote spots. His other major theoretical work, "The Law of Homologous Series in Variation," first published in 1920, was based on the belief that related species tend to vary genetically in similar ways.

When Vavilov first met Lysenko he saw him as a practical agronomist, not as a rival theoretical biologist. Vavilov was not deterred by the fact that Lysenko was different from him both in demeanor and dress. He was impressed with Lysenko's efforts on the farm fields, and complimented him on his work. In a few years Vavilov began to see that Lysenko had grandiose ambitions and nursed grudges against the established geneticists, but Vavilov still hoped to win him over by giving him support. He knew that if an open clash developed between the two of them the Soviet press would side with the peasant agronomist, not with the bourgeois professor. In the countryside at the time of collectivization the "rural correspondents" of the city newspapers often hounded the kulaks (the more prosperous peasants) and praised the poor peasants. Lysenko was often depicted in the press as a poorly educated agronomist who nonetheless was able to give lessons in practical agriculture to diplomaed professors.

Still not realizing how dangerous it was to feed Lysenko's ambitions, Vavilov actually proposed him for membership in the Academy of Sciences of Ukraine in 1934 and, several years later, for corresponding membership in the All-Union, or "big," academy in Moscow. Lysenko responded by sharpening his criticism of the establishment that was beginning to open up to him.

In 1935 at a conference of collective farmers in Moscow where Stalin sat at the head table, Lysenko gave a speech in which he said that disputes over biology were like those over collectivization, with class enemies trying to obstruct Soviet progress:

> You know, comrades, wreckers and kulaks are located not only in your collective farms. . . . They are just as dangerous, just as resolute in science. . . . And whether he is in the academic world or not in the academic world, a class enemy is always a class enemy. . . .[6]

As he often did in later years, Lysenko alternated his attacks on the academic intelligentsia with self-deprecating confessions, apologizing for the fact that he was only an agricultural worker, not an orator or a writer. The combination of ideological aggression with proletarian modesty added up to a message that Stalin liked, and he broke into Lysenko's speech with a cry of "Bravo, Comrade Lysenko, bravo!"[7]

Later the same year Lysenko spoke at yet another conference where Stalin was present and again denounced unnamed scientists who were damaging Soviet agriculture; this time one of Stalin's assistants, Ia. A. Iakovlev, asked Lysenko just who these scientists were, and Lysenko mentioned several names, including that of Vavilov. Attention now began to focus on Vavilov as the most prominent of the geneticists.

In 1935 Lysenko published his first article jointly with I. I. Prezent, who, in contrast to Lysenko, was a member of the Communist Party and a university graduate.[8] Prezent, a lawyer by training, was very adept in fashioning ideological arguments illustrating how Lysenko's form of biology fitted with dialectical materialism. The two men maintained that two kinds of biology were competing, a "socialist" biology represented by Lysenko and Prezent, and a "bourgeois" biology defended by Vavilov and his friends.

From 1935 onward a steady stream of pro-Lysenko propaganda flowed in the meetings of agriculturists, in the popular press, and, increasingly in the journals. Lysenko was by this time receiving significant support from the official bureaucracy, especially in agriculture and education. Vavilov was replaced as president of the Lenin Academy of Agricultural Sciences by A. I. Muralov, who tried to compromise between classical genetics and Lysenkoism; it was a maneuver that did not work, and within a few years Muralov was arrested and shot, as was his successor, G. K. Meister.

Under heavy pressure amounting to mortal threats, some of Vavilov's friends, people who knew that Lysenko was hopelessly ignorant about genetics, began to shift to Lysenko's side in the debates. According to information on the Lysenko period published many years later, during the *glasnost* of the late 1980s, some of Vavilov's associates even wrote false denunciations of their colleagues, hoping to save their skins. One of the worst denunciations, accusing Vavilov of fascist views, was evidently written by Professor G. Shlykov, a department head in the All-Union Institute of Plant Industry, which Vavilov directed. Not until the archives have been thoroughly explored can we be sure how accurate the details of these charges are, but the evidence already available seems strong.[9]

As if to illustrate the old truism that different people act differently under stress, Vavilov now for the first time began to speak openly against Lysenko. Alarmed that the science of genetics itself might be eclipsed, Vavilov abandoned the effort to compromise with Lysenko and pointed out the errors in his biological views. He tried to replicate some of Lysenko's experiments, and announced that he could not do so. But it was already late, and not many people with the power to make a difference listened to him anymore.

A few other brave people continued to speak up against Lysenko. At a conference on genetics in December 1936, A. S. Serebrovskii, an outstanding geneticist and sincere Marxist, called Lysenko's campaign "a fierce attack on the greatest achievements of the twentieth century . . . an attempt to throw us backward a half-century."[10] At another conference in October 1939, Vavilov tried to defend classical genetics against Lysenko by pointing to the importance of hybrid corn in the United States, a direct product of modern genetics. Vavilov even appealed directly to the Central Committee of the Communist Party itself, decrying

Lysenko's "intolerance," his "lack of culture," and his effort to carry out "reprisals against all who intellectually oppose him."[11]

In 1937 and 1938 many scientists fell victim to the purges. David Joravsky has pointed out that the pattern of the purges was arbitrary; not only supporters of classical genetics fell victim, but even some Lysenkoites.[12] No one knew who was safe and who was not. One result of this arbitrariness was to cause most people to lie low, to avoid attracting attention. Such passivity allowed administrators in agriculture and education, many of whom were Lysenkoites, quietly to consolidate their positions and promote their supporters. Only a very courageous person would protest.

On August 6, 1940, while he was leading an expedition in Ukraine, Nikolai Vavilov was arrested. The following July he was sentenced to death by firing squad for espionage and allegedly leading a counterrevolutionary organization. In 1942 the sentence was reduced to imprisonment for twenty years. Vavilov died in a Saratov prison of malnutrition the following year.[13]

After the rehabilitation of Vavilov in the 1960s, Soviet investigators admitted that the charges were false. Indeed, the man in charge of the original interrogation of Vavilov, Aleksandr Khvat, stated in an interview many years after Vavilov's death, "I didn't believe the spying charges, of course. There was no proof." When asked, "Didn't you pity Vavilov? After all, he was going to be shot," Khvat answered, "Ah, there were so many such people!" The interviewer then asked, "Did you think about him after that?" and Khvat replied, "In 1962 I was expelled from the Party in connection with Vavilov's case."[14]

In later years Lysenko always maintained that he was not personally responsible for Vavilov's death. In 1987 Lysenko's son, Iurii Trofimovich Lysenko, wrote a letter to the newspaper *Moscow News* in which he decried the "slander and abuse" that were being heaped on his father's head.[15] The son said that his father had told him that the investigator of Vavilov had come to him and asked, "What can you say in general about the wrecking (spying, counterrevolutionary) activities of N. I. Vavilov?" According to the son, the father replied, "There were and are some differences of opinion on scientific matters between myself and N. I. Vavilov, but I have no knowledge of any wrecking activities of N. I. Vavilov."

If this account is true, it is one more illustration of Lysenko's dishonesty. He had promoted many of the charges that landed on Vavilov, he had spoken of "wreckers" in biology, and on at least one occasion, the conference of 1935, he named Vavilov as being a member of the group of Soviet biologists who were damaging Soviet agriculture. But then he denied being in any way responsible for Vavilov's fate. The most that can be said for Lysenko is that he may not have had a direct line to the secret police; instead, he created circumstances that would cause the police to scrutinize his enemies. In Stalin's Soviet Union such behavior brought grave responsibility, even if Lysenko refused to see it. A respondent to Lysenko's son's 1987 letter put it pithily: "I regard your own letter as amoral. It was read by the survivors of Lysenko's campaign. It will be read by your descendants. . . ."[16]

Vavilov was gone from the scene of Soviet genetics by 1940, but Lysenko's control over biology in the Soviet Union was still not complete. In the research institutes of the Academy of Sciences and on some of the university faculties the teaching of genetics still quietly survived. Soviet scientists, especially physicists, greatly increased their influence during World War II as a result of their work in national defense, and there was hope that after the war Stalin would relax the ideological controls over scientists. And, in fact, in the years 1946–7, such an improvement seemed to be happening. In 1947 the Soviet biologist I. I. Schmalhausen published an article in the main Soviet philosophy journal that was clearly critical of Lysenko's position.[17]

Other events had diminished Lysenko's influence. During the war Lysenko's brother deserted to the occupying German forces in Khar'kov and, after the war, stayed in the West.[18] In the Soviet Union, where the activities of one's family members usually reflected on one's own reputation, this episode caused Lysenko intense embarrassment.

In 1948 Lysenko encountered criticism from a very high level, the office of the Central Committee of the Communist Party.[19] At that time the official in charge of science on the Central Committee staff was Iurii Zhdanov, son of Andrei Zhdanov, one of Stalin's chief lieutenants. Iurii Zhdanov's powerful associations included his marriage to Svetlana Allilueva, Stalin's daughter. Iurii, a chemist by education, had a strong interest in biology, and his knowledge of science was remarkably thorough for a Party bureaucrat. When several Soviet biologists, including V. P. Efroimson and A. A. Liubishchev, complained to the Central Committee about the damage being done to biology by Lysenko, Iurii Zhdanov delved deeply into the question and soon convinced himself that Lysenko was wrong in his denial of Mendelian genetics.

The crisis for Lysenko came to a peak in April 1948. On April 10 Iurii Zhdanov delivered a lecture on science to provincial Communist Party officials gathered at one of the halls of the Polytechnical Museum in Moscow. Lysenko had heard that the lecture would be critical of him, and he arranged to listen by sitting in an adjoining room connected to the lecture hall by a loudspeaker. In his speech Iurii Zhdanov accused Lysenko of monopolizing the field of biology, of preventing anyone from questioning his viewpoints. He continued that Lysenko had failed to improve Soviet agriculture as he had promised, and that he had denied supporters of classical genetics the possibility of showing what they could do to improve the situation. The result was that the Soviet population continued to suffer from poor agricultural yields.

Lysenko never confronted Zhdanov, even though he easily could have done so by walking into the lecture hall. Instead, on April 17, 1948, he wrote a letter to Stalin and Andrei Zhdanov complaining about Iurii Zhdanov's behavior. The existence of the letter was unknown for many years and was first published in 1988 by Valerii Soifer, an ardent foe of Lysenkoism who called the letter "a masterpiece."[20] Lysenko demonstrated his uncanny ability to appeal to all Stalin's prejudices. It is also possible that Stalin was looking for a means to contradict his assistant Andrei Zhdanov, although the historical record here is still unclear.

In the letter, Lysenko adopted the same passive-aggressive tone that he had earlier used. He started out by saying that it had become "very difficult" for him to continue his agricultural work. A struggle was going on between two varieties of biology, the "old, metaphysical" variety and the new "Michurinist" one. ("Michurinism" was the term Lysenko used for his form of biology, in honor of the prerevolutionary selectionist I. V. Michurin.) The way to decide between them should be based on practice. Which one was doing the most to help "socialist agriculture?" But how could he continue to demonstrate practical results when he was being suppressed? An example of attempts to suppress him was the speech given by Iurii Zhdanov a few days earlier. The Party administrators hearing such a speech given by such a high-ranking official would naturally consider it correct, and make it impossible for Lysenko to continue his work. Against this background, all Lysenko asked for was the right to continue his work, to represent "Michurinist biology."

A few weeks later, on May 11, 1948, before receiving any reply, Lysenko wrote I. A. Benediktov, the minister of agriculture of the Soviet Union, requesting that he be allowed to step down as president of the Academy of Agricultural Sciences so that he could devote himself full time to the development of Michurinist biology.

According to Valerii Soifer, Stalin finally answered in July by asking Lysenko to meet with him.[21] At the meeting Stalin expressed concern about the low output of Soviet agriculture and asked Lysenko what could be done about it. Lysenko replied that Michurinist biology could improve agriculture dramatically, and cited as evidence a new form of "branched wheat" he was working on which could give yields five or ten times greater than that of normal wheat. Lysenko was so impressed with this new wheat that he proposed that it be named "Stalin Branched Wheat." Stalin agreed with Lysenko, and, according to the account of Soifer, moved to make Lysenko the leader of Soviet biology. The next month, August 1948, was the time of the infamous meeting of the Academy of Agricultural Sciences at which, with Stalin's approval, formal genetics in the Soviet Union was banned. At the meeting Lysenko announced that his position had been formally approved by the Central Committee of the Communist Party, which meant, of course, Stalin himself. Indeed, a Soviet researcher even reported in 1990 that he had found the 1948 text of Lysenko's speech with corrections made by Stalin. Stalin emphasized the "Malthusian errors of Darwin" and the "reactionary character" of "foreign science."[22]

"Stalin Branched Wheat" was far from a miracle; it was not even new. It had been known since the time of ancient Egypt. Attractive because each stem contains multiple large spikes of wheat, it suffers from enormous disadvantages. It is susceptible to many diseases and its grain contains much less protein than traditional field varieties. Lysenko, however, maintained that he had overcome these problems. He had managed to grow several impressive demonstration fields. Party officials who inspected these fields were not told about extraordinary and expensive measures that Lysenko's workers had to take to protect the wheat from disease, nor were they given information on the low nutritional value of the grain. Independent scientists were not allowed to check Lysenko's claims. Instead, the newspapers and radio launched an enormous advertising campaign for Lysenko's new solution to the Soviet Union's grain problem.

In the months following the 1948 conference research and teaching in standard genetics were eliminated in the Soviet Union. Geneticists were called before Party meetings in their institutions and forced to recant their views. Those who refused were forced out of their positions, and those who agreed were required to work on other subjects. A number of leading genetics laboratories were simply closed down. According to Soifer, over three thousand biologists were fired. Several, such as Professor D. A. Sabinin of Moscow University, committed suicide.

Not until 1965, however, was the agronomist finally overthrown. His demise came as a result of several different factors. First of all, in the fifties and the early sixties, after the death of Stalin, a thaw occurred in Soviet intellectual life, with the emergence of voices deviating, however slightly, from the earlier orthodoxy. A number of biologists, including A. A. Liubishchev, V. N. Sukhachev, I. I. Shmal'gauzen, and I. I. Puzanov, vainly disputed Lysenko's claims and criticized his arrogance.[23] Nikita Khrushchev permitted some criticism of Lysenko to appear even though he himself was a supporter of the agronomist. With Khrushchev's overthrow in 1964 Lysenko lost his most influential protector after Stalin, and the criticism increased. Second, scientists, science administrators, and agricultural experts became ever more aware of the agricultural revolution occurring in the West, a revolution based on the genetics that Lysenko had rejected. The Soviet Union was rapidly falling behind the rest of the world in crop and animal yields. Third, the gap between Lysenko's claims and his achievements on his own farm near Moscow became increasingly clear. The Academy of Sciences appointed a commission to investigate the management of the farm and found fraudulent records. Lysenko was resorting to deception in order to maintain his reputation. No longer could he claim to be an outstanding practical farmer, traditionally his main defense.

Lysenko lost his monopoly position in Soviet biology, but even today genetics in the republics of the former Soviet Union – a field in which Soviet scientists like Vavilov, Kol'tsov, and Chetverikov were pioneers in the twenties – suffers from the effects of the Lysenko years. Young biologists in the seventies, eighties, and nineties, struggling to resurrect the field, had difficulty finding role models among their elders or sustaining traditions in their institutions.

PART III

Science and Soviet society

7

Soviet attitudes toward the social and historical study of science

T H E history of the Soviet Union is a patchwork of dreams and achievements and failures and tragedies. In the previous two chapters we saw in the sciences both the achievements of scientists such as Vygotsky and Fock and the tragedy of Lysenko. In the discipline of the history of science a somewhat similar story can be found, a beginning of great promise and a later story of failure and tragedy, even if not on the same scale as Lysenko in biology.

The Soviet Union was the first country in the world to create an institute or a university department for the study of the history of science and technology.[1] It was also a place where several ideas about the history of science were developed and propagated that had considerable influence on the field worldwide. The history of science soon fell into grave political difficulty, however.

In 1921 the Academy of Sciences created the Commission on the History of Knowledge, a group that took under its purview both the natural and social sciences, and in 1932 this body was converted into the Institute of the History of Science and Technology of the Academy of Sciences of the USSR. Even though the Commission on the History of Knowledge appeared four years after the Revolution, the support for it originally came not from Marxist scholars but from scientists and historians whose intellectual predilections were those of the liberal professoriate during the last years of tsarism. The leader of this group was Vladimir Vernadskii, a geochemist and extraordinarily talented writer on a great variety of subjects.

The idea of promoting the history of science occurred to Vernadskii long before the Revolution. In 1893 he wrote, "I am more and more occupied with the thought of seriously devoting my strength to work on the history of the development of science."[2] From that time until his death in 1945 his interest in the history of science continued, and his papers contain thousands of pages on the subject.[3]

When Vernadskii launched his project the history of science as a discipline was virtually unknown in universities and research institutions throughout the world. At first Vernadskii's ideas about the importance of the history of science amounted merely to his own research agenda, but as early as 1902 he offered a course at Moscow University on "The History of the Modern Scientific Worldview," and after the 1917 revolutions he began to agitate for an organizational center for the history of science, an effort that came to fruition in 1921.

Vernadskii saw the Commission on the History of Knowledge as the foundation for a later institute and museum and as the intellectual center for a new discipline. Under his directorship the Commission launched a publication series of which nine volumes appeared, as well as several other separate publications. The first volume in the series, in 1927, was by Vernadskii himself, and was titled "Thoughts on the Contemporary Significance of the History of Knowledge." Other publications in this series included works on the history of geography, on ancient Oriental science, on embryology, and on physics. Separate works included studies of Russian scientists like the pioneer in structural chemistry A. M. Butlerov as well as West Europeans such as Newton and Berthollet.[4]

Vernadskii's view of the development of science was sophisticated for its time, even though contemporary historians of science would probably say that he overplayed the role of genius and the power of ideas, and underestimated the importance of social context and technology. The sophistication of Vernadskii's interpretation of science was evident in his assertion that genius by itself was only a necessary, not a sufficient, condition for the advancement of science. Also important, he thought, was the presence of nurturing political and social conditions. In his opinion, the great "explosions" of scientific thought that occasionally occur throughout history can be explained by rare "clusterings" of individual geniuses in favorable social environments.[5] As a professor under tsarism he had witnessed the blighting effects of politics and inadequate funding on science, and he hoped for better conditions under the Soviet government. However, he did not agree with the politics of the Bolsheviks, preferring a pluralistic democracy.

Vernadskii advanced special reasons for the need to study the history of science in the early twentieth century. He was convinced that he was living in a unique moment in the history of science, the advent of relativity theory and quantum physics. In his opinion these developments marked the third in three great revolutions in science, the first of which was the birth of science itself in Hellenistic times and the second was the launching of modern science in the seventeenth century.[6] How fortunate we would have been, he mused, if scholars at the time of those two previous great transformations in science had been aware enough of what was happening to engage in self-study. To initiate a center for the study of the history of science at the moment of birth of a new physics was to Vernadskii a historic opportunity.

One of Vernadskii's most interesting conceptions about science and its development was connected with his own work in geology. Vernadskii believed that geologists had previously ignored the importance of plants, animals, and human beings in changing the composition of the earth. In his opinion, a great many mineral deposits either were caused by biological activity or had been affected by biological organisms. He devised the term "biogeochemistry" to convey the importance of combining biology, geology, and chemistry in the study of the earth and he pioneered the investigation of biogeochemical cycles in nature. In 1922 he wrote that of 92 chemical elements currently known to scientists, between 50 and 60 were connected to the history of living organisms.[7]

To Vernadskii, human beings were the latest and the most potent biological organisms affecting the earth. In Paris in the 1920s his lectures were attended by the young French scholars Teilhard de Chardin and Edouard Leroy. According to Vernadskii's biographer Kendall Bailes, "they borrowed from Vernadskii his usage of the term biosphere and he in turn borrowed from Leroy his idea of the noosphere, that is the idea that the biosphere was being transformed into a biological zone controlled by mankind's reason. . . ."[8]

With the advent of the noosphere, caused by human reason, Vernadskii saw science transforming the earth in two different ways. First, technology derived from science had obvious effects by altering the face of the earth. Second, science as a cognitive scheme also changed the earth in an intellectual sense. For example, the world of most eighteenth- and nineteenth-century physicists was made up of material particles, but, after relativity physics, "we are coming to a construction of a world without matter."[9]

By emphasizing the independent power of ideas in determining the nature of the universe we live in, Vernadskii attracted criticism from Marxists for his "idealism." But he was not a consistent idealist any more than he was a reliable materialist; instead, he was thoroughly eclectic in his philosophical tendencies, and he combined flirtations with Kantian epistemology with frequent assertions that he was a "cosmic realist."

Vernadskii's direction of the field of the history of science in Russia came to an end with the takeover of the Academy of Science by the Communist Party in 1929–30, a process in which Vernadskii was the leader of a strenuous and unsuccessful resistance.[10] Even if he had not opposed the Party's assertion of control over the Academy, however, Vernadskii's approach to the history of science would not have been acceptable to the purged leadership of the Academy. According to the new administrators of Soviet research, Vernadskii ignored Marxism in his interpretation of science, he slighted the roles of technology and economic needs as stimuli to scientific progress, and he exaggerated the potency of scientific ideas as forces of social change. Furthermore, he often spoke of the positive effects that religion occasionally had on science. At the end of 1930 Vernadskii was replaced as head of the Commission on the History of Knowledge by Nikolai Bukharin, a leading theoretician of the Communist Party.[11]

Bukharin was a remarkable and highly talented man with a deep interest in the history of science and considerable tolerance for variety in historical interpretation. A committed Marxist, he worked to create a new Marxist tradition in the interpretation of science, both in his own writings and in the scholarship that he promoted. In March 1932 the Academy converted the Commission on the History of Knowledge into a new Institute for the History of Science and Technology, with Bukharin as the first director. The inclusion of the term "technology" in the title indicated that the research of the new organization would be much broader than merely the history of scientific ideas.

The new institute had six sections: the history of technology, headed by Academician V. F. Mitkevich; the history of chemistry, directed by Professors B. N. Menshutkin and T. P. Kravets; the history of physics and mathematics, led by Academician S. I. Vavilov (the brother of the geneticist); the history of biology, headed by Academician B. A. Keller; the history of agriculture, led by Academician N. I. Vavilov; and the history of the Academy itself, led by Academician S. F. Ol'denburg.[12] Bukharin's intellectual flexibility is demonstrated by the fact that none of his section heads was a recognized Marxist and one of them, Ol'denburg, had been a high official in the provisional government that the Bolsheviks overthrew in 1917.

The leaders of the institute made plans for creation of a museum and for an ambitious publication program, including an encyclopedia of the history of technology, and a series entitled "Classics of Science." The library was enriched with the large collection of books on the history of science assembled by Vernadskii.

Bukharin was never personally active in research and writing on the history of science to the degree that his predecessor Vernadskii had been. He was primarily interested in political economy and Marxist philosophy, but he believed that science and technology were closely connected to these fields. Marxist economics was, to him, a science that shared the philosophical assumptions of natural science. Both Marxism and natural science, he often said, were based on the principles of an external material world governed by causal laws. Portions of his writings on political economy are remarkable for the degree to which they draw upon a materialist interpretation of natural science and for the intellectual clarity with which this view is presented.[13]

Bukharin criticized the concepts of "pure science" and "science for its own sake," maintaining that all science is heavily mediated by social, economic, and political factors, and therefore cannot be separated from the society from which it emerges. Science and technology in a socialist society, he thought, would differ from their counterparts in a capitalist one. For example, in a capitalist society new technical ideas are, according to Bukharin, quickly locked up in a system of patents and industrial secrets protected by competitive firms for their individual benefit; on the contrary, he continued, in a socialist economy such technical innovations would be immediately shared by all branches of industry. Here Bukharin sketched a utopian view that would amuse any Soviet engineer or industrial manager of the late twentieth century struggling to keep up with Western or Japanese technology.

In the paper that Bukharin delivered at the Second International Congress of the History of Science in London in 1931 he expanded on his view that science is primarily a social product, not an unmediated intellectual creation.[14] He presented a sociological interpretation of the development of ideas that went beyond the economic determinism displayed in the more famous paper delivered at the same congress by his Soviet colleague Boris Hessen. In this interpretation of science Bukharin anticipated by several decades the work of historians of science in the West who developed the view that science is a social construction. A leading American historian of science, I. Bernard Cohen, observed in 1989 that "Bukharin's piece remains impressive today to a degree that Hessen's is not."[15]

Bukharin observed that scientists sometimes maintain that their findings are objective truth based on pure sensations of nature. However, there is no such thing, Bukharin wrote, as "pure sensation" that is unconnected with society. The "knowing subject," he continued,

> stands on the shoulders of the experience of other people. In his "I" there is *always* contained "we." In the pores of his sensation there already sit the products of "transmitted" knowledge (the external expression of this are speech, language and conceptions adequate to words). In his *individual* experience there are included beforehand society, external nature and history – i.e., social history. . . . Historically there is *no* absolutely unmixed individual sensation, beyond the influence of external nature, beyond the influence of other people, beyond the elements of mediated knowledge, beyond historical development.[16]

This was a remarkably prescient statement of the views historians of science would later develop in Western Europe and America.

Bukharin in the late twenties and early thirties was heavily involved in the administration of industrial research in the Soviet Union. He headed the "Scientific-Research Sector" of the main economic council of the USSR (VSNKh), and was responsible for over one hundred industrial research institutes. In 1931 he convened the First All-Union Conference for the Planning of Scientific-Research Work, where he presented a notable paper in which he tried to distinguish those aspects of research most amenable to planning.[17]

Unfortunately for the history of science in the Soviet Union, Bukharin was falling under Stalin's suspicion. Branded as a member of the "right opposition," Bukharin was expelled in 1929 from the ruling Politburo of the Communist Party even before he became head of the Commission on the History of Knowledge. In subsequent years Bukharin continued to have some influence, especially after he became editor of the government newspaper *Izvestiia* in 1934, but under the conditions of Stalin's autocratic rule his maneuvering room was limited.

In the midthirties Stalin unleashed a terror campaign against the nation and accused many of his erstwhile colleagues of crimes. In 1936 Stalin's chief prosecutor, A. Ia. Vyshinskii, announced that the Institute of the History of Science and Technology, still directed by Bukharin, was a center of an anti-Soviet conspiracy.[18] In April 1937 the main journal of the Academy spoke darkly of "enemies of the people" still hiding in research posts. (All these accusations were many years later recognized by Soviet authorities as complete fabrications.) The following month Bukharin was expelled from membership in the Academy. He was then brought to trial and executed. The same year the Institute of the History of Science and Technology was abolished.

Vernadskii did not give in easily to the destruction of a field that he rightly regarded as his brainchild. In a memorandum to the presidium of the Academy in 1939 he protested against the view that the history of each scientific discipline could be adequately supported by researchers in that discipline. Instead, he wrote,

> The history of technology and science is a special area of science and it makes specific demands on the people who are working in that area. These scholars must have, along with knowledge of their own narrow speciality, broad scientific preparation in economics, history and philosophy. The methods of research of the history of science and technology are essentially different from the methods of research in the institutes of the Academy doing research on technology and natural science. These methods are determined by the essence of the discipline and study of sources, and they do not play an appreciable role in the technical and scientific institutes of the Academy.[19]

Vernadskii's plea was not heeded until the end of World War II, when the Institute of the History of Science and Technology was reinstituted. The earlier charges of treason against Bukharin and the previous institute remained unrefuted, however, until 1988, when Bukharin was restored posthumously to membership in the Academy.[20] In the years between 1945 and 1988 the Institute of the History of Science and Technology promoted some good work, but it displayed its continuing trauma over its predecessor's fate by staying away from discussions of the social and political context of science, concentrating instead on narrow histories filled with technical details. It is ironic, of course, that Soviet history of science, a pioneer in the broad social and economic interpretation of science and technology in the twenties and early thirties, came to be known for its narrow concentration on scientific and technical details and its avoidance of societal issues.[21]

One area where Soviet history of science in the late Stalinist period did make broad claims was an unfortunate one, namely, the unsubstantiated assertion of national priority in scientific discoveries and technical innovations. These claims were a part of the resurgence of nationalism in the Soviet Union during and after World War II. Soviet historians in the forties and fifties credited various Russian and Soviet scientists and engineers with discovering the law of the conservation of mass and energy, the overthrow of the phlogiston theory, the discovery of absolute zero, the establishment of the evolution of species, the creation of the science of structural chemistry, and the invention of the electric light, the telegraph, the radio, the airplane, and many other devices. Most of these claims were abandoned in the sixties and seventies and one or two have been retained that deserve further study. Whether or not they deserve the priority which they have been assigned, such figures as Butlerov in chemistry and Popov and Ladygin in technology merit examination by scholars who are more careful in their work than most historians working under Stalinism.

Today both Vernadskii and Bukharin are great intellectual figures in the former Soviet Union. Vernadskii gradually won a position of respect in the sixties and seventies, especially after the development of an environmental awareness in the Soviet Union that coincided with his emphasis on the impact of humans on the biosphere. By the late seventies and eighties, Vernadskii's works were being reissued in large printings. Streets, squares, and even a subway station were named after him. Bukharin predicted before his execution that "the filter of history will sooner or later wash the dirt from my head,"[22] but could not have dreamed that half a century would be required before that moment came. The full appreciation and evaluation of Bukharin is a process that will take many more years.

HESSEN AND THE 1931 CONFERENCE

While most events that concerned historians of science in the Soviet Union have attracted little attention outside that country, one has had central importance to their colleagues abroad. At the Second International Congress of the History of Science in London in 1931 the Soviet physicist and historian of science Boris Hessen presented one of the most influential reports ever given at a meeting of historians of science.

In fact, two generations after the event it is still being praised and criticized. Arnold Thackray, a leading historian of science in the United States, in 1980 called Hessen's report a "paradigm-setting analysis" and cited its widespread influence in England and North America.[23] On the fiftieth anniversary of the presentation of the original paper, the journal of the History of Science Society, *Isis*, marked the event by picturing the medallion of the 1931 meeting on the front cover of an issue that featured a special discussion of Marxism and science.[24] Also in 1981 the *Dictionary of the History of Science*, a standard reference, cited Hessen's paper as instrumental in the formulation of one of the major interpretative concepts of the history of science, "externalism."[25] In the early 1990s plans were proceeding in the United States to republish Hessen's paper and Western analyses of it.[26]

The word "externalism" brings us to one of the central questions that historians of science ask: What are the main factors influencing the growth of scientific knowledge? In trying to answer this question professional historians of science have often split into two camps, the "internalists" and the "externalists." Internalists are scholars who emphasize the power of scientific ideas and the significance of experimental findings as the major influences on the growth of scientific knowledge. Externalists, who represent a newer trend in the history of science, stress social, economic, and other nonscientific influences on the development of science.

The beginnings of this important debate can be traced back to Hessen's electrifying address in London. It was a moment of high theater as well as pointed intellectual confrontation. In England at a time of deepening economic depression many intellectuals hoped that revolutionary Soviet Russia would present economic and intellectual alternatives to European systems of economy and of thought. The expectations of historians of science were heightened when they learned that the Soviet Union was sending a very distinguished delegation to the congress, headed by the famous government and party leader Nikolai Bukharin.

Most of the Soviet papers were interesting, but it was Hessen's that became the focus of controversy. He chose as his subject Isaac Newton, arguably the greatest figure in the entire history of science. Most previous treatments of Newton had depicted him as a genius whose creativity transcended human understanding. An assessment of the rank of this English cultural hero seemed to call for references to divinity, as epitomized in Pope's famous couplet, "Nature and nature's laws lay hid in night; God said 'Let Newton be!' and all was light." Within the existing literature on the history of science, it was permissible to seek antecedents to Newton in the great works of his illustrious predecessors Galileo, Kepler, and Copernicus, but to indicate that the social and economic conditions of seventeenth-century England might have had something to do with Newton's achievements was considered a debasement of his intellectual nobility.

Boris Hessen announced that Isaac Newton and his work could not be understood outside the context of the rise of mercantile capitalism in England. The English Revolution of the seventeenth century had prepared the way for a new economic order, and this burgeoning society made demands for new technologies. The technologies that England needed at the end of the seventeenth century were for expanding trade, industry, and war. Such technologies called for a new physics, and Newton's three laws of physics and his work on optics would create the applied sciences of ballistics, mechanics, and hydrostatics that the machines and weapons of the age required. Hessen further maintained that the religious and idealistic principles that Newton and many of his followers supported were also tightly connected with economic and political controversies of seventeenth-century England. After a generation of turmoil, England needed stability in order for economic expansion to resume. Newton's affirmation that his system of the universe illustrated the grandeur of God's creation provided English clerics with the raw material for a theology promoting the new establishment. "Newton," Hessen said, "was the typical representative of the rising bourgeoisie, and in his philosophy he embodies the characteristic features of his class. . . . He also was a typical son of the class compromise of 1688."[27]

Hessen presented his theses in an exaggerated and even outrageous form. Few people in London were enamored with his assertion that the proletariat possesses "genuine scientific knowledge of the laws of the historical process."[28] But, aside from these ideological slogans, Hessen had touched on a very sensitive nerve. There was more than a grain of truth in his belief that most historians of science treated scientific concepts "as if they had dropped from the sky," and made little inquiry into the connections between the evolution of science and the evolution of politics and economics.

In the years that followed Hessen's presentation many scholars interested in the history of science acknowledged a debt to Hessen, although few accepted all his political viewpoints. What Hessen forced them to do was to see science as a social product. J. D. Bernal, the brilliant crystallographer and Marxist, wrote that Hessen's essay was "the starting point of a new evaluation of the history of science."[29] Hyman Levy, another British Marxist, commented that the approach of Hessen and his colleagues "crystallized out in remarkable fashion what had been simmering in the minds of many for some time past."[30] But not only Marxists recognized that Hessen had started a crucial debate. The man who was later to become the dean of American sociologists of science, Robert K. Merton, wrote in his now classic work on science in seventeenth-century England that three of the chapters in the book were heavily indebted to Hessen, and that Hessen's mode of analysis, "if carefully checked, provides a very useful basis for determining empirically the relations between economic and scientific development."[31] And Stephen Toulmin, a leading philosopher of science, wrote of Hessen's essay that it scandalized orthodox Newton scholars in a way that could only be compared to the time when "revisionist Biblical scholars had suggested that, after all, the *Song of Songs* was really *meant* to be read as an erotic poem."[32] Such an overturn in ways of looking at science was not easily absorbed; even as late as the 1960s, when the Harvard historian of science Everett Mendelsohn promoted social studies of science he observed that he "felt very much alone."

Historians of science still quarrel about the external history of science, especially about how far it should be taken and what its ultimate implications are; despite the disputes, however, the external history of science is by now well established. Like many other varieties of interpretation of history, it is an approach that can be pursued either subtly or simplistically, and by now a multitude of examples of both kinds of external history of science exists.

Valuable as Hessen's essay was in initiating the development of external history of science, most historians now would agree that it was simplistic. No one today goes to Hessen's original essay for reliable information about Newton. A whole industry of scholarship centered on Newton has arisen since Hessen that regularly produces sophisticated analyses of Newtonian physics of both the internal and external varieties. But Hessen's essay is still remembered as the starting point of one of the most significant controversies in the history of science.

One important facet of the "Hessen Episode" has still not been noticed by the majority of Western historians of science. Hessen was at that moment engaged in a major battle in the Soviet Union in which he was trying to defend relativity physics and quantum mechanics against the attacks of militant Marxist ideologues. He crafted his Newton paper in such a way that it would help him in that struggle. His paper is thus better understood as a result of his peculiar and threatened situation in the Soviet Union than as a model of Marxist analysis of science, either vulgar or sophisticated. In writing his paper Hessen had a strategic goal.[33]

In the late 1920s and early 1930s physicists in the Soviet Union such as Hessen became worried about the attacks being made upon relativity theory and quantum mechanics.[34] Relativity theory was particularly troublesome, for Einstein had recognized the importance in its development of the ideas of the Austrian physicist Ernst Mach, whom Lenin had severely criticized in one of his major books. The suspicions of the Soviet critics of quantum mechanics and relativity physics were heightened when several prominent West European philosophers and scientists con-cluded that the probabilistic approach of quantum mechanics meant the end of determinism as a worldview, while the equivalence of matter and energy postulated by relativity theory marked the end of materialism.[35] Several of them concluded that relativity physics and quantum mechan-ics destroyed the basis of Marxist materialism.

Well-educated Marxists like Boris Hessen, knowledgeable both in poli-tics and in physics, saw the intellectual poverty of these attacks on modern physics. Hessen went to the front line in this battle, simulta-neously defending Marxism and modern physics. Writing in 1927 he insisted that the possibility of drawing conclusions, on the basis of rela-tivity theory and quantum mechanics, that were unacceptable to Marx-ists was no reason for "throwing out the physical contents of the theo-ries."[36] If Soviet Marxists condemned relativity theory as anti-Marxist then what would they do, asked Hessen, if relativity turns out to be correct as a physical theory? The only way to avoid the conclusion that Marxism was in error, he continued, was to see the difference between the physical core of science and its philosophical interpretation, a theme to which he would return in his paper on Newton.[37]

Hessen noted in articles written before his trip to London that this problem of linking science too rigidly to ideology did not arise with relativity theory. Newtonian physics, so hotly defended in the name of materialism by such Russian physicists as A. K. Timiriazev, had also been used for philosophical purposes that Marxists could not accept. Newtonian physics easily lent itself to an ideology of a "divine first impulse" that set the solar system in motion. Newton himself found this view attractive. Yet Hessen obviously thought it ridiculous for atheists and Marxists to reject Newtonian mechanics for this reason.[38]

But defending modern physics at a time when revolutionary tensions were high was no easy task. In the process of doing so Hessen incurred serious criticism. His social and ethnic background did not help him. Son of a bank employee (a profession particularly despised by militant revolutionaries), he was also a Jew. He was a cosmopolitan scholar, educated in the West, fluent in German, French, and English. To the radical young students being pushed to the top of the Soviet educational system Hessen was a typical member of the old Jewish intelligentsia, perhaps "progressive" at the time of the Russian Revolution, but dis-tinctly falling behind the times as Stalin called for proletarian militance.

In 1930 and 1931 relativity physics was under heavier criticism than at any time in the five years before or after this date. The early attacks on relativity led by old-fashioned physicists continued, but they were now more than matched by newer threats. Beginning in 1930 a worrisome danger arose with the appearance of the "Bolshevizers" of philosophy and science, younger militants taking advantage of the Cultural Revolution then in progress and calling for the "reconstruction" of physics on the basis of dialectical materialism.[39]

Hessen and his views on physics came under very heavy criticism at a conference on the state of Soviet philosophy that was held October 17–20, 1930. Although present, he was not permitted to speak in his own defense.[40] He was denounced as a "metaphysicist of the worst sort,"[41] a "pure idealist,"[42] and as a deserter of the cause of materialism who interpreted relativity physics in the same spirit as the Western mystic astronomer Arthur Stanley Eddington.[43] He was criticized for paying insufficient attention to the ideas of Engels and Lenin.[44] Particularly mistaken, said his detractors, was his definition of matter as a "synthesis of space and time," a wording that came from one of his defenses of relativity theory.[45] In the final resolution of the conference Hessen was censured by name twice, once for his philosophical views on relativity theory and again for his opinions based on quantum mechanics.[46]

Attacks in the Soviet Union on relativity attained a new level after November and December 1930, when Einstein published articles in the *New York Times Magazine* and the *Berliner Tageblatt* under the titles "Science and Religion" and "What I Believe," in which he defended a form of deism similar to that of Spinoza. One of Einstein's Soviet critics responded that deism was logically inherent in the concept of a four-dimensional space–time continuum and that therefore relativity must be rejected. He noted Hessen's defense of relativity theory, a doctrine that he condemned as "a rotten swamp."[47]

One of Hessen's bitterest critics was Ernst Kol'man, a Czech Marxist who had emigrated to the Soviet Union. In an article published in January 1931, Kol'man maintained that "wreckers" were trying to corrupt Soviet physics just as wreckers had earlier tried to disrupt Soviet industry. The implication was serious, since the engineering "wreckers" had been brought to trial and many of them imprisoned. Kol'man in the same article tried to illustrate how the wreckers in physics were trying to discredit materialism:

> "Matter disappears, only equations remain" – this Leninist description of academic papism in modern physics gives the clue to the understanding of the wrecker's predilection for the mathematization of every science. The wreckers do not dare to say directly that they want to restore capitalism, they have to hide behind a convenient mask. And there is no more impenetrable mask to hide behind than a curtain of mathematical abstraction.[48]

Kol'man asserted that it was time for Marxists to reject Hessen's view that relativity theory was inherently Marxist and to recognize that the "most harmful and dangerous of all things is empty, naked theoretization."[49] He continued that Marxist philosophers should notice that Stalin had announced that "technology in the current stage decides everything" and therefore they should turn from analyses of theoretical science to analyses of the practical tasks of industrialization.[50]

In an article published only three months before the London conference, Kol'man issued a direct challenge to Hessen, calling on him to change his ways, to correct his political mistakes:

> Comrade Hessen is making some progress, although with great difficulty, toward correcting the enormous errors which he, together with other members of our scientific leadership, have committed. Nonetheless, he still has not been able to pose the issue in a correct fashion, in line with the Party's policy. . . . One must speak directly here, and say that there is no Bolshevism in Hessen's science, nor in that of his comrades. This has to be said forthrightly. Comrade Hessen now has the possibility of showing in his practical work that he really wants to correct his mistakes.[51]

Many years later Kol'man defected to the West, disillusioned by the Soviet Union; he confessed to me in emigration that the Communist Party had sent Hessen to London as a test of his political orthodoxy and that the Party had dispatched Kol'man as a member of the delegation with the duty of keeping an eye on Hessen.[52]

Hessen's London paper satisfied the requirements that Kol'man laid down. It eschewed theoretical physics and mathematics, contrary to most of his previous papers. It strongly emphasized the role of practice in determining theory, and it obeyed Stalin's command to stress technology. Hessen wrote that although Newton's *Principia* is "expounded in abstract language," its "earthy core" is actually technical problems arising out of industry and trade in the seventeenth century.[53] Throughout the paper Hessen copiously quoted Marx, Engels, and Lenin.

However, Hessen had not abandoned his major goal of protecting the core of modern science from ideological attack. He embedded in his Newton paper a subtle message about the relationship of science to ideology. He defended science against ideological perversion by pointing to the need to separate the great merit of Newton's accomplishments in physics from both the economic order in which they arose and the philosophical and religious conclusions that Newton and many other people drew from them. Hessen knew that not even the most radical critics of relativity physics in the Soviet Union questioned Newtonian physics; if he could show that the same contextual critique could be made of Newton that some Marxists in the Soviet Union were making of Einstein, then the lesson seemed clear. Hessen was illustrating that Marxists should simultaneously recognize the value of Newton's physics, even though it developed in mercantilist England and was used as a tool to support religion, and the value of Einstein's and Bohr's physics, while acknowledging that they arose in imperialist Europe and are often used to counter Marxism.

When it came to discussing the relationship of physics to economics, Hessen pulled out a textbook Marxism that he employed to great effect. After years of trying to warn his colleagues in Moscow about the damage that could be done with an unrestrained Marxism, he must have found satisfaction in realizing that in London he could only help his cause and that of Soviet physics by letting fly full force. His performance carried the implicit message to Bolshevik critics of relativity physics: "What you do to Einstein and Bohr, I can do to Newton, so let's leave the physics alone."

This message became almost explicit when Hessen in the paper praised the "great results" and "elements of healthy materialism" contained in Newton's *Principia* while criticizing his "general religio-theoretical conception of the universe."[54] Hessen believed that even though Newton linked his system of the universe with the idea of a "divine first impulse" his system could equally well be accepted without this religious assumption. Back in the Soviet Union Hessen had also been saying the same thing about Einstein, separating his physics from his religious views.

We see, then, that despite all his concessions to his critics, Hessen left room for the defense of theoretical physics and its differentiation from ideology. By placing emphasis on technology and practice in determining theoretical physics he freed physics from being condemned merely by the philosophical or theological interpretations that may be placed on it. He believed that the development of twentieth-century physics could be analyzed in the same way that he explained Newtonian physics, and thought there was no more reason to accept attacks on materialism in the name of twentieth-century physics than there had been to accept such attacks in the name of Newton, whose religious views were merely a "product of his time and class."[55] The unwritten final line was that when Einstein wrote on religion or philosophy he also merely expressed his social context and therefore these views should not be held against his physics.

It is ironic, but not contradictory, that Hessen was doing to Newton something rather similar to what Hessen's ideological critics in the Soviet Union were doing at the same moment to Einstein. Hessen was maintaining that Newton's physics was based on the ideological assumptions and promoted by the economic interests of bourgeois England in the seventeenth century. Hessen's foes in the Soviet Union were maintaining that Einstein's physics was based on the ideological presuppositions and buoyed by the economic interests of imperialistic Europe of the late nineteenth and early twentieth centuries. But Hessen differed sharply with his critics on the conclusion that should be drawn from these analyses. His critics believed that the ideological association of a theory was an important factor in judging that theory's validity. Hessen, on the contrary, wished to differentiate between the social origins of science and its cognitive value. He knew that he would have an easier time convincing militant Soviet Marxists that Newtonian physics had enduring value despite its bourgeois social origins than he would demonstrating that the still little understood relativity theory also must be valued despite its social origins in capitalistic central Europe.

It is difficult to say just how successful Hessen was in achieving the goals of his brilliant paper. Relativity physics survived in the Soviet Union, although it continued to have many critics. Externalism in the history of science eventually prospered, especially outside the Soviet Union. In an intellectual sense, then, one might contend that Hessen's effort succeeded. However, it is doubtful that Hessen's paper was a turning point in the battle over relativity physics. The available evidence points to the contrary, since the paper did not attract nearly as much attention back in the Soviet Union as it did in the West. In a personal sense, Hessen failed. He died in prison in 1938, a victim of the purges, along with six members of the eight-man delegation to London in 1931, including Bukharin.

"SCIENCE STUDIES," OR *NAUKOVEDENIE*

A strong interest among Soviet scholars studying the history and social dimensions of science in the 1920s and again in the 1960s and 1970s was "science studies," often called *naukovedenie*.[56] *Naukovedenie* is the social study of science as an institution, and encompasses sociology of science, science management, and science organization. In its narrowest form of practice it has the utilitarian goal of improving the performance of scientific researchers; in its broader forms it is a cognitive effort to understand science better by bringing to bear upon it all the relevant social science disciplines. One of the first persons to use the term *naukovedenie* was I. Borichevskii, who defined it in 1926 in terms of its broader, cognitive goal:

> On the one hand, it (*naukovedenie* lrg) is a study of the inherent nature of science, a general *theory* of scientific *cognition*. On the other hand, it is a study of the social purpose of science, of its relations with other types of social creativity. It is something we could call a *sociology of science*. This area of knowledge does not yet exist; but it must exist: It is required by the very dignity of its object, i.e., of the revolutionary power of exact knowledge.[57]

The birth of *naukovedenie* in the twenties was closely connected with the effort to plan science in accordance with the principles of a socialist state; before one could try to plan science one must have empirical information about the science establishment. An early leader in this effort was S. F. Ol'denburg, permanent secretary of the Academy of

Sciences, who promoted non-Marxist work in this area in the twenties; in the late twenties and early thirties, Nikolai Bukharin stimulated further research with his vision of a uniquely socialist science. Between 1921 and 1934 workers at the Academy of Sciences made statistical and organizational surveys of the scientific personnel and institutions of the USSR that were remarkable in their detail and in their awareness of the potential of science and scientific institutions as natural resources. Anticipating similar work in other countries by more than a decade, the multivolume surveys, not to mention less ambitious handbooks and outlines, provided an enormous amount of data for the study of the growth of scientific disciplines and institutions in Russia.[58]

Soviet authors in the twenties attempted to improve scientific research by suggesting changes in research techniques and the use of laboratory equipment; by proposing reforms in publication and indexing operations; by calling for information-retrieval systems, including primitive computers; and by developing quantitative criteria for evaluating the effectiveness of scientific research. A few even launched psychological and sociological studies of the nature of scientific creativity.[59] The assumption underlying all this activity was that within the framework of socialism it would be easier than elsewhere to submit scientific research to analysis, to ascertain its principles, and to improve its conduct.

This research was largely abandoned in the early 1930s as the social sciences withered under Stalin's authoritarian policies. By the time a revival began in the 1960s Western scholars were far ahead in a field in which Soviet authors had been pioneers. The writings of J. D. Bernal in England and Derek Price in the United States sparked new interest in science studies in many countries. Bernal's 1939 book *The Social Function of Science* and Price's 1961 and 1963 books *Science since Babylon* and *Little Science, Big Science* were influential early works, even though most later specialists in science studies saw them as elementary and even naive.[60] Bernal emphasized the need for the multidisciplinary study of the nature and patterns of scientific development; Price urged statistical analysis of scientific behavior and publication trends. They were soon followed by a host of Western sociologists, economists, political scientists, historians, and ethnographers who applied far more sophisticated methods to the study of science as a social phenomenon. All this scholarship created a sense of urgency in the Soviet Union by the late 1960s that it was necessary to catch up with the West in science studies.

Most of the original leaders of the rebirth of science studies in the Soviet Union in the 1960s were closely connected to the hard sciences. The first centers were at the Department of Mathematical Statistics at Moscow University, the Institute of Mining in Novosibirsk, and the Institute of Cybernetics in Kiev. As Yakov Rabkin has noted, part of the reason the scientists were so interested in the field was that they were trying to get control of science policy after years of being submitted to Communist Party regulation.[61] One important goal of their analyses of the flows of scientific information, for example, was to show how important it was for Soviet scientists to have closer contact with Western scientists, both through more access to Western publications and through international exchanges that would allow person-to-person contacts.

By the late sixties or early seventies, professional historians of science in the Soviet Union became interested in science studies; a gradual shift occurred in the location of most research in the field, from centers of natural science and engineering to institutes on history of science and social science. The Institute of the History of Science and Technology of the Academy of Sciences of the USSR, located in Moscow, became a leading focus of *naukovedenie* under the directorship of S. R. Mikulinskii.

The assertion of control over the field by social scientists eventually led to tensions with the natural scientists, who were not any more interested in having historians or sociologists trying to "improve" science than they were in the efforts of Party leaders to achieve the same goal by constantly meddling in their research programs. As a result, by the late seventies and early eighties "science studies" had lost much of its allure to natural scientists and had become more and more just another type of social science research. It was now located in a variety of institutes pursuing work in history of science, economics, and sociology. As the studies became more and more specialized and embedded in the distinct terminologies of the individual disciplines, the aspiration of *naukovedenie* in the 1920s to show how planned science in a socialist state would be distinct from science elsewhere was largely lost.

NEW VIEWS IN THE HISTORY OF SCIENCE

In the last half of the 1980s, when all intellectuals in the Soviet Union were discussing reform, historians of science and scientists began reexamining the history of Soviet science. Many aspects of that history had been distorted or suppressed during the Stalinist period; even thirty-five years after Stalin's death, many topics still awaited adequate treatment. Most dramatic was genetics. In the 1920s and 1930s Soviet Russia was home to a vibrant school of geneticists, including Nikolai Vavilov, N. V. Kol'tsov, S. S. Chetverikov, A. S. Serebrovskii, and others.[62] In some areas, such as population genetics, the Soviet group was among the world's best. This school was destroyed by Lysenko and his supporters. Only after several decades was genetics restored in the Soviet Union. This long history contains at least three separate phases that Soviet historians by the late 1980s had still not properly studied: the development of a thriving science of genetics before Lysenko; the roots, causes, and consequences of the destruction of that school by the Lysenkoites; and the painful recovery of Soviet genetics in the sixties, seventies, and eighties.

In the late 1980s Soviet historians and biologists began to discuss these topics much more directly than ever before. They pointed out that Soviet biology had still not recovered entirely from Lysenkoism. Other scientists and historians noted that the weight of Stalinist history pressed on many other disciplines, too, not just biology. In 1987 and 1988 the journal *Problems of the History of Science and Technology* organized roundtable discussions "to restore the truth about various events in the history of Soviet science, which have until recently been either hushed up or glossed over."[63] The participants in the discussions pointed to Stalinist intrusions in physiology, cybernetics, physics, psychology, and other fields. Soviet journals published articles describing the work of prominent scientists who had suffered under Stalin. The journal *Nature*, for example, devoted its entire October 1987 issue to Nikolai Vavilov.

At the beginning of the last decade of this century young Soviet historians of science declared their intention to write "the social history of Soviet science," and they held several conferences devoted to that theme. Two of these historians wrote what they called "An Experimental Guide to an Unknown Land: A Preliminary Outline of a Social History of Soviet Science from 1917 to the 1950's."[64] It was, of course, sad and ironic that the social history of Soviet science was, to Soviet historians, an "unknown land" in 1990; after all, Soviet historians in the 1920s and early 1930s had pioneered the idea of such a history of science.

In their preliminary guide the two historians, D. A. Aleksandrov and N. L. Krementsov, indicated that they would develop a sophisticated interpretation of the evolution of Soviet science. They were not satisfied, for example, to blame all the ills of Soviet science on Stalin personally, although he would receive much blame. Some of the problems of Soviet science, they observed, began before Stalin achieved complete control, especially the progressive takeover by the government of all scientific institutions and funding sources. Even some scientists who later became heroes in their resistance to ideological incursions in Soviet science could not resist the temptation to seize the many levers of administrative power offered to them by the Soviet state system. Nikolai Vavilov, for example, was simultaneously president of the Agricultural Academy, director of the Institute of Plants, director of the Institute of Genetics, president of the Geographical Society, and the head of a number of other organizations. Aleksandrov and Krementsov correctly saw that such a concentration of posts later "eased the seizure of power" by Vavilov's rival Lysenko. Such a criticism of one of the icons of anti-Stalinists was rare among reformers of the late 1980s and it showed that the coming generation of historians of science in the former Soviet Union may go beyond the mere identification of heroes and villains and instead look for institutional and social reasons for the emergence of such individuals.

This reexamination of the history of Soviet science, still in process, means that our knowledge of many important events in the history of twentieth-century Soviet science is rapidly changing. A truly definitive history of science will probably never be written in any country, but the distance between what we know and what we need to know is very great in the case of the history of Soviet science. Many important archives have still not been examined by anyone, including historians in the former Soviet Union. The fact that so much of the written history of what became the world's largest scientific establishment is incomplete and biased is both a warning and a challenge to scholars currently working in the field.

8

Knowledge and power in Russian and Soviet society

ONE of the most revealing aspects of Russian and Soviet history has been the changing and contradictory relationship of science and the state, or knowledge and political power. Scientists obviously have needed the state for support, and the state has equally clearly wished to have the benefits of science, yet despite this link of mutual dependence, and possibly even because of it, the relationship between science and political authority in Russia has been filled with dramatic conflict.

Readers familiar with the political problems of science in other countries may question whether Russia and the Soviet Union represent a special case. The history of science in the United States, for example, is replete with political, religious, and financial struggles. A few illustrative examples in America are the rejection by religious fundamentalists of Darwinism in the nineteenth century and the continuing struggle over "creationism" in this century; the stripping of J. Robert Oppenheimer's security clearance during the early years of the Cold War; controversies over the financing and administration of federally sponsored research at the time of the creation of the National Science Foundation and later; and a multitude of recent political problems surrounding such issues as nuclear power, genetic engineering, and environmental degradation. It is clear that science and technology become involved in intense political disputes everywhere.

Nonetheless, the intensity of the conflicts between science and the state in Russian history has been far greater and more dramatic than in any other of the European and North American powers with which Russia has been most frequently compared. Indeed, at moments these confrontations have been so intense that an outside observer might wonder if the coexistence of science and the tsarist or Soviet state was possible. If in other countries the reputations of scientists and the degree of support for individual projects have often been in balance during the conflicts, in Russia and the Soviet Union scientists have faced dismissal from their positions, imprisonment, and even execution, and the very existence of scientific disciplines has been threatened. In recent decades, these dire outcomes no longer were prevalent, but Soviet scientists continued to struggle for rights that were assumed in most other countries, such as free travel abroad and looser bureaucratic control. The troubled relationship between knowledge and power did not begin in the Soviet period, but can easily be found in tsarist history. The root of the problem was the inherent contradiction that arises when a state tries both to modernize and to remain authoritarian.

The life of Russia's first significant scientist, Mikhail Lomonosov, clearly reveals the ambiguity of the relationship of science and the state in eighteenth-century Russia. As we saw in Chapter 1, Lomonosov benefited enormously from state educational policies that allowed him to make the transition from a peasant boy in the far Arctic to full member of the Academy of Sciences of St. Petersburg. He returned the compliment by writing obsequious odes to the rulers of Russia, Peter I, Anna, and Elizabeth I. Nonetheless, he protested vociferously against the dictatorial administration of the Germans who controlled the Academy, and he became involved in squabbles that on several occasions became physical. For a period of eight months he was actually placed under house arrest. He was honored more after his death than during his life.

A somewhat similar pattern, albeit with different specific causes, was seen in the life of another Russian scientist discussed earlier, Nikolai Lobachevskii, the great mathematician of the nineteenth century. Like Lomonosov, Lobachevskii was buoyed educationally by the modernizing policies of the state, and rose high above the primitive education of his mother. At the university level, however, both as a student and a young faculty member Lobachevskii collided with the efforts of the administration to enforce religious orthodoxy on the university community. The inconsistency and vacillation of tsarist educational policies were revealed a few years later, however, when the former iconoclast was chosen to be rector of Kazan' University, a position he could never have assumed without the government's approval.

Later in the same century the eminent chemist Dmitrii Mendeleev lived through a variety of political clashes while benefiting from the effort of the government to produce a technical intelligentsia. Educated in Western Europe on a government scholarship, Mendeleev also acquired an admiration of European economic and political policies. The government during Mendeleev's most creative years was suspicious of his political inclinations, moderate though they were. As described earlier, he was fired from his university position when he was fifty-six years old in retaliation for his support of a student petition calling for political reform. He was never allowed to resume teaching. Once again, however, the inherent inconsistency of the tsarist government's attitudes toward the technical intelligentsia was revealed some years later, when Count Witte appointed Mendeleev to a high government position and made him his technical advisor. Witte valued Mendeleev's insistence on the necessity for Russia to become an industrially advanced nation, and in order to make progress toward that goal he was more willing to accept political unorthodoxy than his predecessors.

Already we see in the lives of these three prominent scientists – Lomonosov, Lobachevskii, Mendeleev – a pattern of alternating rejection and embrace by the state that is typical of much of Russian and Soviet history. The pattern will appear many times, even in recent years, as the lives of such prominent Soviet scientists as Peter Kapitsa and Andrei Sakharov illustrate. Both experienced long periods of house arrest and persecution, and both received high honors before and after their periods of disgrace. Such ambiguity was visible in the attitude of the Soviet government toward all intellectuals, not just scientists and engineers, but it was particularly graphic with regard to the latter because of the obvious necessity for a modernizing state to rely on the technical intelligentsia to achieve its goals. But neither the tsarist nor the Soviet government was willing to accept the full political implications of this reliance.

Some Russian and Soviet scientists experienced the alternating pattern of support and repression under both the prerevolutionary and postrevolutionary regimes. An excellent example here is the outstanding geologist and geochemist Vladimir Vernadskii.[1] At the time of the Russian Revolution Vernadskii was already fifty-four, a distinguished geologist and a member of the Imperial Academy of Sciences, a man whose personality and political views were fully formed. Yet he lived on as an active scientist under the new Soviet government for almost another entire generation, dying at the end of World War II. As a liberal in politics and an eclectic in philosophy, Vernadskii never felt comfortable with either of the two systems of government that occupied almost all of his life, the tsarist empire that fell in early 1917 and the Soviet government that arose toward the end of that year. Under both systems Vernadskii feared for the fate of science and learning, and under both he was threatened with dismissal and arrest. And yet under both he also received high honors.

After the Bolshevik Revolution of 1917 the attitude of the government toward scientists and engineers became even more problematic than under the tsarist government. Despite the fact that the tsarist regime was oppressive, it still allowed considerable working room for independent-minded people. There is no known case under tsarist rule of the execution of a scientist or engineer merely because of his or her political views, although several student terrorists in the late nineteenth century who were executed had science or engineering backgrounds. In the latter case the punishment was for acts of violence rather than political deviance. Stalin and his secret police did not make the distinction.

The years of "War Communism," from 1918 to 1921, were the first period of a series of oscillations in the relationships of the Soviet state and the technical intelligentsia. In subsequent periods the state and the scientific community would sometimes move closer together; in others, they would move apart. As earlier discussed the more militant Bolsheviks were deeply suspicious of the "bourgeois specialists" in the years immediately after the Revolution. Lenin, however, insisted on a policy of watchful reliance on and cooperation with the scientists and engineers inherited from the previous regime. With the coming of the New Economic Policy in 1921 the bonds between the regime and the technical intelligentsia became closer. For a few years it appeared that an actual alliance between the technical community and the Party-controlled government might develop. However, at the end of the twenties came the Cultural Revolution and Stalinism, when the government and science moved apart. Then, in the late fifties, after the death of Stalin, when the scientists were basking in the successes of the atomic weapons and space programs, the two began to move toward each other once again, only to separate after the tightening of political controls in the midsixties and throughout the Brezhnev period of the seventies. Under Gorbachev (after 1985) a dramatic closing of the gap again occurred. Just as in the midtwenties, people began to speak of the possibility of an alliance between the state and the technical intelligentsia, perhaps even of the development of a technocracy; the prospect of a state in which leading scientists and engineers played influential roles was obviously far more pleasant to the scientific community than the terror it had experienced under Stalin or the tight controls it encountered under Brezhnev, but it was one that possessed its own potential dangers and soon provoked criticism from the nonscientific community and some members of the Party.

One of the most fascinating periods in the history of the relationship of Soviet power to the technical intelligentsia was the period after the Revolution and, especially, the 1920s, when very few of the technical specialists were sympathetic to the new Bolshevik regime. In fact, as late as 1928, only 138 of approximately 10,000 engineers in the Soviet Union were members of the Communist Party.[2] The great majority of the technical specialists would have preferred the Western-style provisional government that lasted for a few months in 1917 after the overthrow of the tsarist regime. Immediately after the Bolshevik Revolution in the fall of that year many engineers and technicians joined the strikes protesting the Bolsheviks.

In retrospect, however, it seems surprising how quickly these acts of defiance melted away and were replaced by, at first, implicit compromise with the new government, and, soon thereafter, explicit and even eager cooperation. Engineers and technicians who in private, among family and friends, continued to say that they considered the Bolsheviks socially and ideologically repugnant, nonetheless often found certain aspects of the new economic and political system beckoning. In the 1920s an intersection briefly occurred in the Soviet Union between technocratic tendencies visible in many countries and a vision of a unique possibility for technocracy in the new centrally planned Soviet economy.[3] The fact that this developing alliance failed dramatically and even violently at the end of the twenties does not erase its significance.

Historians like Edwin Layton have chronicled the rise of a technocratic ideology among American engineers in the first decades of this century.[4] World War I, the Russian Revolution, and the Great Depression were events that gave the movement special impetus. During World War I all the governments of the participating industrialized nations took control of their economies and assigned to specialized central committees responsibilities for producing and distributing the necessary munitions and supplies. Engineers and other technical specialists often played major roles in these planning committees. Some of them saw here a model for a more rational way of running an economy in peacetime. In the United States, Thorstein Veblen called in his 1921 book *The Engineers and the Price System* for the creation of committees he called "soviets" (the influence here of the Russian Revolution was unmistakable), which would run the economy in a "rational" way he distinguished from the "irrationality" of capitalism. When the Great Depression came at the end of the 1920s, some engineers were convinced that the appropriate response was to create a new kind of industrial economy, one that was centrally planned in accordance with the rational criteria best utilized by engineers. Howard Scott enjoyed a brief period of fame when he founded Technocracy, Inc., and called for the transfer of political power to engineers as a way out of the depression.[5]

Many of these events found echoes in Russia and the Soviet Union. There, too, the technical specialists had for the first time been given governmental planning responsibilities during World War I. In contrast to other countries, however, this central apparatus was retained after the Revolution and became the foundation for the Bolshevik planned economy. Many of the same engineers and technically trained specialists who helped the tsarist government during the war served the new Bolshevik government of Russia in the twenties. A number of them began to believe that by background and knowledge they were uniquely qualified to run a command economy based on the sophisticated technologies of industry, transport, communication, and agriculture.

A prominent leader of the new technocratic thought was Peter Pal'-chinskii (1875–1929), a remarkable Russian mining engineer whose biography has been insufficiently explored in either Soviet or Western literature.[6] Most mentions of Pal'chinskii in the existing literature note that he was later accused by the police of being the leader of the infamous

Industrial Party, that in 1929 he was shot in secrecy by the Soviet authorities, and that he was accused of conspiring with emigré and Western capitalists to overthrow the Soviet government. An examination of the record shows that Pal'chinskii was entirely innocent of the charges (an innocence later acknowledged in Soviet histories) but also that his approach to technology was one that was simply incompatible with the Stalinist mode of industrialization.

Pal'chinskii's program for industrial planning emphasized rationality. The planning engineer, he said, could not perform "miracles," but could make impressive contributions to the economy if he were allowed to assess each problem in an open and rational manner.[7] Local conditions, such as availability and costs of coal, water transport, educated workers, and construction materials, would dictate different solutions in different places to problems that at first glance appeared similar. The single most important factor influencing engineering decisions, Pal'chinskii maintained, was "human material."[8] Success in industrialization and high productivity in production were not possible, he repeatedly emphasized, without highly trained workers and adequate provision for their social and economic needs.

In order for the Soviet engineer to be able to apply this new form of rational analysis to problems of industrialization, Pal'chinskii believed that the engineer's role in society must change. Earlier, he observed, the engineer had been assigned a "passive" role by society, and had been asked to find solutions to technical problems which were assigned to him by higher authorities. Now, he continued, the engineer must emerge as an "active" economic and industrial planner, suggesting where economic development should occur and what form it should take.[9]

Pal'chinskii's vision of the new Soviet engineer was based on several different motivations. There is no question that an important source of his opinions was his conviction that this broad approach to engineering would result in more efficient industrial enterprises. But it is also clear that the new model engineer was important to Pal'chinskii's sense of professional pride. Edwin Layton observed that engineers in the United States in the same period had an "obsessive concern for social status." Pal'chinskii and his colleagues were eager to promote the engineer to a new prominence in society, and they believed that the new Soviet state, with its emphasis on centrally planned industrialization, provided unusual opportunities for this promotion.

The historian with the advantage of hindsight sees an almost inevitable collision between Pal'chinskii's program of industrialization and Bolshevism. Many tensions existed between Pal'chinskii's vision and that of the Communist Party, and these tensions grew explosively once Stalin gained absolute control over the Party. The basic tension was one of political authority. The Communists had never allowed professional groups to have the kind of autonomy or broad concerns that Pal'chinskii wanted for engineers. Pal'chinskii emphasized a rational approach to modernization, never promising more than what could be achieved, and stressed the necessity for fulfilling workers' full social, economic, and educational needs. Stalin, on the other hand, promoted an ideological campaign for economic advancement that included wildly unrealistic goals; he was quite willing to sacrifice large numbers of workers in the effort toward those goals. He insisted on the construction of gigantic hydroelectric power stations, which he found impressive in scale and revolutionary symbolism, ignoring local conditions that indicated that thermoelectric plants would be more economical in many instances. Stalin pulled in poorly educated peasants from the countryside and placed them in industrial plants where they were not qualified to perform their new tasks. The high accident rates that resulted were for Stalin acceptable costs, while for Pal'chinskii they were signs of irrationality, inefficiency, and injustice.

Stalin had always mistrusted the engineers educated before the Revolution, much more than Lenin. Stalin had been a member of a commission investigating the strikes of the university faculties and engineers immediately after the Revolution, and he considered the technical intelligentsia potential saboteurs. Pal'chinskii not only had a differing conception of how the Soviet Union should industrialize but also was a person with dangerous ambitions. In an interview with H. G. Wells in 1934, Stalin voiced his own opinion on the social function of engineers: "The engineer, the organizer of production, does not work as he would like to, but as he is ordered. . . . It must not be thought that the technical intelligentsia can play an independent role."[10]

While Pal'chinskii was the most outspoken engineer in developing an alternative vision of industrialization and the role of the engineering profession, other engineers pursued similar concerns. Kendall Bailes has analyzed the fuller dimensions of the "technocratic tendency" in Soviet politics in the 1920s.[11]

One center of technocratic thought in Russia in the twenties was the journal *Engineers' Herald* (*Vestnik inzhenerov*), whose main editor was I. A. Kalinnikov, who had held many responsible positions in engineering education, including the presidency of the famous Moscow Higher Technical School. Kalinnikov would later be accused in court, along with the already executed Pal'chinskii, of being one of the leaders of the Industrial Party. In 1927 Kalinnikov helped organize a discussion group called the Circle on General Questions of Technology that announced its intention to develop "a whole new worldview, fully adapted to contemporary technical culture."[12] One of the spokesmen of the Circle, the engineer P. K. Engelmeier, called for engineers "uniting, not only along trade union lines, but on the basis, so to speak, of ideology. . . ." Engelmeier's failure to mention Marxism in this new ideology immediately brought criticism from Communist Party ideologists.

Yet another focus of the Soviet technocratic movement was among the technical advisors to the central Soviet economic planning apparatus. Under the "Supreme Council of the National Economy" (*Vesenkha*) existed a Scientific-Technical Administration that was responsible for developing policies guiding industrial research and development. Among its leaders were other engineers who would later be brought to trial, including N. F. Charnovskii, S. D. Shein, and V. I. Ochkin. These men advanced a clearly technocratic program calling for the scientization of not only Soviet economic development, but also such fields as industrial psychology and management. According to one of their documents, "the future belongs to the managing-engineers and engineering managers." This was a phrase that Party critics later used against the engineers with great effect, maintaining that it showed that they considered themselves superior to the working class even though the Revolution had been based on the proletariat.

From the standpoint of Stalin's besting of his rivals in his struggle for political power in the Soviet Union it was very convenient that one of his main critics, Nikolai Bukharin, was associated with the technocratic tendency. This association was not a conspiracy, as Stalin later maintained, but merely a congeniality of viewpoint as well as bureaucratic association. Bukharin and his like-minded Party colleague A. I. Rykov often defended the engineers and praised the sort of technocratic planning which they favored. Bukharin even repeated at one point the phrase "The future belongs to the managing-engineers and the engineering-managers."[13] Furthermore, for a brief period of time he was the head of the Scientific-Technical Administration, which was a center of technocratic thought. Stalin thus had the opportunity to strike against both one of his main rivals and what he saw as the arrogant aspirations of the engineers.

And strike he did, with tragic effects. Pal'chinskii was seized and shot without any pretense of a trial. Then in November 1930 eight leading Soviet engineers were brought to trial on charges of conspiring to overthrow the Soviet government. This event, however, was only the beginning of a reign of terror among Soviet engineers, several thousand of whom were arrested. When one considers that only about ten thousand engineers existed in the Soviet Union at the time, the effect was obviously calamitous. The arrested engineers were thrown into camps, a privileged few in special research and development prisons where they were asked to work on tasks stipulated by the government. Alexander Solzhenitsyn's famous novel *The First Circle* describes one of these prison laboratories, created in the wake of the Industrial Party Trial. The result of this purge was that in the following decades Soviet engineers were expected to concentrate on the narrow technical tasks assigned to them by the Party leaders, and larger societal and economic issues that Pal'chinskii had considered intrinsic to the engineering task became the exclusive concern of the Communist Party leaders, and, under Stalin, of the supreme leader. One might assume that this turn of events meant the end of all possibility for the emergence of a technocracy in Soviet society. Under Stalin, such a possibility was, indeed, excluded. However, after Stalin's death the issue was to reemerge, although in a very different form.

The primary issue concerning the technocrats in the 1920s had been what attitude the regime should adopt toward those technical specialists whose experience and talents were needed but who were educated before the Revolution and whose loyalty was thus questioned. At the beginning of the 1930s a vast expansion of the Soviet technical educational system began. The education that the new engineers received was very narrow, with little attention to the economic and social issues that Pal'chinskii had stressed. Many of these engineers were from the lower classes, all of them benefited significantly from the Soviet system, and their loyalty was not questionable in the way their predecessors' had been. By the post–World War II period they occupied leading positions throughout the Soviet economy. Thus, by the 1950s the question of the role of engineers in the Soviet Union had been transformed. Industrial managers with engineering backgrounds were often the most stalwart supporters of the Communist Party's policies. An implicit agreement had been struck between the Party and a new generation of engineers: The engineers would support the Party's policies and the Party would promote the engineers in industry, agriculture, and the military forces. Throughout this period, however, the principle laid down at the end of the twenties remained: The engineers would not raise basic political questions, but instead carry out the Party's orders.

The question of technocracy came up again, however, from within the Party itself. As the generation of Old Bolsheviks who created the Revolution died off, or were killed in the Great Purges, where were the new Party leaders to come from? The answer was overwhelmingly from among the graduates of technical institutions, the new engineers. Indeed, by the 1960s and 1970s so many of the top Party and government functionaries had engineering backgrounds that American specialists on the Soviet Union began to comment that an engineering education played the role of preparing people for high political offices in the Soviet Union that law or business education did for future political leaders in the United States.[14] A few statistics will show how striking this trend was and how unparalleled it was among the industrialized countries of the world. Between 1956 and 1986 the percentage of members of the

Politburo, the top political body in the Soviet Union at that time, who had received their educations in technical areas rose from 59 percent to 89 percent.[15] If one defines "technocracy" as "rule by people who were educated in technical areas," the Soviet Union in its last decades was clearly a technocracy.

However, such a definition is inadequate. Most of the members of the Politburo were professional Party activists who had spent years in full-time Party work. How significant was it that many years before they had received technical educations? Leonid Brezhnev, the top leader of the Soviet Union for eighteen years, will serve as an example. Educated as a metallurgical engineer, he spent forty years in full-time Party work where the professional ethos was far more ideological and political than it was technical. Was he better described as a "technocrat" or as a "Party functionary"? Clearly, the latter title was more accurate. If "technocrats" are defined not only as "people who were educated in technical areas" but also as "people whose professional work in those areas was the formative experience in their lives" Brezhnev probably did not qualify as a technocrat.

However, even if we apply a much more rigorous definition of "technocrat," requiring that such a person must not only receive a technical education but actually have worked for a considerable portion of time in a technical specialty, the Soviet Union still was a country with a remarkably high percentage of its leaders who could be called "technocrats." Between 1956 and 1986 the percentage of the members of the Politburo who had not only received technical educations but had actually worked for at least seven years in positions requiring this education rose from 24 percent to 53 percent. And the average length of time the members of this 53 percent had worked in such positions was 19.7 years.[16] A strong case can be made that an education and two decades of work experience in a technical area qualifies a person as a technocrat.

But the new generation of technocrats who came to rule the Party and the economy in the post-Stalin period were not in the same mold as Pal'chinskii and his colleagues of the 1920s, who had wanted the creation of a technocracy in the Soviet Union. The vision of the members of that older generation was no doubt also flawed in its heavy concentration on science as the key to all human activity, but they had much broader social concerns than the following generation of technocratic functionaries produced by Soviet engineering institutions after the 1930s. Pal'chinskii had vainly called for much more emphasis in the education of engineers on economics, politics, and sociology. The professors of the engineering institutions who taught the new generation of engineers under Stalin understood very well what fates such concerns had brought Pal'chinskii and his colleagues; they therefore stayed as far away as possible from issues involving politics and social justice, and concentrated on science and technology. The humanities and social sciences education of Soviet engineers was supposedly served by obligatory courses on historical and dialectical materialism. Thus, a new type of technocrat was produced by Soviet engineering schools, a person with a strikingly narrow vision. And the top leaders of the nation, with few exceptions, came from this stratum.

Did the narrow engineering education of the great majority of the Soviet Union's top administrators influence their management style and policy preferences?[17] For decades Soviet leaders emphasized enormous construction projects that were seriously flawed from the standpoint of investment choices, environmental considerations, and social costs. Many of the top administrators were former engineers who admired mammoth construction projects but who knew little about economics and cost-benefit analysis, not to speak of sociology and human psychology.

The large-scale Soviet construction projects included the most ambitious programs in hydroelectric power and canal-building in the twentieth century, as well as the largest nuclear power plants ever built. Even more breathtaking projects were discussed, such as the Northern Rivers Project, which would have reversed the flow of several of Siberia's largest rivers in order to provide irrigation water for Central Asian agricultural regions. Called the largest civil engineering project in history and favored by land reclamation engineers and Central Asian political leaders, the project was vehemently opposed by environmentalists, Russian nationalists, and several leading economists. Soon after Gorbachev came to power, the Northern Rivers Project was shelved.[18]

Soviet agricultural policies also displayed a search for a technological fix for what was essentially an economic and social problem. The original preference for collectivized agriculture was based not only on the principle of socialist ownership of the land but also on a conviction that the full potential of modern agricultural machinery, such as tractors and combines, could not be fulfilled as long as the land was divided into small private plots. There was a certain justification for this belief, as the growth of the average size of farms all over the world since the 1940s indicates, but it was too narrowly grounded on reliance on technology and not sufficiently attuned to the economic and psychological factors that make the difference between a hard-working private farmer and a listless state employee. When it became clear to Khrushchev in the 1950s and early 1960s that Soviet agriculture was in deep trouble he reached once again for a technocratic solution: the extension of massive, mechanized state farms (which he called "agricultural cities") to virgin lands previously not cultivated. This program soon ran into trouble because it united flaws of the original technocratic vision of collectivized agriculture with the difficulties of raising crops on arid lands. Even after abandoning Khrushchev's utopian schemes for agriculture the Soviet government under Brezhnev continued to emphasize large mechanized collective and state farms; in the 1970s the Soviet Union produced more tractors and combines than any other country in the world. All this equipment, however, could not solve the motivational problem that lay at the base of low Soviet productivity in agriculture.[19]

The beginning of the former Soviet Union's recovery from narrow technocratic visions probably dated from the catastrophe of the Chernobyl nuclear power plant in 1986. That event was followed by a series of other less spectacular but equally significant technical failures in nuclear submarines, transportation (ship and train wrecks with large losses of lives), and environmental degradation. Mikhail Gorbachev identified the cause as "the human factor," and began calling for new approaches to technology, with much more attention to contextual and social issues such as economics, safety, workers' benefits, environmental risks, and managerial practices that took psychological and sociological factors into account. He began to turn more and more to economists and sociologists for advice and less and less to engineers. The ghost of Peter Pal'chinskii, who had warned of the effects of narrow technical education, was haunting the Soviet Union.

THE ROLE OF THE NATURAL SCIENTISTS

The natural scientists considered themselves part of the "creative intelligentsia," and tended to look down on the engineers as narrowly educated and concerned only with applied and managerial problems. After Stalin's death the natural scientists emerged as spokespersons on larger social issues in a way in which the engineers usually did not. Perhaps the engineers still remembered the stern treatment they had received a generation earlier when they advanced broader aspirations, or perhaps by the 1950s the engineers had been thoroughly socialized into the industrial establishment, with its emphasis on orthodoxy and loyalty.

By the late 1950s and early 1960s the importance of natural science to the Soviet government was incontestable, and the prestige of the scientists rose immeasurably. Nuclear physicists had given the Soviet government awesome weapons, and, just as in the United States, these physicists were enjoying an unprecedented prominence. Space scientists gained similar prestige after the Soviet Union launched the world's first artificial satellite in 1957 and put the first human being into orbit in 1961. In the intellectual thaw of the late fifties and early sixties these scientists began to speak out on topics outside their own special realms, a clear indication of their growing influence and ambitions.

In the Khrushchev period, later often termed a harbinger of the *glasnost* of Gorbachev, leading scientists attempted to change state policies, a marked departure from the glacial silence and repression in the scientific community which had prevailed only a few years earlier under Stalin. The Soviet scientists met Western colleagues in conferences such as the Pugwash series, where they discussed international problems of peace and security; they helped battle obscurantism at home, opposing the followers of Lysenko in genetics and giving shelter to geneticists in radiation laboratories who were trying to develop their field under difficult conditions; they helped to reorganize the Academy of Sciences, placing more emphasis on fundamental research and establishing new centers, such as *Akademgorodok* in Novosibirsk; and they called for reforms in Soviet education that would present gifted science students with special opportunities for rapid advancement. Many intellectuals, and especially scientists, hoped for a new Soviet political order in which they would play influential roles.

Unfortunately, this hope died after the fall of Khrushchev in 1964 and especially after the invasion of Czechoslovakia by Soviet troops in 1968. Much tighter controls over Soviet intellectuals and their contacts with Western colleagues soon followed. The dissident movement that had emerged in the early sixties was almost wiped out by trials and imprisonments. Anti-Semitism grew in the Soviet Union and often was directed against Jewish intellectuals.

A particularly revealing way to examine these transitions is to look at the biography of the Soviet Union's most famous scientist, Andrei Sakharov, who enjoyed great prominence in the early Khrushchev period, fell under increasing suspicion under Brezhnev, and was denounced, punished, and exiled at the end of the seventies, only to return to prominence and elected office under Gorbachev. It is a dramatic history, unprecedented in twentieth-century science, a story that tells much about the relationship of knowledge and power in the Soviet Union.

ANDREI SAKHAROV

Andrei Sakharov was the scientist who led the Soviet attempt to create a hydrogen bomb.[20] For his obvious brilliance in successfully carrying out this task much earlier than any Western observer thought possible he was elected to a full membership of the Academy of Sciences in 1953 when he was thirty-two years of age; he was the youngest scientist in the Soviet Union to hold such a rank. Together with Igor Tamm, he also worked out the Tokamak design for controlled nuclear fusion, the approach that still dominates research on nuclear fusion reactors today, in the United States as well as in Russia. The Soviet government showered him with honors, including the Order of Lenin and a Stalin Prize.

Before 1957, Sakharov seems to have had few or no qualms about his social role as a scientist. He believed that his work on the hydrogen bomb was justified, since he thought the world would be safer if both superpowers possessed such weapons, rather than only one. In his work on atomic weapons Sakharov differed from at least one of his fellow Soviet physicists, Peter Kapitsa, who refused to join the atomic project and who, as a result, was placed under house arrest for several years. However, Kapitsa later told me that his real objection was not to weapons, but to Lavrenty Beria, the secret police chief in charge of the atomic project.[21] Sakharov's path to the same destiny of house arrest was different from Kapitsa's, but when he finally arrived he found the conditions of his confinement even more severe.

Sakharov first came to public prominence in the Khrushchev period, that time in the late fifties and early sixties when scientists emerged as prominent spokespeople on many topics. The first steps in Sakharov's dramatic decline in official circles were closely connected with his scientific work. Intimately familiar with nuclear weapons research, he was deeply disturbed by the problem of radioactive fallout from continuing above-ground weapons testing. A sense of guilt, or at least responsibility, played a role here. Sakharov later wrote, "I felt myself responsible for the problem of radioactive contamination from nuclear explosions."[22] In the late fifties he began at conferences to speak against atomic testing, and wrote letters and memoranda to high government leaders. In 1961 Sakharov wrote a note to Khrushchev himself, and passed it to him at a dinner party. The note called for Khrushchev not to resume atomic testing after a three-year moratorium. Sakharov later reported that Khrushchev's response was, "[Sakharov has] moved beyond science into politics. He's poking his nose where it doesn't belong. . . . Leave the politics to us – we're the specialists."[23] Despite the reprimand, at this point Sakharov did not lose the tolerance of the higher authorities.

But he pressed on, raising other sensitive issues. He opposed Khrushchev's system of compulsory practical work for all Soviet students, maintaining that an interruption of the educations of promising students would be damaging to their development. He joined with other Soviet scientists who were trying to overcome Lysenkoism in biology, and helped block the candidacy of one of Lysenko's close associates for membership in the Soviet Academy. Khrushchev, who had given his personal backing to the candidate, became infuriated.

The three major problems that occupied Sakharov's attention in this interim period were all closely connected with science: atomic testing, scientific education, and the troubles of Soviet genetics. But Sakharov's involvement in protest efforts, even around such technical issues, automatically brought him into contact with Soviet dissidents with other concerns, who naturally looked to him for leadership and protection. Under their influence Sakharov's concerns became broader.

In 1966 he joined with twenty-four other prominent Soviet intellectuals to protest what they feared would be the rehabilitation of Stalin. Here was an issue that was of a different sort from the others; science was not its defining ingredient. The political horror of Stalinism did impinge on science, to be sure, but certainly no more than on many other facets of Soviet culture and politics. Sakharov was now approaching that invisible line in Soviet politics that separated permissible from impermissible behavior, a line that shifted with the times. In the next two years Sakharov moved toward the line at the same time that the line moved toward him. His famous 1968 manifesto, *Progress, Coexistence, and Intellectual Freedom,* went far beyond issues of science to sketch a vision of world peace based on cooperation between the Soviet Union and the United States and an eventual convergence of the two systems of government. This was a goal that Sakharov would later abandon, as he found the Soviet system less and less acceptable, and later still would resuscitate when Gorbachev's reforms brought back portions of his old dream.

By 1968 Sakharov had taken on issues the Soviet government considered to be entirely within its own preserve. In his memorandum of that year he hailed the reforms going on in Czechoslovakia, the attempt to "give socialism a human face" by Dubček and his supporters. According to Sakharov, the Prague Spring was a development of great promise. The degree to which Sakharov had contradicted official views became dramatically clear a few weeks later when Soviet military forces invaded Czechoslovakia and overthrew the Dubček government.

Sakharov's views had now crossed the line and gone quite beyond what Soviet authorities were willing to allow. He was removed from all classified research and was dramatically demoted in status and pay, although he retained his rank of "academician," a status the Soviet Academy of Sciences refused to remove even when Sakharov was in total official disgrace and under what amounted to imprisonment.

Following his demotion Sakharov turned more and more from scientific problems to social problems, and even within the category of social problems, he turned from general and abstract issues to deeply personal and moral ones. He became a defender of dissidents, attending their court trials whenever possible. He publicized the plight of persecuted religious believers and oppressed nationalities. He called on the Soviet government to allow citizens to exercise freedoms guaranteed by the Soviet Constitution but denied in practice. He helped organize a Committee on Human Rights. He protested a variety of Soviet policies, including the invasion of Afghanistan by Soviet troops in 1979.

Shortly after the last event he was seized by the secret police and illegally removed to Gorky, a city closed to foreigners, and told that he could not leave. He was constantly harassed, and on several occasions abducted to medical or penal institutions, where he was forcibly fed, bound, and even tortured. On his return to his Gorky apartment he was under constant surveillance. His telephone was removed and a special jamming transmitter was placed on the roof of his apartment building so that he could not listen to foreign broadcasts. He was constantly reviled as a traitor in the Soviet press, and even many of his colleagues in the Academy of Sciences (although not all) signed letters condemning him.

After Mikhail Gorbachev came to power, Sakharov's first hint that his situation was going to change came when a telephone crew arrived and reinstalled his telephone. The reason for the telephone became clear the next day when it rang and Sakharov found himself talking to none other than Gorbachev himself, who invited him to return to Moscow, imposing no conditions. Sakharov accepted and soon occupied a prominent position in Soviet society once again. The transition, however, was not easy. When scientific researchers in the Academy of Sciences nominated him as a delegate to the Congress of People's Deputies, Gorbachev's new national assembly, the administrators of the Academy removed him from the list. Only a massive demonstration in front of the Presidium of the Academy on February 2, 1989, persuaded the administrators to allow the nomination to go forward.

Sakharov was duly elected, and when the assembly met, Gorbachev called on Sakharov as the first speaker. It was a stunning turnaround. Yet the uneasiness of Soviet authorities with this outspoken scientist was still clear; when a speaker at the assembly later denounced Sakharov for criticizing the actions of the Soviet army in Afghanistan, Gorbachev stood and joined others in applause for Sakharov's critic. Just before Sakharov's death in December 1989, he was planning new political struggles aimed toward broadening democracy in the Soviet Union. The position of vital but disruptive scientists in Soviet society was still uncertain, just as it had been before the Revolution. One suspects that Lobachevskii or Mendeleev would have understood Sakharov's situation well.

Not enough time has yet gone by since the demise of the Soviet Union to know what sorts of relationships will develop between the scientific communities and the new governments of the independent republics. A crucial issue will be, of course, the degree to which those republics become truly democratic. If this transition is successfully made, for the first time in the history of the countries of the former USSR one of the essential tensions between science and government there – the effort to promote science within an authoritarian political system – will have dissipated. In such a situation the frictions between science and government will look much more like those occurring in Western countries, where the issues are often hotly argued but almost never have repressive outcomes. However, it would be much too early to conclude that this situation has already come about in the states of the former Soviet Union. Even in Russia, where the call for democracy is currently vocal, the results of the first elections to the new Russian Academy of Sciences, as described in the next chapter, show that political influence on science is still very heavy.

9

The organizational features of Soviet science

THE Bolsheviks who took over Russia in 1917 were committed to the creation of a modern, industrialized state and were enthusiastic about science and technology. Indeed, no group of governmental leaders in previous history ever placed science and technology in such a prominent place on their agenda. It is true that the more radical revolutionaries were suspicious of prerevolutionary scientists and engineers, seeing them as members of the despised bourgeoisie (attitudes that found violent expression later, especially when abetted by Stalin) but Lenin in the early years gave strong support to the technical specialists, and the Communist Party promised from the start the patronage of science on an unprecedented scale.

Much progress in science was made. In a period of sixty years the Soviet Union made the transition from being a nation of minor significance in international science to being a great scientific center. By the 1960s Russian was a more important scientific language in a number of fields than French or German, a dramatic change from a half-century earlier.

The first characteristic of science in the Soviet Union that the observer from Western Europe or North America was likely to notice was the uncommonly large role played by the central government. This feature was inherited by the Soviet government from its tsarist predecessor. However, at the very end of the empire, in the early twentieth century, a few important exceptions to the government's monopoly over the administration and support of research and development began to emerge, supported by industrialists and private philanthropists. Even a few private educational organizations emerged, such as the Shaniavskii University and the Moscow Women's University. The Ledentsov Society was a philanthropic organization that, allowed to develop freely, might well have become a private foundation similar to those emerging in Western Europe and North America in the early twentieth century. The Moscow Scientific Research Institute Society, organized in 1914 by a group of biologists with business support, was devoted to the raising of private funding for research. One of its early ventures was the organization of an Institute of Experimental Biology headed by N. K. Kol'tsov, a very successful institution in the years immediately after the Revolution, albeit under different financial and administrative control than that envisioned by its founders.

Private efforts to support research were cut off soon after the Revolution. The tradition of state control, still strong even at the very end of the tsarist period, passed intact to the new Soviet government in 1917 and was a characteristic of Soviet science and technology until the last decade of the century, when decentralizing reforms began. While the Soviet government in the twenties and thirties supported science and technology on a scale unmatched by that of any other government in the world, at the same time it reduced diversity and hindered initiative by creating a system in which all research organizations were parts of huge state bureaucracies.

The organizational structure of science and technology created by early Soviet administrators and planners was based both on what they inherited from the tsarist regime and on their vision of a socialist economy different from, and, in their opinions, superior to the capitalist economies of the West. Some of the flaws of Western science and technology, they thought, were their inefficiency due to competition among secretive independent industries, lack of centralized planning, and inadequate financial support from the government.

Perhaps the most significant reform of science which the Soviet government enacted in the twenties was the establishment of the idea of the research institute, and the creation of a system of research based on this idea.[1] There are, of course, research institutes in all scientific nations today, and there were quite a few even in the 1920s; one might doubt, therefore, that the Soviet idea of a research institute was in any way extraordinary. In the Soviet Union, however, the term "scientific-research institute" (*nauchno-issledovatel'skii institut*) came to carry a stature and a meaning that it did not have in any Western country. After World War II almost all outstanding scientists and engineers in the Soviet Union were members of an institute or had connections with one. (The main exception was university scientists, but faculty members without institute connections played a remarkably small role in Soviet research.) By 1990 there were several thousand research institutes in the Soviet Union, the majority falling under the jurisdiction of the industrial ministries. Many were quite large organizations, some employing thousands of researchers. The most prestigious were under the Academy of Sciences of the USSR and were usually in the basic sciences. In the entire academy system (including the republic academies) there were about six hundred institutes. By the last decade of the century certain streets in Moscow, Leningrad (St. Petersburg), Novosibirsk, and Kiev were lined on both sides with institutes. In Moscow alone, in the area of the city south of October Square and extending to Moscow University and beyond were dozens of research institutes. This geographical region contained probably the largest concentration of research talent in the world.

Several factors influenced the decision of administrators and Communist Party leaders in the 1920s to make the institute the basic organizational kernel of Soviet research. At that time the concept of research in integrated institutes, as distinguished from research in universities or academies, was still relatively new in all countries. The Soviet planners looked over their shoulders at the new types of research organizations developing in the United States, Germany, Britain, and France and attempted not only to catch up but actually to anticipate Western trends. At the same time, they found no one Western model entirely satisfactory. In the end, they combined foreign models with their own socialist innovations.[2]

Every year during the twenties Soviet scientists journeyed to Western Europe and America where they discussed not only their fields of specialization, but also questions of science organization. In 1923 the permanent secretary of the Academy of Sciences, S. F. Ol'denburg, went to France, England, and Germany to examine the organization of scientific research abroad. He went again in 1926, and upon his return home he wrote: "If the eighteenth century was the century of academies, while the nineteenth century was the century of universities, then the twentieth century is becoming the century of research institutes."[3]

Certain Soviet journals of the twenties were literally filled with reports on the best ways to organize science. The journal *The Scientific Worker*, for example, contained in almost every issue between 1925 and 1930 one or more reports on science in foreign countries; the total number of such "foreign country reports" in this one journal during this six-year period was over 50. The leading countries treated in *The Scientific Worker* were Germany (approximately 20 articles); the United States (approximately 10); France (8) and England (5).

Germany clearly emerged as the country most appropriate as a model for the organization of science. Each leading nation, however, had its Soviet analysts and even emulators. The United States was admired for the strength of its industrial research, the scale of its educational and scientific effort, and its "cult of efficiency." Many Soviet critics thought, however, that the rampant capitalism of the United States, the decentralization of science organization, and the emphasis on commercial applications of science made American science inappropriate for Soviet replication. The Soviet physicist A. F. Ioffe, for example, complained after a 1926 visit to the United States that antiintellectualism and crass commercialism were distorting American science.[4] Other Soviet critics extended similar analysis to Great Britain, where it was thought there was much less antiintellectualism than in the United States, but still a great deal of commercialism; furthermore, British scientists seemed to be particularly hostile to the idea of the planning of science.[5]

French institutions also had their Soviet admirers, especially such organizations as the *Institute Pasteur*, but Soviet observers seemed to believe that the organization of science in France was too heavily conditioned by a long and unique history to be reproducible. In addition, in the 1920s France seemed too static in its population and its institutions to provide many examples of the latest models of the organization of science.

Germany seemed to Soviet visitors both more familiar and exciting, despite its postwar economic difficulties. Germany, like Russia, had industrialized later than France and England, and its institutions in science and education were heavily conditioned by this upsurge of the late nineteenth and early twentieth century. Furthermore, in the 1920s Germany was attempting to adapt the institutions of a recently overthrown empire to the needs of a new government. Academic relations between Russia and Germany had always been close, particularly in science.

The appropriateness of the German model for science was, to be sure, ambiguous. Germany, like all the West, was in Russian eyes of the twenties a capitalist domain where the organization of science heavily reflected class interests. The Soviet critics perceived in the organization of science and education in Weimar Germany the influence of a philosophy of education and knowledge that Soviet radical critics, in particular, found unacceptable. Yet without question the Germans were creating new organizational forms in scientific research that provoked intense interest among Soviet scientists. Many of them liked the new institutes of the Kaiser Wilhelm Society.[6]

The points of difference between the German and the Soviet approaches concerned the relationship of teaching and research; of applied to pure research; of institutes and the Academy; and of creativity to individuality. When the Germans created the Kaiser Wilhelm Society in the first years of the twentieth century, a primary motivation was the belief that the German universities were losing their status as elite institutions and were being flooded by students, especially premedical students.[7] The professors called for relief from teaching duties. The planners of the Kaiser Wilhelm Society also wished to utilize the growing interests of industrialists in scientific research while preserving fundamental research from utilitarian encroachment. In addition, the Germans were seeking to keep the old Prussian Academy of Sciences a largely honorific organization free from new administrative burdens. And, finally, the Germans placed great emphasis on the "the free reign of a great personality," the scientific geniuses who would head the new institutes.[8]

The Soviet planners and science administrators could not accept many of these assumptions underlying the German science reform, which they saw as elitist and derivative of idealistic German philosophy. Yet they accepted many of the organizational principles of the German reforms while buttressing them with a different philosophy and set of motivations.

To committed socialists and communists, for example, the German effort to divorce teaching from research sounded very strange. To create citadels of pure thought, untainted by teaching responsibilities or concerns with educating a new generation of technical specialists, seemed to them to be a reinforcement of the "caste-like secludedness" they thought was an unfortunate characteristic of Russian science inherited from the tsarist regime.[9] Yet eventually the Soviet government separated advanced research from teaching in the Soviet Union to a higher degree than even in Germany, but for a very different reason. The Germans, as we have seen, wanted to promote research in separate institutes because they feared the effects of mass education upon the quality of scientific research. There were some Russian scientists who had similar fears – particularly among the old intelligentsia – but most of them were silent about these worries as the Soviet mass education campaign gathered speed. To the leaders of the Communist Party and its security organizations, however, there was an entirely separate, and somewhat ironic, reason for wanting to keep research and pedagogy fairly separate. The leading scientific researchers in the early years after the Revolution were not sympathetic to Soviet power and were feared as a bad influence upon Soviet youth. Since there was no way of quickly replacing the old intelligentsia (how does one transform a good Communist factory worker into an internationally known physicist?), the logical solution was to convert the universities into mass institutions where the spirit of socialism was carefully observed while maintaining the advanced research institutes on a separate level.[10] Thus, while the Germans feared the effects of mass education on science, the Soviet authorities feared the effects of bourgeois scientists on education.

The division between research and pedagogy that developed in the Soviet Union by the early thirties was never absolute. Many members of the Academy of Sciences taught in the universities, while the universities also developed laboratories. Research in some fields – for example, mathematics – remained strong in the universities. Furthermore, after 1930, when Communist influence within Academy institutes had become more secure, the Academy developed a system of graduate study (*aspirantura*).[11] Nonetheless, the resulting pattern of research and education was based on a degree of separation of the two that was much greater than in Western states, particularly the United States and England. Some Soviet scientists were worried even in the twenties about the effects of depressing the place of research in the universities.[12] Under Stalin, however, the pattern set in place in the twenties became even more entrenched. Many years later, under the leadership of Gorbachev and Yeltsin, it was necessary to make special efforts to revive university research.

On the question of the relationship of science to industry a curious affinity of views occurred between German mandarin intellectuals, who viewed the development of industry with mixed feelings, and looked nostalgically back to the days when German learning had been unsullied by industrial concerns, and the opinions of Soviet socialist critics, who perceived the "perversion of science" by capitalist industry. But the Soviet critics of capitalism did not believe that similar distortions of scientific research could occur in a socialist economy. Their reason for separating research from individual plants was not fear of perversion, but desire for the advantages of centralization. Once again, the Soviets supported the German model, but for a different reason.

A genuine distinction arose between the approach of the German science administrators to the Prussian Academy of Sciences and the Soviet science administrators. By drawing a line between the Prussian Academy and the new Kaiser Wilhelm Institutes the German leaders continued the trend that was already well established by the late nineteenth century: The Academy was a largely honorific organization, important as a learned body facilitating the publication of scholarly works and serving as an advisory council for the new research institutions, but not directly responsible for the organization and administration of the new complex institutes.

In the Soviet Union, however, the Academy of Sciences increased its status and activity with respect to other research organizations, particularly the universities. Many new institutes were created within its system. Although a large number of research institutions were also created outside the framework of the Soviet Academy, it became the prestigious center of Soviet science. From the 1930s onward the Academy of Sciences was the only one of the eighteenth-century academies of science of Europe that continued to dominate intellectually the scientific research of its nation in the twentieth century.

The basic principle behind the administration of the Kaiser Wilhelm Institutes was the free reign of the creative personality of the institute director, a principle fully in line with traditional German idealism; the first principle of the new institutes being created in the Soviet Union was, at least officially, that of collectivism. The new institutes were conceived as giant coordinating centers for the expression of cooperative endeavors in the exploration of nature and in the development of technology. The prominent plant geneticist N. I. Vavilov, discussed earlier, a sincere socialist, wrote in 1919: "From the work of solitary scientists we are shifting to collectivism. Modern institutes and laboratories – they are, so to speak, 'factories of scientific thought.' "[13]

Despite the Soviet praise of the collective principle in the administration of research, the actual management of the institutes was soon entrusted to powerful directors, just as in Germany. Because of political and economic pressures, the directors' creative personalities were never quite given "free reign." Political and economic pressures were always present. But in the Academy system in particular the directors of the institutes usually exercised great authority. In fact, one of the criticisms of the Soviet Academy voiced in later years was the view that senior distinguished scientists refused to step down from their authoritative positions early enough to make room for younger scientists.

The fact that the early planners and administrators of Soviet science and technology extended the system of centralized research institutes to industrial technology as well as fundamental science is important for understanding the later strengths and weaknesses of Soviet research. The early industrial managers believed that Western technology was hampered by competition between capitalist companies that concealed their research results from their rivals. In the United States, for example, Soviet critics castigated the "wasteful" competition in research on topics like synthetic fibers of the laboratories of Dow, Du Pont, Monsanto, and Union Carbide; these critics cited the "superior" example of the Soviet Union, where one centralized synthetic fiber research institute in a large city would, they thought, make its results available to the entire Soviet chemical industry.[14]

This desire to have large centralized research institutes located in capital cities rather than on industrial sites coincided with the general Marxist penchant for central planning that was so strong in the early decades of Soviet history. Many years later it would become clear that this prejudice for centralization harmed Soviet industrial research by the creation of unwieldy bureaucracies and by the distancing of research and industry from each other. Even in the 1920s a few Soviet administrators worried about this possibility. Iu. L. Piatakov, for example, proposed in 1925 that research institutes be attached directly to industrial enterprises.[15] This suggestion was sharply opposed by the leading Bolshevik F. E. Dzerzhinskii, a police official and also an industrial administrator. Dzerzhinskii maintained,

> In the attaching of institutes to factories or trusts I see a great danger, since this would mean the restricting of the scale of these institutes and their intellectual achievements. . . . It seems to me that these institutes must be independent . . . because their goals and interests must not be tied to those of trusts and factories (but) . . . must be connected to those of the whole country. Perhaps science will show that we need to abolish a whole series of trusts. This requires the independence of institutes from factories.[16]

In this statement we see both Dzerzhinskii's belief in the superiority of planning from above and also his suspicion that many of the private trusts being permitted to operate in the Soviet Union in the twenties as a part of the New Economic Policy (NEP) might soon be dismantled or nationalized (as they, in fact, were). Dzerzhinskii's view prevailed, and on-site industrial research remained weakly developed in the Soviet Union. Compared to other industrial nations, a strikingly small percentage of Soviet research scientists and engineers were employed directly in industry. Even as late as 1982, only 3 percent of Soviet researchers with the degree of *kandidat* (roughly equivalent to the American Ph.D.) were employed by industrial plants.[17]

The system of research and development that emerged from these considerations proved very capable at some tasks, not so capable at others. The centralized control permitted Soviet planners quickly to marshall resources for a few high-priority tasks, such as building hydroelectric power plants, creating atomic weapons, or promoting a space program. It also supported impressive fundamental research in a few areas, such as mathematics and theoretical physics. The system proved much less adept in providing industrial research support across the whole spectrum of high-technology industry, especially for the consumer economy. The centralized approach favored in the formative years of the Soviet Union created a chasm between industrial research and industrial production, with the industrial research institutes and the plants where production was actually taking place separated geographically, organizationally, and even philosophically. The researchers working in the centralized institutes in the large cities gained a reputation for being little interested in production at local sites. As later Soviet critics would say, they simply "threw their published articles over the transom," assuming that it was the responsibility of the industrial plants to put their ideas into practice. The factory managers, worried about meeting output quotas, were reluctant to interrupt production in order to try out new ideas that existed only on paper.

The Soviet discussions in the twenties over the proper organization of scientific research and the relevance of foreign models still influence debate today, even after the breakup of the Soviet Union. It is striking how many of the issues about the organization of science that were reopened in later years, especially after 1985, have their roots in the decisions of the late twenties and early thirties.

By the mid-1930s, Soviet science and technology had taken on the basic organizational features that remained constant until the end of the Soviet period, despite numerous small reforms. Research was concentrated in three distinct pyramids: (1) the academy system, headed by the USSR Academy of Sciences, but including a number of other specialized academies and the academies of each of the republics as well; (2) the institutions of higher education, such as the universities and technical institutes; (3) the ministerial research establishments, usually industrial research institutes. Above all three was the State Planning Commission (GOSPLAN) of the Council of Ministers, which determined the overall budget of each of the pyramids. And above the Council of Ministers was the Communist Party, acting through the Central Committee or the Politburo. These highest organs were, in theory, responsible for the actual research work of the pyramids, but on all but high-priority or ideologically troubled topics these central bodies did not interfere with research; the individual pyramid – academy, industry, education – controlled its own work within the assigned budget.

After the 1920s there were many different attempts in the Soviet Union to exercise better control over science and technology by creating one central organ assigned that function. Various committees and agencies were created with the legal power to perform this duty, but none was able to carry out such a universal function in a truly effective way. An attempt was made in the twenties to create an All-Union Committee for Science, but disputes among the Academy, the industrial commissariats (antecedents of the later ministries), and the Soviet government resulted in only fragmented control. In the thirties and forties a variety of agencies attempted to fulfill the same function (for example, the Department of Science of GOSPLAN), but actual supervision of research by these agencies was limited. The industrial ministries and the Academy of Sciences continued to perform the most important research functions, working within the framework of general goals laid down by the Party and government. After World War II a decree of the Party Central Committee established the State Committee for the Introduction of New Technology into the National Economy (GOSTEKHNIKA), which went through several permutations and name changes before being transformed in 1961 into the State Committee of the USSR Council of Ministers for the Coordination of Scientific Research Work (GKKNIR), which was given broad powers but which, again, was not able to use them productively. In fact, the GKKNIR was even less successful than some of its predecessors, in part because its existence coincided with the decentralization of the economy by Khrushchev on the basis of the "regional economic councils." With the reemergence of the centralized ministerial system of directing the economy in 1965, a new research coordinating committee, the State Committee for Science and Technology (GKNT), emerged, one that existed until the end of the Soviet Union in late 1991.

According to its statute, the State Committee for Science and Technology was responsible for the coordination of science and technology policy in the entire USSR. The State Committee was therefore often seen as a serious rival to the Academy of Sciences as the most influential science body in the Soviet Union. The actual practice, however, was that the State Committee concentrated its attention on trying to coordinate industrial research and on obtaining foreign technology, not always with a great deal of success. After jostling for power, the Academy and the State Committee worked out a modus vivendi by which the Academy continued to be the primary organization in fundamental science, while the State Committee concerned itself with technology policy and technology transfer.

These two organizations, the most important science and technology bodies in the Soviet Union, were quite different. The State Committee had no research laboratories (although it did have some policy and information centers) and was primarily an administrative body. The Academy of Sciences was the largest single research organization in the world, with hundreds of institutes and tens of thousands of researchers.

As already noted, the greatest problem embedded in the organizational scheme of Soviet science and technology was the separation of research and industry. Again and again the Communist Party and the government called for this gap to be narrowed. An examination of the record of resolutions passed by top Communist Party and government organs shows that in 1926, 1929, 1947, 1961, 1965, 1966, 1975, and 1985 a major goal of the directives was bringing science and production into closer contact.[18] It is too early to say whether the breakup of the Soviet Union and the devolution of power to the individual republics that came in 1991 will eventually help with this problem. So far the problem has remained.

The Academy of Sciences before 1929 did not have engineers among its members and did not engage in engineering research. The first engineers were elected in that year; between 1929 and the late 1950s their numbers grew until the department of technical sciences (engineering) had more members than the three departments for social sciences and humanities combined.[19] The Party leaders hoped that by placing engineers in the Academy the division between pure and applied research would be overcome, to the benefit of industry. The engineers in the Academy, however, found their research homes in large centralized institutes similar to those of their colleagues in the basic sciences, and often had only distant contacts with industry. As a result, the government and Party continued to pester the Academy with requests for industrial assistance, which was often rendered late and unenthusiastically. Until Stalin's death in 1953 it was not possible for the fundamental scientists in the Academy to question publicly the policy of flooding the Academy with engineers, although many of them feared that the strength of the Academy in basic sciences was being undermined without the problem of assistance to industry being solved. In the 1950s, however, talk of reform began to spread in the Academy. In 1955 the president of the Academy, A. N. Nesmeianov, observed that there were too many engineers among the members and that government and Party officials continually interrupted the research of the Academy institutes with requests for solutions to narrow production problems. According to Nesmeianov, the Academy should concentrate on what it does best, fundamental research, and leave industrial research to other agencies, presumably the research arms of the industrial ministries.[20]

Soon a great debate over science and industrial policy erupted in the Soviet Union. Dozens of articles appeared in the scholarly and popular press, with a variety of proposals for improving the quality of Soviet research in both fundamental and applied science.[21] The most aggressive proponent of the fundamental scientists' position was the Nobel laureate chemist N. N. Semenov, who maintained that there was no necessity for engineers in the Academy. He argued that science is not an appendage of industry, as many Stalinists maintained, but has its own independent assignment, which is the "thorough study of nature." He disagreed with Engels's emphasis on practical needs as influences on the development of scientific theory. Did industry ever hint at the possibility of unleashing atomic energy, he asked? On the contrary, Semenov maintained that atomic energy was the fruit of pure science. Semenov dodged here the question of how much influence military requirements, a practical need, had on the time and place in which atomic energy appeared. But Semenov certainly had a point in calling for a sophisticated understanding of science that would encompass two rather independent sources of scientific advance: the demands of production, which he thought Soviet planners never failed to emphasize, and the internal logic of science itself, which he thought Soviet planners ignored.

The leading spokesman for the engineers – and the chief opponent of the point of view expressed by Semenov – was Ivan Bardin, the head of the department of the Academy that contained most of the engineering institutes. Bardin had been one of the first engineers ever elected to the Academy, and he flatly accused Semenov of trying to erase the past thirty years by advocating a return to the Academy of prerevolutionary times, which had been an ivory tower of theoretical research. Bardin pointedly asked, "Just why must the USSR Academy of Sciences, which was awakened to the need for contact with life by V. I. Lenin, constrict the range of its work and retreat to the position of the ill-remembered Imperial Academy of Sciences?"[22]

The most important reason for the emotional content of Bardin's reply was Semenov's proposal to oust the engineers from the Academy, thereby depriving them of the title of "academician," which carried greater prestige than any other professional title in the Soviet Union. Underlying the whole debate, of course, was the tension between theorists and practical engineers that has appeared in scientific organizations all over the world. Engineers have often suspected that theorists consider them intellectually and socially inferior. Evidence for this prejudice can be found almost everywhere, from the greater prestige usually awarded to universities over technical institutes, to the fact that engineers are usually excluded from awards like the Nobel Prize. In the Soviet Union this same tension existed, despite the professed Marxist preference for labor over "empty" theorization and despite the fact that in the first decades of Soviet history the practical life of the engineer was constantly praised in official writings, novels, and films. It is interesting that the ethos of the pure scientist was not only preserved during this period, but actually grew in strength.

The decline of the prestige of engineers in the Soviet Union was particularly evident in the post-1960 decades. It seems the fundamental scientists won this debate, in large part, because they gained the support of Nikita Khrushchev. Khrushchev was poorly educated and on many occasions he displayed a lack of appreciation of theoretical science. Nonetheless, he somehow grasped that it was time to shift the priorities of the Academy from industrialization to the further expansion of knowledge. He observed in one of his speeches, "I consider it unwise for the Academy of Sciences to take on questions of metallurgy and coal mining. After all, these areas were not within the Academy's domain earlier. . . ."[23]

In the first half of the 1960s, the system of science administration in the Soviet Union underwent several reforms. About half of the institutes in the Academy, most of them involved in industrial research, were removed from the Academy structure and assigned to industrial ministries. With the Academy no longer directly involved in engineering research, it was necessary to create a new body responsible for coordinating Soviet science and engineering policy. This was the aforementioned State Committee on Science and Technology, which was created in 1965 on the basis of an earlier committee established in 1961. Some of the members of the Academy agreed with Semenov that these reforms allowed the Academy to concentrate on what it did best. Others, however, worried that the restricting of the role of the Academy to basic research would diminish the relative standing of the Academy in the entire Soviet scientific establishment. Furthermore, a strong case could be made by the late sixties and early seventies that the old dichotomy between fundamental and applied science was no longer valid. In some of the most exciting areas of high technology, such as computer science and molecular biology, the distance between fundamental science and application was so short that often the same people were involved in both stages. (Military and space research also often involved both scientists and engineers.) During the years after the reforms of the sixties, therefore, the Academy gradually again became more and more involved in technologies, not old technologies like coal and steel, but new ones in biology, computers, automation, space, and defense. Some members of the Academy, adjusting to the new world of high technology, welcomed this development, while others, still worried that the Academy would become the handmaiden of industrial and military interests, continued to resist it. The top administration of the Academy tried to strike a balance by moving into the new areas of technology while prohibiting Academy institutes from deriving more than 25 percent of their budget from industrial contracts.

In the seventies and eighties Soviet administrators continued to make many attempts – with varying degrees of success – to bridge the gap between basic research and industrial application. They formed a great variety of "associations," "technological centers," and "complexes," each including under one umbrella all stages of the R and D cycle from basic research to production. Most widespread were the science-production associations (NPOs), which numbered over three thousand by the late seventies. One of the best-known NPOs was the Svetlana Electronic Instrument Manufacturing Association in Leningrad, which introduced a range of improved scientific instruments.

Although the science-production associations had some success in introducing innovations into Soviet industry, they often failed to gain the close involvement of the best fundamental researchers. The NPOs were usually under the control of the production ministries (government departments responsible for research and production in specific fields), and rarely worked closely with institutes in the Academy of Sciences. Furthermore, despite their common affiliation with the NPOs, the factories and the research institutes often continued to follow their own independent plans and goals.

By the late seventies many Soviet scientists and research administrators were beginning to believe that further reforms were necessary. Soviet science and technology had achieved many successes, such as the space program and the creation of the second most powerful industrial establishment in the world, but new worries were emerging. Although the Soviet space program remained a leader, it no longer enjoyed the preeminent position that it held in the late fifties and early sixties, and the position of the Soviet Union as the second largest industrial economy in the world was being overtaken by Japan. In some areas of technology, especially computers and genetic engineering, the Soviet Union had clearly slipped behind a number of other nations. It was apparent the productivity of Soviet research was falling; the Soviet Union had by the late seventies the largest research establishment in the world, but it seemed to be receiving inadequate return on this enormous investment. Thane Gustafson, a Soviet studies specialist at Georgetown University, observed, "By any measure – whether Nobel prizes, frequency of citation by fellow specialists, origin of major breakthroughs, or simple quantity of publications – U.S. scientists lead their Soviet colleagues in most disciplines, and in many there is simply no competition."[24] In 1977 the noted Soviet economist L. V. Kantorovich and his colleague A. G.

Kruglikov calculated that science had boosted the USSR's national income by 2.2 percent in 1966, but that the figure had fallen to only 0.8 percent by 1976.[25] Furthermore, the Soviet scientific establishment was visibly aging; between 1976 and 1986 the percentage of researchers in the Academy of Sciences under age forty holding the degree of doktor declined by 67 percent, and the percentage of full members (academicians) under age fifty declined by 66 percent.[26] The morale of Soviet scientists had also been damaged by the repression of dissidents like the prominent physicist Andrei Sakharov, anti-Semitism in research institutions, and the increasing conservatism of the late Brezhnev period.

These concerns were present at the time of the assumption of the leadership of the Communist Party by Mikhail Gorbachev in 1985. Gorbachev launched the most fateful series of reforms in the entire history of the Soviet Union, a radical program that finally, despite his efforts, brought an end to the country he was trying to revive and the political party he had been chosen to lead. Despite his disappearance from the political scene after December 1991, many of the reforms that Gorbachev initiated continued to influence society in the following era in the former Soviet Union. Science and technology felt the full impact of these changes.

Gorbachev's more general political and economic reforms, such as the policy of *glasnost* (openness), liberalization of political controls, and decentralization of economic activities were aimed at Soviet society as a whole, rather than at science and technology per se. Nonetheless, these broad reforms were part of an effort to modernize Soviet society and to improve the working conditions of creative people, especially scientists, engineers, and skilled laborers.[27] The release of the famous physicist-dissident Andrei Sakharov from forced exile in the city of Gor'kii was an example of *glasnost* extended to the scientific sphere. The allowing of factory workers to participate in meaningful elections of factory directors was a recognition that highly skilled workers must be given more autonomy than was necessary during the basic industrialization drive. And the decentralization of economic activities, including price formation, was designed to foster the energizing and innovative effects of competition.[28]

Reforms not specifically aimed at science and technology but which had effects in this area included the Law on Individual Labor Activity, launched in May 1987,[29] and the Law on the Cooperative System, adopted in May 1988.[30] Under these laws private citizens, operating under a system of registration and control, were allowed to engage in profitable small-scale trade and household services. Most of these cooperatives were small restaurants, repair shops, taxi services, and studios in the arts and crafts. Soon, however, scientists and engineers began using them to provide high-tech consulting services to local industries, to establish scientific supply houses, to create innovative software programming centers, and to build scientific instruments. Here was a belated recognition by Soviet administrators that giant enterprises are often less creative than individuals or small groups. The growth of private initiative in the area of science and technology was quite remarkable. By 1990 there were more than 10,000 science and technology (S & T) cooperatives in the Soviet Union involving about 250,000 people and handling business of approximately 3.5 billion rubles a year.[31] The largest single area of activity of the new cooperatives was computers (both hardware and software), but the movement rapidly spread throughout the communication, transport, chemical, medical, and power industries.

The work of the new S & T cooperatives, even though impressive, was hampered by both official regulations and public resentment. According to the regulations, the cooperatives were basically moonlighting organizations whose employees worked at regular jobs during the workday. Often the cooperatives were "attached" to official scientific research institutes and used the equipment of those institutes in the evenings and on the weekends. Many disputes naturally arose between official administrators and private entrepreneurs over compensation and diversion of resources. In some cases the cooperatives were so successful that they actually came to dominate their parent institutes; in other cases the official organizations tried to absorb the cooperatives. Some bureaucrats tried to squelch the cooperatives with a multitude of regulations, including punitive taxation. Others simply levied shakedown payments. Criminal organizations demanded protection money from the cooperatives. Ordinary citizens often resented the enrichment of entrepreneurs and wrote letters to the newspapers asking that the cooperatives be closed down. By the early nineties the situation of the S & T cooperatives was a very mixed one. With the move toward more radical economic reforms that came with the end of Communist rule in late 1991, however, private initiatives of many different types joined the older cooperatives.

While private groups sprang up in Soviet science and technology the official administrators continued to seek a more effective system for stimulating innovation. In the middle and late eighties the favored organization for high technology was the Interbranch Scientific-Technical Complexes, known as MNTKs.[32] These conglomerates combined both research and production facilities and were usually under the control of institutes of the Academy of Sciences, instead of production ministries. Within a few years about two dozen of the new MNTKs were created. They included complexes working on biotechnology, machine tools, computers, robotics, fiber optics, chemical catalysis, lasers, welding, automation, and petrochemicals. All had acronyms for titles, such as ROBOT, KATALIZATOR, BIOGEN, and PEVM (personal computers).

Despite this promising start, by the beginning of the 1990s the movement toward MNTKs slowed. More and more, private initiatives in S & T cooperatives drew away the creative energy of the official organizations like the Academy institutes. The industrial ministries resisted the Academy's encroachment on their prerogatives and managed to prevent the Academy from entirely taking over the MNTK movement. Furthermore, radical critics began to attack the Academy itself as an example of centralized research. Thus criticized both from the right (the bureaucrats of the centralized ministries) and from the left (reformers who, wanting a Western style scientific establishment, questioned the need for Academy institutes) the MNTK movement with its principle of research complexes dominated by Academy institutes lost much of its steam.

Under Gurii Marchuk, who was president of the Academy of Sciences of the USSR from 1986 until its disappearance in December 1991, the Academy's leadership, in response to pressure from younger researchers, rather reluctantly enacted several reforms. The goals of the new measures were to decentralize administrative controls, liberalize travel regulations, revitalize the subsidiary academies in the Soviet republics, democratize the system of choosing the top administrators, and enforce retirement of older administrators.

The last reform caused considerable controversy. Under Brezhnev some institute directors and senior administrators remained in their jobs into their eighties or even their nineties. The new policy of the reform era required directors of institutes to retire at age seventy, thus opening the way for younger, more creative scientists and overcoming the lethargy that prevailed in such institutes. Not surprisingly, senior Soviet scientists complained that these policies limited their privileges. In consolation for older scientists losing their administrative posts, the Academy created the ranks of "honorary director" and "advisor to the presidium," positions with prestige but little power, and also allowed these senior scientists to retain their perquisites, such as limousines and offices.[33]

The traditional method of funding research in the Soviet Union was through block funding of large institutes. Every year each institute usually received an incremental increase in its budget; that sum would then be split up among the various departments of the institute. This method of funding gave great authority to the institute director. Individual researchers within the institute were not free to apply for funds to outside organizations.

This system had never been absolute. It was supplemented by contracts between institutes and various other government organizations, civilian and military, for task-directed research. These contracts also went through the central administrations and were under the control of the institute directors. As mentioned, within the Academy of Sciences, the sum of such contracts was normally restricted to 25 percent of any institute's total budget. This policy did not normally apply to institutes and universities outside the central Academy; some of them became heavily dependent on industrial and military contracts.

One of the many reform suggestions of the late eighties was a move toward a system of peer-reviewed support for research similar to that in many Western countries. Several central funding organizations were established, and principal investigators were encouraged to submit applications for peer review. This reform increased the authority of individual researchers and diminished that of institute directors, even though directors still possessed administrative review powers over the individual proposals (as is usually the case in the United States and other countries).

An unprecedented protest against the senior administration of the Academy broke out in the spring of 1989. The demonstration was provoked by the Academy presidium's refusal to nominate several reformers to the new Soviet legislature, including Andrei Sakharov and Roald Sagdeev, the long-time head of the Academy's institute of space research. Several thousand Academy workers gathered in the driveway before the presidium building just off Lenin Prospect in Moscow and jeered the decision, waving signs calling for President Marchuk and the members of the presidium to resign, and urging democratic reforms within the Academy structure. The presidium eventually relented to the pressure and allowed the members to nominate Sagdeev and Sakharov to the legislature.

In the last years and months of the USSR the Soviet scientific and industrial establishment began to open up to the influence of the world marketplace. The number of ministries and enterprises licensed to conduct foreign trade greatly increased. Through imports, this measure increased the flow of Western technology into the Soviet Union. And through exports, it allowed the Soviets to use the marketplace as a laboratory for testing and refining their products and technologies.

Soviet enterprises were also permitted to set up joint ventures with Western partners inside the USSR, with the chief aim of improving access to foreign technology. Soviet economic administrators expressed strong interest in joint ventures in fields such as the textiles, machine building, petrochemicals, pulp and paper, agriculture, electronics, and communications industries. They ran large advertisements in such places as the *International Herald Tribune* and *The Wall Street Journal* promoting themselves as business partners. For the first time in Soviet–West relations, a joint task force was established by the Soviet Chamber of Commerce and the International Chamber of Commerce for the purpose of studying the legal and administrative framework of joint ventures in the USSR. A rather large number of American, Japanese, and West European companies initiated agreements. Among the leaders were Monsanto, Occidental Petroleum, Archer-Daniels-Midland, Combustion Engineering, and Siemens. After the Soviet Union supported the move toward a unified Germany in 1990, the German government and industries became active, rendering direct technical aid to the Soviet Union. After the end of Communist rule in 1991, and the dissolution of the Soviet Union, republican leaders such as Boris Yeltsin pushed for a great expansion in foreign investment in industry.

RUSSIAN SCIENCE AFTER THE END OF COMMUNISM

The end of communism and the breakup of the Soviet Union in 1991 brought major changes to science and technology in that country and its successor states. From 1925 until 1991 the Academy of Sciences, home of most of the leading work in the fundamental sciences, was an "all-union" organization, meaning that it belonged to the entire Soviet state, not to any one of the republics within it. Of the fifteen republics, all but Russia, by far the largest, had their own republican academies of sciences, over which the all-union, or "big" Academy exercised intellectual leadership. Since most of the members, research institutes, and property of the big Academy were located in Russia (and, in fact, were overwhelmingly in or near a few large cities such as Moscow, Leningrad, and Novosibirsk) many members of the old Academy believed for years that creation of a separate Russian academy was unnecessary, and would, in fact, be redundant.

This situation began to change dramatically with the growth of nationalism in the late 1980s and early 1990s, as each republic strove for its own identity, and, ultimately, its independence. Since the big Academy was dependent for its funding on the Soviet state, the disappearance of that state meant the Academy's demise unless all the newly independent republics would agree to support it through some sort of vestigial central coordinating organ, or the Russian republic would itself pick up the burden for the entire Academy. Among scientists and science administrators much of 1990 and 1991 was devoted to a strenuous debate of these issues.[34] In the end, the hope for retention of some central organs responsible for the big Academy died, and Russia assumed control of a new Russian Academy of Sciences made up of institutions from the old Soviet one. In return for assuming financial responsibilities, Russia forced a major expansion of the membership of the Academy in an extraordinary series of events, to be briefly described below. However, the expansion of the membership was not accompanied by a growth in financial support; on the contrary, scientific institutions in the former Soviet Union were caught in a severe financial crisis.

The fate of the old Soviet Academy of Sciences was also influenced by the fact that it was increasingly seen by reformers as a politically conservative pillar of the discredited Soviet regime.[35] Actually, in recent decades the Academy had occasionally displayed a degree of independence, such as its refusal under Khrushchev to elect the Lysenkoist biologist N. I. Nuzhdin to full membership despite Khrushchev's insistence, and its unwillingness under Brezhnev to expel Andrei Sakharov from membership, despite calls for doing so by Communist Party organizations. But these small acts of unorthodoxy in earlier years could not counteract both the appearance and the reality that the Academy had become a part of the centralized apparatus of the Soviet state. It cooperated fully with the secret police in controlling the travel abroad of its scientists who wished to attend scientific meetings, it was dominated by senior scientists who often ignored the needs of younger colleagues, it condoned the anti-Semitism that infected some of its research institutes, and it suffered from the bureaucratic rigidity endemic to all Soviet organizations.

The link between the Academy and the old Soviet order was solidified by the Academy's behavior during the abortive coup of August 1991. At a moment when many intellectuals rose to the defense of democratic government and continuing reforms the leaders of the Academy by their silence indicated their willingness to live with the right-wing coup leaders and their hope for a return to the old order. For many intellectuals and scientists the Academy's apparent complicity with the military and police leaders of the attempted *putsch* was evidence that it must undergo a radical restructuring, along with most other Soviet institutions.

Even before the August 1991 coup attempt the future of the Academy was increasingly uncertain. The previous year, on August 23, 1990, Gorbachev had issued a decree that granted the Academy of Sciences of the USSR autonomy from the state and had given it ownership of all the state property over which it had previously made use. The leadership of the Russian Republic saw this move as a challenge to Russia's right to control all the property within its borders, and the Russian Supreme Soviet therefore refused to recognize the validity of Gorbachev's decree. Furthermore, Boris Yeltsin, president of Russia, and a number of influential members of the Russian Supreme Soviet indicated that they favored the creation of a new organization, a "Russian Academy of Sciences," which would obviously be a rival to the old Soviet Academy, vying for most of the same institutions and resources. On March 25, 1991, the presidium of the Russian Supreme Soviet created a special organ, the Central Organizational Committee, to work out a plan for a new Russian Academy of Sciences.[36] The committee was chaired by Academician Iurii Osipov, an applied mathematician from Yeltsin's home city of Ekaterinburg (Sverdlovsk).

Osipov's committee elaborated a complicated system for the election of members to the new academy. Since one of the criticisms of the old Soviet Academy had been that it was made up of members living mostly in the large cities of Moscow, Leningrad (now St. Petersburg), and

Novosibirsk, the new electoral system favored provincial cities by being based on twelve regional committees. Of all the electors, 70 percent were appointed by the twelve regional committees, 20 percent were appointed by the Academy of Sciences of the USSR, and the remaining 10 percent were appointed by Osipov's committee itself.

The final vote took place on December 6, 1991, and a total of 39 candidates were chosen as full members of the new academy and 108 were chosen as corresponding members. An examination of the slate of elected members quickly showed that the elections were not free of political influence, as the reformers had claimed would be the case, but instead had been shaped by a constellation of new political forces. While the old Academy of Sciences of the USSR had often been under pressure from the Communist Party and the Soviet government, the newfledged Russian Academy of Sciences was clearly under the influence of the emergent non-Communist government of Russia. One of the first-elected corresponding members of the Russian Academy was Ruslan Khasbulatov, chairman of the Russian Supreme Soviet. Khasbulatov, an economist, had never been considered for membership in the old Academy of Sciences of the USSR. Another corresponding member of the new Russian Academy of Sciences was Ernst Obminskii, the Deputy Minister of Foreign Affairs of Russia, a person who had earlier been voted down as a candidate for the all-union Academy. Yet another example of politics at work in this new scientific institution was the election to full membership of Vladimir Shorin, the chairman of the committee on science and education of the Russian parliament. And several other examples of candidates being chosen for reasons that had more to do with politics than intellectual achievements could be identified. And, finally, twenty candidates with Jewish names were rejected, which probably indicates continuing anti-Semitism. Nonetheless, it would be a mistake to believe that all the members of the new academy were chosen on a political or prejudicial basis. A number of outstanding scientists who had been ignored by the old Academy were elected to the new one. In mathematics, in particular, the elections of such candidates as Iu. L. Ershov, O. A. Oleinik, Ia. G. Sinai, and I. R. Shafarevich were overdue recognitions of talent (although Shafarevich's published anti-Semitic views sullied his fine reputation in mathematics).[37]

During the final months and days of the Soviet Union there were actually two rival academies of science, the old all-union one and the new Russian one. For a while they even had the same name, since the Academy of Sciences of the USSR in October 1991 changed its name to the Russian Academy of Sciences. However, the two academies, claiming the same research establishment, obviously could not both continue to exist. Furthermore, there was no way that the old "big" Academy could force its will on the new one, since the Soviet government which had been its traditional source of support was disappearing as the discussions progressed. Yet the newborn Russian Academy of Sciences needed the prestige and members of the traditional Academy. A way out of this dilemma was found by fusing the two academies; the approximately 250 full members and 450 corresponding members of the "big" Academy were combined with the newly elected 39 full members and 108 corresponding members of the just-born one to form a single "Russian Academy of Sciences" (RAN) that took over the research establishment of the old Academy.[38]

In early December 1991 it was still unclear just what kind of institution the new Russian Academy of Sciences would be. In order to help answer this question, from December 10 to December 12, 1991, a "Conference of Scientists of Academic Institutions" was held in Moscow, attended by 700 elected representatives from research institutes of the Academy, 300 full and corresponding members of the Academy, and invited participants from overseas, including the United States, the United Kingdom, France, Italy, Germany, and the Netherlands.[39] (I was among the invited participants from the United States.) The primary issues that emerged at the conference were: How would the new Russian Academy of Sciences be organized and governed? Who would control the property of the old academy, the individual research institutes within the new academy, or its central administration?[40] The answers to these questions would determine just how different (and how decentralized and democratic) the new academy would be from the old one. The questions aroused the emotions of the attending scientists, and many disputes quickly appeared.

During the debates four different groups struggled for influence: The first one, the "radicals," called for a great decrease in the significance of the Academy, and wanted to build a system of science organization in Russia that would be similar to that in Western countries, particularly the United States. In other words, most fundamental science would in the future be done in the universities, and the Academy would either be abolished or turned into an honorific organization like the National Academy of Sciences in Washington or the Royal Society in London. Such an honorific organization might perform some advisory functions for the government, but it would not actually control research laboratories, all of which would be located outside its framework.

The second group, the "deep reformers," wanted to preserve the academy system, including its network of research institutes, but aimed for a separation of the Academy's role as a learned society from its function as controller of research institutes. In other words, the members of the Academy (full and corresponding) would remain an elite group but would not be the primary administrators of the system of research institutes, contrary to the old system. The research institutes would be put together in an "association of institutes" governed by an elected "coordinating council." This proposal was favored by many, probably the majority, of younger research workers in the academy system and even had fairly heavy support from the two chairpersons of the conference, especially Aleksei Zakharov, but even, to a certain extent, Academician Evgenii Velikhov. Zakharov was elected a board member by the "Voters' Club," which was dominated by younger researchers. Velikhov was director of the Kurchatov Institute of Atomic Energy, which was not a part of the academy network, and he had often expressed the view that the relative weight of the Academy system in Russian science should decrease.[41] The position of the "deep reformers" was opposed by many senior members of the academy whose power stemmed from the fact they were directors or senior administrators in the research institutes.

The third group, the "mild reformers," wanted to maintain the present system of the Academy, including the leadership of the institutes by its senior members, but would introduce an element of democracy into the administration of the Academy. The mild reformers proposed revamping the senior governing body of the Academy, the General Assembly, by including in its membership not only the full and corresponding members of the Academy, but also elected members from the lower research institutions. They suggested that up to 50 percent of the members of the General Assembly could be elected in this way. The mild reformers eventually prevailed, although their proposals for reform were watered down.

The fourth group, the defenders of the system existing at the time, were people who believed that the old Soviet Academy of Sciences had performed well and had only minor flaws. They saw the whole reform movement as an attack on the integrity of science. The leader of this group was Gurii Marchuk, the president of the Academy of Sciences of the USSR. Marchuk delivered a bitter speech in which he said that all of science in the Soviet Union was being threatened by dark forces similar to those which had arisen shortly after the Russian Revolution in the "Proletarian Culture" movement.[42] Marchuk received surprisingly heavy applause after the speech, but the feeling was widespread that Marchuk's time had passed, just as Gorbachev's had. Both were continuing to defend the old "Soviet center" after that task had become hopeless.

Although group three, the "mild reformers," won, as time goes on the new Russian Academy of Sciences (RAN) looks more and more like the old Soviet Academy of Sciences. Truly meaningful reforms did not take place. Members of the RAN continued to be the major administrators of Russian science. Control over property went to the Russian Academy of Sciences as a whole, not the individual institutes. The main change was the provision for representatives of the institutes in the General Assembly, but just how this system will work is still not clear. Many critics of the new system believe that these representatives will in most cases be scientists who desperately want to become members of the RAN themselves and who therefore will do little that might displease their seniors. In the end, the system of privilege, perquisites, and authority that the old Academy represented turned out to be too powerful to permit genuine reform. The election of Iu. S. Osipov instead of Velikhov as president of the Russian Academy of Sciences was the choice of a more conservative candidate (although Velikhov was no radical). Osipov, a very able man, is nonetheless a representative of the old military-industrial complex and a defender of the system of a dominant central academy, even if now only of Russia. Many of the young research scientists in the academy system were deeply disappointed in the outcome of the debate over reform. The possibility of building a research establishment in Russia similar to those in most other developed countries (i.e., one in which the universities would be the dominant force) seemed small.

Nonetheless, Osipov called for continuing reform. His first priority was the creation of an equivalent to the National Science Foundation in the United States and an accompanying system of peer review. These changes would be very healthy. The financial difficulties of Russia in the early nineties, however, made financial innovations of this sort problematic.

The end of the Soviet Union meant changes in other top science policy bodies inherited by Russia. The State Committee on Science and Technology of the USSR (GKNT), last chaired by Academician Nikolai Laverov, was abolished, and its functions were transferred to a new Ministry of Science, Higher Education, and Technological Policy of Russia, headed by Boris Saltykov, formerly the deputy director of the Analytical Center of the USSR Academy of Sciences. Saltykov, an economist and specialist in science and technology policy, was in charge not only of the budget of the Russian Academy of Sciences but of the universities as well. He was reportedly a strong advocate of a market-oriented economy.

In 1992 science in the former Soviet Union was in deep crisis. In the face of economic inflation the governments of the successor republics were not able to maintain science budgets in real terms at their earlier levels. "Hard currency" was almost unavailable, and, as a result, foreign equipment, chemical reagents, and foreign periodical subscriptions were increasingly difficult to obtain. Research scientists found that their salaries were in many cases lower than those of taxi drivers or industrial workers. Some research scientists emigrated abroad and many more were considering such a step. Israel alone had, by late 1991, accepted over 6,000 basic researchers from the former Soviet Union.[43] By this time 20 percent of the staff of the famous Lebedev Institute of Physics were working abroad, either permanently or temporarily. About a third of the staff of the equally famous Steklov Institute of Mathematics had also left.[44] Many of the scientists who departed did not announce whether they were emigrating permanently or were intending eventually to return. Their positions at institutes such as the Lebedev and Steklov were usually left open, in hopes that they would return.

A great fear to Western governments was that scientists and engineers from the former Soviet Union who had been engaged in military research might, out of desperation, sell their knowledge and talents to third world countries. The United States led an effort by several of these governments to establish an institute near Moscow that would provide employment for some of these scientists, especially those who had been involved in nuclear weapons.[45] This support obviously did not help the hundreds of thousands of scientists and engineers who were not experts in weapons and who were in even greater need. Western foundations, learned societies, and private organizations provided some help, but it was clear that the survival of culture and science in the former Soviet Union was dependent on a stabilization of the economies of the various new states and the development of new means of support.

The history of the organization of Russian and Soviet science and technology stretches like an arc through four stages: a tsarist system that, while somewhat different from Western models, was clearly becoming more similar to the organizations of other industrialized nations; an early Soviet system in which administrators proudly sought to create a distinct system superior to those of other nations while selectively drawing on the latest foreign models; a late Stalinist and Brezhnevite period in which the disadvantages of the unique Soviet research system, despite its accomplishments in a few high-priority tasks, became increasingly evident; and a new reform era after 1986 in which administrators concentrated on trying to create a system similar to those in the capitalist nations their predecessors scorned. Thus the Soviet Union gave up its exceptionalist aspirations in science and technology based on the assumption that socialism and centralization were inherently superior foundations for scientific development. As they have done in many other areas, former Soviet science administrators joined with the rest of the industrialized world in their organizational aspirations. Even today, however, there remain many characteristics of the long period in which they and their predecessors emphasized the uniqueness of the organization of Soviet science. An understanding of the organizational characteristics of science and technology today cannot be gained without taking into account each of these four periods and the marks they left on the scientific establishment of the former USSR.

Conclusions

WHAT a remarkable story the history of Soviet science and technology is! Just to start with one of its most striking and sobering features, let us remind ourselves of the role that political persecution played in its development. Soviet scientists who were arrested by the police and accused of grave crimes included people who, before or after their arrests, were the designers of the Soviet Union's most famous airplanes, the main theoretician of the Soviet hydrogen bomb, the head of the Soviet space program (who directed the launching of the world's first artificial satellites), three Nobel Prize–winning physicists, several of the giants of the development of population genetics, two successive presidents of the Agricultural Academy, the director of world-famous Pulkovo Observatory, a founder of the Moscow school of mathematics, the director of the Leningrad Astronomical Institute, a physicist who was a pioneer in the development of "externalism" in the history of science, the director of Tashkent Observatory, two different directors of the Khar'kov Physics Institute, a forerunner of animal and plant ecology, the rector of the Moscow Technical School, the head of the trade union organization for engineers and technicians, the director of the Institute of Physics of Moscow University, the director of the Microbiological Institute, the dean of biology at Moscow University, the director of the Institute of Medical Genetics, the director of the Institute of Hybridization, the director of the Institute of the History of Science, the director of the Institute of Plants, and many, many more. Many of these scientists either were shot or died in labor camps.

And these are only a few of the most famous. The persecution ran through the ranks of average researchers and designers as well. Probably half of the engineers in the Soviet Union in the late 1920s were eventually arrested. The percentage of scientists arrested may never be known, but it was probably lower than the engineers. We do know that in just a few weeks during autumn 1929 approximately 650 members of the staff of the Academy of Sciences were purged. According to the official figures published by the secret police, 19 percent of the staff personnel in the departments surveyed were fired or seized at this time.

And the peak of the purges did not come until 1937! Approximately 20 percent of all Soviet astronomers were arrested in 1936 and 1937. Even long after World War II some of the Soviet Union's most eminent scientists and engineers worked in prison laboratories. Details on these tragic events can be found scattered throughout this book, including Appendix Chapters A and B.

There is nothing in the annals of the history of science to parallel this record. The forced recantation of Galileo, the burning at the stake of Giordano Bruno, the hounding of J. Robert Oppenheimer by U.S. security organs – all these pale in comparison to the story of persecution in the history of Soviet science. And most amazing of all, Soviet science survived. But it did not merely survive; in some areas it even flourished. When in the 1970s I served on a panel of the National Academy of Sciences of the United States that evaluated the quality of Soviet science, leading American specialists in the fields of theoretical physics and mathematics wrote in their official reports that Soviet scientists in their fields were as good as any in the world. One can only agree with the Soviet historian of science Aleksei Kozhevnikov who wrote in the Gorbachev period, "One of the main paradoxes in the history of Soviet science and an interesting problem is not why this science works badly – it is quite natural – but just the opposite: why despite all unfavorable conditions it still works and works sometimes better than one would expect." Kozhevnikov pointed out that five Nobel Prizes were awarded to Soviet physicists for work done in the 1930s and 1940s, a period of terror and tyranny. One of those physicists, Peter Kapitsa, had been kidnapped on Stalin's orders just three years before he did his most important research.

Some progress can be made in explaining this paradox, but at the very beginning of the effort one should observe that science is simply much hardier than most people have thought. Since organized scientific research appeared in world history only recently and occurs only in advanced nations, most people have compared it to a fragile flower, perhaps the most delicate product of civilization. Scientists have contributed to this image when they have pleaded for nurturing conditions for their work. On the contrary, once established, and so long as it is supported financially (which was definitely the case in the USSR even in the worst periods), it will survive. Modern science has become increasingly resilient as it has become ever more intertwined with industrialized and militarized societies. And governments have become increasingly willing to finance it even when they may abuse it in other ways.

Many years ago the sociologist of science Bernard Barber asked, "How long does it take to 'kill' science? Indeed, can it really be extinguished in modern industrial society? Probably not, and probably it cannot even be weakened beyond a certain point in such a society." The offspring of civilization possesses an inner inertia of its own. Scientists have become as adept in arguing that their support is in the nation's interests as have farmers, military officers, or industrialists. In time, the question seems to be not whether governments will destroy science but whether science will destroy governments.

Of course, governments can harm science, hobble its progress, even eliminate certain areas temporarily, as the Soviet Union destroyed genetics. Genetics came back to the Soviet Union in part because a genetics revolution was sweeping the rest of the world, helping other nations to become stronger by raising better crops, breeding better animals, producing superior medicines and pharmaceutical materials. A large nation that did not join this revolution could not remain a world power.

Science is more robust than most observers in the past have believed, but the achievement of Soviet science has additional explanations. One reason that Soviet science was able, despite oppression, to do as well as it did in the thirties, forties, and early fifties was the momentum it had built up before this time. A remarkable generation of scientists in the USSR, educated just before the Revolution or in the Revolution's most idealistic period, the twenties, was thoroughly in step with world science. These scientists were the friends and sometimes the former students of international scientific leaders like Rutherford, Bohr, Einstein, Planck, Dirac, Bateson, and Muller. Others were the disciples of world-rank prerevolutionary scientists in Russia, such as the chemist Mendeleev, the physicist Lebedev, the physiologist Pavlov, the biologist Mechnikov, the mathematician Luzin, the astronomer Struve, the geologist Vernadskii, and the soil scientist Dokuchaev. Combining their close international connections with native strengths, early Soviet scientists wrote some brilliant pages in the history of science. In mathematics, population genetics, psychology, animal and plant ecology, and physics, early Soviet Russia moved toward world leadership. And without any question Soviet science benefited from the largesse of the Soviet government, which strongly supported science financially and institutionally.

Like a well-built ship with a skilled crew that has entered a terrible storm, Soviet science was not easily destroyed by Stalinism. It suffered incredible damage, several parts of it were completely destroyed, there were moments when even the crew thought it would go under, but it survived. The metaphor of a good ship in a storm, however, conveys only part of the story of Soviet science; it fails to represent adequately the drama and paradox of the situation. Some scientists in the Stalinist period took refuge from the political turmoil surrounding them by burying themselves in their work to a degree unlikely in normal circumstances. In a few cases, repression may even have contributed to Soviet science, although of course only temporarily. In the Soviet Union of Stalin's time not only was the world outside the laboratory or study dangerous, it also offered few attractive diversions. As a result, science almost totally absorbed many researchers' lives. In "blackboard science" in particular, where the only tools required were chalk and blackboard, or paper and pencil, talented Soviet researchers who had been thoroughly initiated into world science before the whirlwind hit continued to work intensively. Even if arrested, they sometimes continued their efforts in prison camps. Only physical annihilation could, and tragically often did, stop them. No fragile flower, this Soviet science!

If part of the reason that Soviet scientists managed to do as well as they did under Stalinism is that they were so much a part of world science before the storm hit, one can begin to understand yet another paradox: After Stalinism had passed, Soviet science seemed to falter. Many physicists and mathematicians of the 1980s and 1990s actually spoke of the 1930s and 1940s as the "golden days" of their fields in the Soviet Union, despite the political horrors of those times. Those same scientists expressed fears that science in the former Soviet Union was declining just as it was becoming free. Can political freedom be harmful to science? Not at all. The intellectual costs of Stalinism were delayed, becoming most visible long after Stalin's death. The eminent scientists of Stalin's time were not products of that time, but of the preceding generation, when they were educated in Western Europe or in the best Russian and Soviet universities before Stalinism. By the 1970s that generation was gone. Science students in the Soviet Union in the seventies and early eighties had few role models at home and little opportunity to study abroad. They could not acquire that initiation into the ethos of path breaking science that an outstanding graduate student most needs. To return for a moment to the metaphor of a good ship in a storm, the ship began to leak most badly from the damage it had incurred after it reached port.

One of the reforms of the 1980s and 1990s was to permit vastly expanded contacts between science in the former Soviet Union and the rest of the world. Not enough time has yet passed to know how adequate an answer to the problems of Soviet science this broadening of contacts will be. It unfortunately comes at a time of financial crisis. But a knowledge of the history of Russian science leads one to be hopeful. In the nineteenth century Tsar Nicholas I curtailed the foreign travel of Russian students, but his successor Alexander II restored the contacts. Let us hope that the recent reforms will have the effect of Alexander II's in science, and that a young equivalent of Dmitrii Mendeleev is now studying abroad, later to return home to one of the republics of the former Soviet Union where he or she will push forward the frontier of his or her field.

The history of Russian and Soviet science is an excellent illustration of the influence of social, economic, and political factors on the development of science. Poignantly, the earliest and best-known exponent of this viewpoint was a Soviet scholar who became one of Stalin's victims, Boris Hessen, whose work and influence are discussed in Chapter 7. Hessen applied the externalist methodology too simply and literally, for his own reasons, but his early effort has helped later historians to develop a more sophisticated form of externalism.

The contemporary historian of science who studies Russian science has no difficulty in seeing that many of its visible features relate to the environment in which it developed. This observation will stand even though it is quite possible to imagine a scientist from, say, France or the United States who might, upon returning from a visit with colleagues in the former Soviet republics, affirm that "science is the same everywhere, and I was perfectly at home in the laboratory where I worked." Science in the former Soviet Union is certainly more similar to science in other industrialized countries than literature or art or philosophy. However, a broad view of its development of the type I have attempted to give in this book shows clearly that its development is closely connected to its social environment.

Applying an externalist interpretation to Russian and Soviet science, we see that its strengths and weaknesses, as detailed especially in Appendix Chapters A and B, its moments of progress and regress, and many of its characteristics can be explained in social terms. The strength of mathematics and theoretical physics has many times here been described as, at least in part, a natural product of talented researchers in a repressive political atmosphere and a tightly controlled economic one. Gifted young people gravitated to fields where achievement was possible despite the political and economic barriers of tsarist Russia and the Soviet Union. Conversely, the weakness of industrial research and innovation was, again in part, a result of the absence of a free economy. The cyclical pattern of advance and retreat in Russian technology over several centuries, as discussed in Appendix Chapter B, has been produced by a combination of talented native engineers, periodic foreign stimulus, and an unsupportive domestic environment. Breakthroughs in technology have not usually been sustained. On the other hand, Soviet strengths in "big science and technology" in areas such as atomic weapons and space technology came from centralized governmental control over resources and personnel, a degree of control possessed by few other governments; this ability to focus on a few high-priority projects helped the Soviet government in areas important to its security. This same centralized control harmed innovation in the Soviet Union in many other fields, of which computer technology is a primary example. Computer hardware and software development has benefited in all countries from governmental support, but some of the most innovative work has been done by individual entrepreneurs and eccentric geniuses of a sort who would have had trouble striking out on their own paths in the centralized economy and research establishment of the Soviet Union. And, last, some even of the intellectual characteristics of Soviet science can be traced to external influences.

As I described in Chapter 5, Marxist philosophy played a role not only in the repression of Soviet science but also in some of its prouder moments. Marxism was an important influence in the views of some of the Soviet Union's best scientists, including the work on psychology of Vygotsky, the ideas about origin of life of Oparin, the interpretation of quantum mechanics of Fock, the formulation of the foundations of mathematics of Kolmogorov, and in the conceptions of a number of other outstanding Soviet scientists. This cognitive role of Marxism in some eminent scientific work in the Soviet Union is the least understood and appreciated of the characteristics of Soviet science.

Because of the somewhat exotic and extreme character of Russian and Soviet history, at least compared to Western Europe and America, it may be that social factors have had a more visible impact on Soviet science than on science in those countries. As a historian of science, however, I resist the conclusion that externalism "works" only in exotic locales, and not in "normal" ones.

The history of science in the United States, for example, has also been heavily influenced by external factors. Just think of the importance of atomic weapons research, the advent of the National Science Foundation, and the Cold War on the development of American physics. Any careful examination of the history of American science would show how external factors in the United States were of a different variety from those in the Soviet Union. The consumer culture in America has had a great impact on the development of technology, not only of the domestic type, but even of fairly sophisticated electronics, such as that utilized in computer games. The individualistic nature of American culture has placed a premium on creativity by the researchers, and on their ability to apply for and receive funds from foundations and the government. At the same time, this intense competition has sometimes resulted in fraud and premature claims. The strength of the American research university system has resulted in the United States receiving far more Nobel Prizes than any other nation, and has attracted researchers from all over the world. The weakness of the American primary and secondary educational systems makes us wonder if the tradition of excellence in science can be maintained. The success of white male researchers from elite institutions in winning grants more often than others and those from less prestigious institutions has raised problems of justice in the dispensing of research funds. The importance of military interests in the post–World War II period has had a major impact on research and development in the United States, attracting some of the best researchers and funding some of the most expensive endeavors, allowing foreign competitors such as Japan to make major inroads in nonmilitary sectors, such as consumer electronics. The profitability of the pharmaceutical and scientific instrumentation industries has made them the envy of the world. Meanwhile, the increasing involvement of universities with industry in fields such as genetic engineering has raised serious ethical issues. These are only a few of the many ways in which American culture, economy, and political traditions have influenced U.S. science and technology.

The influence of external factors may be more obvious at first glance in the case of Russian and Soviet science than in science elsewhere, but that influence is universal. Demonstrating that influence in other countries, however, is the task for other historians. I have tried to trace these patterns in Russia and the Soviet Union.

APPENDIX

The strengths and weaknesses of Russian and Soviet science

Appendix Chapter A: The physical and mathematical sciences

THE Soviet Union was traditionally very strong in the theoretical foundations of physics. Western observers of Soviet science often spoke of the "blackboard rule," meaning that their Soviet colleagues could be expected to excel on those topics where world-rank work could be done with tools no more complicated than blackboard and chalk. While there were many exceptions to the blackboard rule, it nonetheless had considerable accuracy.

Although physics was not distinguished in nineteenth-century Russia, at the end of that century and during the first years of the twentieth the foundations were laid for the later impressive growth of Soviet physics. The most important figure here was probably P. N. Lebedev (1866–1912), who as a young man studied under Kundt in Strassburg. Lebedev in 1900 presented a paper at the World Congress of Physicists in Paris in which he demonstrated the existence of light pressure, a phenomenon predicted earlier by Maxwell, but not previously confirmed. A year later he published his classical study *Experimental Research on Light Pressure*. Buoyed by the international reputation that this work brought him, Lebedev launched a program to create a school of Russian physicists capable of participating in research at a world level. He modernized the physics laboratory at Moscow University, and he created a colloquium that eventually served as the kernel for the Moscow Physical Society. When Moscow University was closed by political turmoil in 1911 Lebedev was a leader among a group of Russian scientists who sought a more pluralistic organizational framework for science. Critical of the dominating role of the government in science, Lebedev and his colleagues worked to energize private and philanthropic support for research, such as that offered by the Ledentsov Society, the Shaniavskii University, and the Moscow Society for a Scientific Institute. As noted in Chapter 9 these efforts were aborted by the advent of the Soviet government with its penchant for centralization. The name of Lebedev lived on prominently in Soviet physics, however, since the main physics institute of the Soviet Academy of Sciences was named after him. Pluralism in science, as embodied in Lebedev's ideas, received a new boost decades later during the reforms at the time of the fall of communism.

Important as Lebedev was in the formation of twentieth-century Russian and Soviet physics, he did not sympathize with or participate in the revolutionary development of physics in the early twentieth century marked by the development of relativity and quantum theory. Two physicists of the prerevolutionary period who did recognize the importance of these new trends were O. D. Khvol'son and N. A. Umov. Khvol'son published widely in West European languages and enjoyed an international reputation because of his insightful presentation of the concepts of the new physics. In Russia he was also known as a defender of religion and its compatibility with science, especially the new physics. Umov was a polymath who worked in a great variety of fields: thermodynamics, light dispersion, magnetism, philosophy of physics, and the organization of science and education. He did not leave a major mark in physics research itself, but as a leader of the scientific community in the late nineteenth and early twentieth centuries he is still remembered today.

An important influence in Russian and Soviet physics was Paul Ehrenfest, an Austrian Jew who married a Russian mathematician, T. A. Afanasieva, and who spent much time in Russia, both before and after the Revolution. While living in St. Petersburg and other cities of the Russian Empire from 1907 to 1912 Ehrenfest and his wife wrote an important monograph on statistical mechanics. Paul Ehrenfest was also a pioneer in exploring the significance of Max Planck's concept of energy quanta. A colleague of Abram Ioffe, Ehrenfest participated in the development of a major school of Russian and Soviet physics even though he was never a citizen of Russia or the Soviet Union.

The true blossoming of physics in Russia came after the Revolution. In general relativity theory, A. A. Fridman (Friedmann) in 1922 produced a brilliant mathematical approach that showed that Einstein was wrong in his opinion that his equations of 1915 could lead only to a static universe, an error that Einstein graciously admitted after seeing Fridman's work.[1] Beginning in the early 1930s Vladimir Fock, Lev Landau, and Igor Tamm made contributions to quantum field theory that attracted attention from leading physicists throughout the world.[2] At about the same time, P. A. Cherenkov began his work under the supervision of S. I. Vavilov on the action of radiation on liquids that later came to be known as the Cherenkov effect, for which, in 1958, he received the Nobel Prize, along with I. M. Frank and Tamm.

Ia. I. Frenkel' was well known in the thirties and forties for his work on electrodynamics and especially for his two-volume text on the subject, published first in German.[3] In the thirties and forties, L. A. Artsimovich, I. Ia. Pomeranchuk, and D. D. Ivanenko did important work on quantum electrodynamics. After World War II the textbooks on theoretical physics by L. D. Landau and E. M. Lifshits were known to physicists everywhere.

In the 1930s Landau and B. I. Davydov established a strong tradition in plasma physics that has continued to the present day. In later years some of the most influential workers in this field were V. L. Ginzburg, A. A. Vlasov, R. Z. Sagdeev, E. P. Velikhov, L. A. Artsimovich, M. A. Leontovich, A. D. Sakharov, and I. E. Tamm. Sakharov (later famous in the West for his protests against Soviet violations of civil rights) and Tamm suggested the Tokamak toroidal model for controlled fusion that was widely accepted throughout the world as a basis for continuing experimentation. In the 1960s and early 1970s Soviet scientists were at the very forefront in theory and device construction in plasma physics, but in the late seventies their position weakened because of deficiencies in experimental diagnostics and computer analysis.[4] Indeed, the leading research on the Tokamak idea, Soviet in conception, was transferred to the United States, especially to Princeton.

The founder of Soviet work in solid state physics was A. F. Ioffe, a great figure in the history of Soviet science. Ioffe (1880–1960) established in 1918 the Leningrad Physico-Technical Institute (LFTI), which became the cradle of Soviet physics. The Western historian of this institute, Paul Josephson, has described the 1920s as the "flowering of Soviet physics," a time when a whole group of talented young Soviet physicists flourished under Ioffe's tutelage.[5] Among the members of Ioffe's school were A. P. Aleksandrov (future president of the Academy of Sciences), A. I. Alikhanov, L. A. Artsimovich, P. L. Kapitsa (Nobel laureate), I. K. Kikoin, V. N. Kondrat'ev, B. P. Konstantinovich, I. V. Kurchatov (later leader of the Soviet atomic weapon project), L. D. Landau (Nobel laureate), P. I. Lukirskii, M. A. Mikheev, I. V. Obreimov, N. N. Semenov (Nobel laureate), D. V. Skobel'tsyn, Iu. B. Khariton, L. V. Shubnikov, Ia. I. Frenkel', and many others.

In addition to serving as the birthplace of the Leningrad School of Physicists, the LFTI (later known as the Ioffe Institute) organized other centers of physics research, providing the initiative for fifteen scientific-research institutes, and over one hundred factory physics laboratories.[6] The LFTI tried to foster research equal to the world's best and to combine fundamental and applied concerns. At the home institute Ioffe was largely successful in the twenties but in later years had difficulty overcoming political and economic barriers. Particularly troublesome was the creation of close links between fundamental research and industrial application; the Soviet economic and political system did not provide sufficient incentive for the practical exploitation of innovative (and temporarily disruptive) theoretical breakthroughs.

One important offshoot of Ioffe's LFTI met a tragic fate.[7] In 1928 Ioffe was asked by the government to organize a physicotechnical institute in the city of Khar'kov, at that time the capital of the Ukrainian republic. The new institute, known as the Ukrainian Physico-technical Institute (UPTI) soon attracted a constellation of brilliant physicists, many of them from the home institute in Leningrad. The group included I. V. Obreimov (who became director), L. V. Shubnikov, A. Leipunskii, L. D. Landau, K. D. Sinel'nikov, M. Ruhemann, L. V. Rosenkevich, and B. Podolskii. Most of these men had studied in Western Europe and were closely connected to the international physics community. They soon invited foreign physicists to visit and work in Khar'kov, and many came, including Bohr, Dirac, Ehrenfest, Houtermans, and Weisskopf. Khar'-kov quickly became the third leading center of physics in the USSR, after Leningrad and Moscow. The scientists at UPTI built the first Soviet linear accelerator and confirmed in 1932 the experiments of Cockroft and Walton on splitting the nucleus.[8] Lev Landau created at Khar'kov his famous school of theoretical physics, and acquired as a student E. M. Lifshits, who later would write with him the noted textbooks in physics.

The great terror of 1937–8 hit UPTI so hard that it never recovered: Shubnikov, Gorskii, and Rosenkevich were arrested and shot, Weissberg and Houtermans were arrested and later extradited to Germany, Obreimov and Leipunskii were arrested but subsequently released, Ruhemann was forced to leave the Soviet Union, and Landau was seized in Moscow, to which he had earlier moved. Peter Kapitsa and Niels Bohr pleaded with Stalin for Landau's release, and, after a year in prison, he was allowed to work in Kapitsa's Institute of Physical Problems.[9]

In the thirties and forties Ioffe and his coworkers did important work on the properties of semiconductors, on photovoltaic cells, on thermoelectric effects, and, starting in the fifties, on microminiaturization. All of this was important both for Soviet physics and Soviet industry, but it must be admitted that in almost no cases did the Soviet Union lead in industrial applications in these areas. The transistor revolution in world electronics was powered largely by economic and entrepreneurial driving forces that the Soviet Union could not match. As a result, Soviet industrial managers tried unsuccessfully to catch up in this field for several decades. One of the major stimuli to the Soviet microelectronics industry in the sixties came from an American who defected to the USSR during the McCarthy period, the electrical engineer Alfred Sarant, who took the name in the Soviet Union of Filip Staros. In the Soviet Union Staros was given official honors for "significant contributions to the establishment and development of national microelectronics."[10]

In theoretical investigations Soviet physicists remained very strong. Starting in the forties they rapidly developed the mathematical foundations of quantum electrodynamics. Applying this theory to the study of condensed matter, they became world leaders. In the sixties and seventies Soviet work on condensed matter theory was described by an official report of the National Academy of Sciences of the United States as "some of the most innovative and important work in the world."[11] The leading institution in the Soviet Union in condensed matter theory was the L. D. Landau Institute of Theoretical Physics. Other Soviet institutions with important research in condensed matter theory were the Ioffe Physico-Technical Institute in Leningrad, the Institute of Physical Problems in Moscow, and the Lebedev Physics Institute in Moscow. In the seventies and eighties cooperative work between the United States and the Soviet Union in condensed matter theory was strong, despite political vicissitudes. The leaders of a joint research group called Frontiers in Condensed Matter Physics were, on the Soviet side, L. P. Gor'kov, and, on the American side, J. R. Schrieffer, a Nobel laureate.

Another bright page in the recent history of Soviet physics was quantum electronics, where new methods were found for the generation and intensification of electromagnetic waves. In 1964 the Soviet physicists N. G. Basov and A. M. Prokhorov, together with the American physicist C. Townes, received the Nobel Prize for fundamental research in quantum electronics leading to the development of lasers and masers. In subsequent years Soviet researchers continued to devote a great deal of attention to this area, but encountered familiar problems in instrumentation, computer analysis, and industrial applications. Because of the great significance of this topic for military applications, however, Soviet researchers in quantum electronics received top priorities in access to scarce instruments and computing facilities.

As a result of its unusual properties and anomalies, the physics of liquid helium has been an especially attractive problem for researchers. Furthermore, liquid helium is invaluable for attaining extreme cold. Peter Kapitsa, one of the Soviet Union's best-known scientists, began research on low temperatures in England where he lived in the twenties, and continued to work on the topic in the Soviet Union, where he was forced to stay by Stalin in 1934 when he came home to visit relatives. Kapitsa placed a great premium upon scientific experiment, an art he regarded as most highly developed in Rutherford's laboratory in Cambridge, England (where Kapitsa worked for thirteen years), and insufficiently appreciated in his native Soviet Union. Working together with the theoretician Landau, Kapitsa combined exquisite experimentation with refined theoretical analysis. Landau in 1962 received the Nobel Prize for his theory of condensed phases, especially of liquid helium, and in 1976 Kapitsa won the same high honor for his work in the superconductivity of liquid helium at temperatures near absolute zero.

Kapitsa's independence and bravery are evident in a number of his actions: As mentioned earlier, he successfully won in 1938 the release from prison of Landau, who had been placed there by Stalin's secret police on the ridiculous charge that Landau was a "German spy"; after

World War II Kapitsa refused to work on atomic weapons because of his antipathy toward L. Beria, the secret police chief who also directed the weapons project; as a result, Kapitsa was placed under house arrest for eight years; and, in the seventies and early eighties, he objected strenuously to growing anti-Semitism in Soviet science.[12] On his death in 1984 at the age of eighty-nine he was the only member of the presidium of the Soviet Academy of Sciences who was not a member of the Communist Party.

Soviet scientists have done impressive work in relativistic physics and astrophysics, continuing the tradition established by Fridman. Throughout the fifties and sixties the leading Soviet interpreters of relativity physics were A. D. Aleksandrov and V. A. Fock. Fock was particularly well known abroad for his penetrating analysis of the principles of general covariance and the equivalence of inertial and gravitational mass.[13] In the seventies and eighties the most prominent Soviet relativists and astrophysicists were A. L. Zel'manov, I. S. Shklovskii, V. L. Ginzburg, I. M. Khalatnikov, A. D. Sakharov, Ia. B. Zel'dovich, and I. D. Novikov. The reputation of the latter two grew markedly in the West after the translation into English in 1983 of two of their books, *The Structure and Evolution of the Universe* and *Evolution of the Universe*. Zel'dovich, Novikov, and Shklovskii all worked at the famous Space Research Institute in Moscow, directed by Roald Sagdeev, a strong proponent of international scientific cooperation.

In atomic and high-energy physics the Soviet Union had a strong record. Authorities in the United States were startled in 1949 when the Soviet Union exploded an atomic device years before many of them had predicted it possible. The scientist in charge of the Soviet atomic project was Ivan Kurchatov (1903–1960), an outstanding physicist who in the 1920s had been a researcher on dielectrics in Ioffe's Leningrad Physico-Technical Institute. In 1953 the Soviet Union exploded a hydrogen bomb ten months after the United States exploded their hydrogen device; subsequent investigation showed, however, that the Soviet Union actually possessed a deliverable hydrogen bomb before the United States. The Soviet physicist responsible for the design of the hydrogen bomb was A. D. Sakharov, the brilliant scientist already mentioned several times. In both the atomic and hydrogen bomb projects prison camps and prison labor played large roles. Sakharov was offended by the use of prison labor in his scientific projects, and later sharply criticized the practice.[14]

In high-energy physics the Soviet Union possessed strong theorists and powerful experimental facilities. V. I. Veksler and I. V. Kurchatov were Soviet leaders in the 1940s in developing accelerator technology. In 1949 the Soviets introduced the most powerful accelerator in the world at that time, the synchrocyclotron at Dubna, operating at first at 480 million electron volts (MEV), and, in 1953, at 680 MEV. Four years later the Soviets completed at Dubna a 10 billion electron volts (GEV) proton synchro-phasotron. The Dubna facility became the center of the Joint Institute for Nuclear Research, where physicists from the Soviet bloc engaged in co-operative research; it was, in part, a response to cooperative efforts by West European and American scientists at CERN in Switzerland.

Other leading Soviet centers in high-energy physics have been the I. V. Kurchatov Institute, the Institute of Nuclear Physics at Gatchina, the Lebedev Physics Institute in Moscow, the Institute of High Energy Physics at Serpukhov, and the Institute of Nuclear Physics in Novosibirsk. In the 1960s the Soviet Union possessed the world's most powerful proton accelerator at Serpukhov, and planned to recapture this position after losing out to competitors in the United States and Western Europe. Academician Gersh Budker, director of the Institute of Nuclear Physics in Novosibirsk, was a figure among the world's atomic physicists in the sixties for his direction of research on the institute's colliding beam accel-erator. Soviet physicists not only were strong in atomic theory, but also advanced new ideas in accelerator technology (for example, collective acceleration and electron cooling). However, Western studies of Soviet work in nuclear and elementary particle physics have indicated that rarely did Soviet scientists make breakthroughs even at moments when they possessed the most powerful facilities. Their scientific productivity was somewhat low, the work frequently undisciplined, and computa-tional facilities lagged. To some extent they overcame their backward-ness in computer technology by using applied and analytic mathematics in unusually creative ways, but mathematics, no matter how innovative, could not entirely make up for the lack of the latest computers.

MATHEMATICS

Of all fields of knowledge, it was mathematics to which Russia and the Soviet Union made the greatest contributions. The Soviet Union became a world power in mathematics. Indeed, Moscow is probably today the city of the greatest concentration of mathematical talent anywhere. The main competitor is no doubt Paris, since mathematicians in the United States, another leader in mathematics in the last generation, are more widely distributed geographically than in France or the Soviet Union.

Russia has, of course, a great tradition in mathematics, dating back to Euler and the Bernouillis in the early eighteenth century. N. I. Lobachev-skii (see Chapter 2), M. V. Ostrogradskii, and P. L. Chebyshev in the nineteenth century solidified the reputation of Russia in mathematics. By the early twentieth century Russian mathematicians were working at the leading edge of mathematics in many areas: Chebyshev and A. A. Markov in the theory of numbers and probability; V. A. Steklov and A.

N. Krylov in differential equations; D. F. Egorov, K. A. Andreev, and A. K. Vlasov in geometry; D. A. Grave, S. O. Shatunovskii, and F. E. Molin in algebra; N. N. Luzin in the theory of functions; and many others.

Soviet mathematicians worked across the entire spectrum of pure and applied mathematics, and the National Academy of Sciences of the United States in the 1970s rated their work "second to none."[15] It is true that in the late seventies some weaknesses began to show up in Soviet mathematics because of lags in computer technology and discrimination against Jewish mathematicians. Nonetheless, the strength of the Soviet Union in mathematics was so great, and the tradition of outstanding achievement so deeply rooted, that even serious blows such as these did not dislodge Soviet mathematics from its eminent position.

While Moscow was the focus of Soviet mathematics, leading mathematicians could also be found in cities such as Leningrad, Novosibirsk, Odessa, Khar'kov, and in some of the Baltic republics. Outstanding Soviet mathematicians often created schools consisting of their pupils, such as the school of geometry specialists that clustered around A. D. Aleksandrov at Leningrad University, or the group involved in functional analysis centered around, first, N. N. Luzin, and, later, I. M. Gel'fand, at Moscow University.

The glory of Soviet mathematics appears a bit unusual when one views it against the traditional aims of Soviet science as enunciated by government leaders. From the time of the Revolution onward science administrators in the Soviet Union called for the creation of a new science that would be collective in nature, centered in large research institutes of the Academy of Sciences, and one that would serve Soviet industry by uniting theory and practice. Although one should be careful not to exaggerate the degree to which Soviet mathematics deviated from these aims, the fact remains that most of its significant achievements were not collective works, but the creations of one, or at most, two individuals; that, alone among the fields of Soviet science, it continued to draw much of its strength from the universities, not the centralized institutes of the Academy; and that, last, some of its greatest achievements were not in areas closely tied to industrial applications, but, instead in pure theory. When Soviet mathematicians made great advances in areas with significant potential for applications, such as information theory and computer algorithms, often their work received its most advanced exploitation in other countries. (An exception here may be the space program, in which mathematicians such as M. V. Keldysh played leading roles, and developed ingenious mathematical means of overcoming Soviet lags in computer technology.).

The dominance of Moscow in Soviet mathematics was relatively recent. Before the twentieth century, the major center in Russian mathematics was St. Petersburg (later Leningrad, now St. Petersburg again).

Most of the work was done in the University of St. Petersburg and the Academy of Sciences, where Euler and the Bernouillis had established the tradition. In the late eighteenth and early nineteenth centuries Russian mathematicians had not been able to continue that tradition at its earlier heights, but in the last half of the nineteenth century an independent and powerful "St. Petersburg school" of mathematics emerged. It became one of the leading forces in world mathematics. Early pioneers were V. Ia. Buniakovskii and M. V. Ostrogradskii, but the real creator of the school was P. L. Chebyshev (1821–1894).[16] Chebyshev worked on many topics in mathematics, including the theory of numbers and integration of algebraic functions, but it was his work on probability that had the most influence on the St. Petersburg school. The tradition in probability that Chebyshev established is visible in the former Soviet Union still today, where probability has remained a central concern.

At first the St. Petersburg school of mathematics followed mainly Chebyshev's interests. Over time, however, it became much broader and gradually became a cluster of subschools occupied with a variety of topics, including the theory of the best approximation of functions, the theory of differential equations, mathematical physics, as well as the theory of numbers and probability. One characteristic of the school was attention to both theoretical and applied problems. Chebyshev himself was interested in the theory of mechanics and ballistics, and helped Russian army artillery specialists. He also designed a calculating device and was interested in steam engines and hinge-lever driving gears. If the St. Petersburg mathematics school had relatively little influence on Russian industry, the fault lay more with the lack of native industrialists with the necessary interests and funds than with the school itself.

One of the founders of the Moscow school of mathematics was N. N. Luzin (1883–1950), whose most important work was on the theory of functions.[17] He had been introduced to the importance and inherent interest of the theory of functions by his Moscow teacher, D. F. Egorov. While studying in Göttingen and Paris from 1910 to 1914, Luzin began writing a work that would become a classic in the history of mathematics, his *The Integral and Trigonometric Series*. In 1916 he presented the manuscript as his master's degree dissertation. His professors were so struck with the work that they recommended skipping the master's degree and directly awarding the doctorate, something that had not been done at Moscow for over sixty years. One year later Luzin was made professor in the faculty of pure mathematics at Moscow, and he immediately established Moscow University as a world center in mathematical research on the theory of functions. His students included some of the most famous mathematicians in the Soviet Union. Among the most noted of them were P. S. Aleksandrov, L. A. Liusternik, P. S. Novikov, M. A. Lavrent'ev, and A. N. Kolmogorov. In the late twenties and early thirties Moscow University produced an explosion of mathematical research. Luzin published numerous works on the theory of real functions, A. Ia. Khinchin and A. N. Kolmogorov were creating the noted Moscow school of probability, M. A. Lavrent'ev and M. V. Keldysh were applying analytical functions to hydrodynamics and aerodynamics, P. S. Aleksandrov was doing outstanding work on topology, and I. G. Petrovskii was working on the theory of partial derivatives.

In the heyday of Luzin's influence in Moscow, from roughly 1914 to 1930, his famous seminar at the university was jokingly called "Luzitaniia," and its members were "Luzitaniians." These young mathematicians were critical of the form of classical analysis that reigned in St. Petersburg (after 1924 Leningrad), deriding it as inherently conservative. On the other hand, the St. Petersburg mathematicians, most of them disciples of Chebyshev, were suspicious of the new ideas about functions and sets with which the Luzitaniians were so entranced. The distinguished academician V. A. Steklov, for example, downplayed the significance of Luzin's dissertation, considering it "Göttingen chatter."[18]

The rivalries among Soviet mathematicians on occasion led to downright hostility, and may have set the stage for the outbreak of verbal warfare in Soviet mathematics in the thirties, a time of ideological militance that often provided a pretext for the venting of old personal disputes, as well as attacks by younger mathematicians on older ones. Luzin's teacher, D. F. Egorov, was arrested in 1930 and died in prison exile the following year. Egorov was a classmate and friend of Pavel Florenskii, a Russian Orthodox priest and mathematician who carried on extensive correspondence with Luzin about mathematics. This link between mathematics and religion was criticized by the ideological militants. Florenskii was arrested and murdered while in prison.[19] A few of the younger mathematicians, pushed onto the educational ladder in the wake of the Revolution without adequate preparation, made use of these rivalries for their own purposes. The Luzitaniians were a highly talented group known for their intuitive, almost philosophical, approach to mathematics and for their devotion to abstract theory to the detriment of practical concerns. Their philosophical inspiration was West European, and sharply differed with the Marxist philosophy being urged on scholars by the Bolshevik ideologists of the thirties. The established Leningrad mathematicians were, at least at first, no more kindly disposed toward Marxist philosophy than the Moscow ones, but Chebyshev's tradition of interest in mechanical and engineering problems came closer to the Marxist slogan of the "unity of theory and practice" than that of the early Luzitaniians. And in the figure of A. D. Aleksandrov, later rector of Leningrad University, the old St. Petersburg group produced a distinguished mathematician who was also a sincere Marxist.[20]

In the middle and late thirties, a time of purge and terror in the Soviet Union, Luzin was sharply criticized by Communist Party militants for publishing many of his most important works abroad and for not displaying sufficient ideological awareness.[21] Unnamed authors in leading Soviet mathematics journals accused him of "servility toward foreign science," as demonstrated by the fact that he usually published in West European journals, and of "enmity toward anything Soviet." As a result of these attacks, Luzin was expelled from Moscow University, but not from the Academy of Sciences, although he was demoted there. Paradoxically, the venerable Academy, which earlier had not been immediately receptive of Luzin's mathematics, provided a refuge for him when he needed it. The reason is not difficult to see: Soviet universities were far more vulnerable to political pressure than the Academy, which retained its old method of governance by secret ballot of the members. At the same time, Luzin changed his behavior in a way better to protect himself against the charge of "subservience to foreign science." After 1935 every one of his publications was in the Russian language, while before that date over two-thirds of them had been in French. The chastened Luzin survived, living on until 1950. His students continued to prosper; a few of them embraced aspects of Marxist philosophy, a development that would have surprised Luzin in his younger years. As we saw in Chapter 5, the great mathematician Kolmogorov wrote articles on the foundations of mathematics for the *Large Soviet Encyclopedia* that are based on Marxist epistemological realism and which differ markedly from corresponding articles in Western encyclopedias such as the *Britannica*.

The list of Soviet mathematicians who made major advances in their field is long, and to enumerate them all would be tedious. Soviet mathematicians worked in all major fields. One of the early leaders of the school of algebra was the colorful radical O. Iu. Shmidt, a student of Grave's.[22] Other leading Soviet algebraists included N. G. Chebotarev, E. B. Dynkin (Lie algebra), D. I. Fadeev, A. G. Kurosh, and A. I. Mal'tsev. A. N. Kolmogorov, already mentioned, was probably the best-known Soviet researcher on probability, but note should also be made of Iu. V. Linnik. The group involved in functional analysis has also been famous, including, in addition to those already named, L. V. Kontorovich, A. G. Vitushkin, M. G. Krein (leader of the Odessa school), I. N. Vekua, and V. I. Arnold. In differential equations, leaders have been N. N. Bogoliubov, O. A. Oleinik, S. L. Sobolev, and M. I. Vishik. And, finally, in theory of numbers, outstanding work has come from Iu. I. Manin, I. M. Vinogradov, and I. I. Piatetskii-Shapiro.

How does one account for the outstanding strength of Soviet mathematics? First, both the tsarist and Soviet governments favored mathematics and supported it strongly. The advent to St. Petersburg in the eighteenth century of such mathematical luminaries as Euler and Bernouilli came with strong government approval. Part of this support derived, no doubt, from the fact that mathematics was far from politics and presented no threat to the regime. Such conservative ministers of education as Count Uvarov in the nineteenth century favored fields such as classical studies and mathematics because they brought international laurels, were politically acceptable, and were not very expensive to support. Even in the Soviet Union some of these same elements remained constant. (It should be noted that several outstanding Soviet mathematicians either became involved in difficulties with Soviet political authorities in the post-Stalin period, or emigrated, or both, as did I. G. Shafarevich, D. M. Kazhdan, A. S. Esenin-Vol'pin, and A. A. Zinov'ev.)

Once a great tradition was established in Russian mathematics in the eighteenth century the inferiority complex that plagued Russian scientists in many other areas no longer threatened. To a large degree, mathematics became a "Russian science," one in which Russians expected to do well, just as they assumed they would excel in chess and ballet. The fact that Russia was backward in the eighteenth and nineteenth centuries in industry was no obstacle to excellence in mathematics. After the Revolution, mathematics could be maintained even in moments of great social turmoil. Although Soviet mathematicians lost many of their personal contacts with mathematicians in other countries after the late 1930s, they maintained international traditions in mathematics. Several of the leaders of Soviet schools of mathematics, such as N. N. Luzin and S. N. Bernshtein, studied abroad before the Revolution and continued for decades to teach mathematics in the spirit in which they had been educated in both Western Europe and Russia.

A further insight into the strength of Soviet mathematics can be gained by placing oneself in the position of a talented young Soviet student trying to decide which area he or she (there have been outstanding Russian female mathematicians, from S. V. Kovalevskaia to O. A. Ladyzhenskaia) wished to pursue professionally. Practical occupations such as business or law were not often attractive in the Soviet Union, since it was almost impossible to make a great independent career in these areas. The humanities and social sciences beckoned primarily to people with political interests or ambitions. Some of the natural sciences, such as biology and even branches of physics (for example, atomic physics and weapons research), were too close to social and political issues for some fiercely independent intellectuals. For the person who had natural talent in mathematics and who was looking for a career where he or she would encounter the fewest obstacles, either of a material or a political sort, mathematics seemed the logical choice.

A. Ia. Khinchin, a leading Soviet specialist in probability theory, was quite explicit on why young idealists in the USSR found mathematics superior to other fields of professional activity:

In common lawsuits each of the disputing sides usually aims toward the solution of the issue that would be desirable or profitable from its standpoint; with the greatest inventiveness possible, each side devises arguments for the resolution of the case in its favor. Depending on the epoch, the social environment and the contents of the dispute, the sides appeal to various higher authorities – common morals, "natural" law, holy writ, the juridical code, the current rules of internal order, and often even the opinions of authoritative scholars or recognized political leaders. All of us have many times observed the passion with which such disputes are waged. . . .

Only mathematics is fully spared from this. . . . Every mathematician learns early that in his science any attempt, whatever the reason, to act in a tendentious way, inclining toward a certain solution of the problem, favoring only those arguments which speak in favor of that solution – any such attempt is doomed to fail and to bring nothing but disappointment. Therefore, the mathematician quickly learns that in his science only correct, objective arguments free from all tendentiousness have benefit for his cause. . . .[23]

There is, of course, a great irony in the opinion of Khinchin and his colleagues that mathematics as a field is the farthest of all areas of knowledge from social influence. Young Soviet students who followed Khinchin's advice and went into mathematics were not illustrating the lack of social influence on mathematics but instead were demonstrating the strength of such influence. It was because of the particular characteristics of Soviet society, a society in which political, economic, and ideological factors impinged on so many areas of knowledge, that they found mathematics so attractive. It is true that once they were at work in mathematics they were rather distant from social influences, but their entry into the field can be explained, at least in part, by the social environment in which they lived and worked.

It also must be noticed that Soviet mathematics was not free of political and personal strife, as we have already seen. Foreign visitors to the Steklov Institute of Mathematics in Moscow in the 1970s and early 1980s found many examples of tensions. L. S. Pontriagin, a member of the institute, was not only an outstanding topologist, but a rabid anti-Semite who attempted to keep Jews out of the journal that he edited, *Mathematicheskii Sbornik* (Mathematics Collection). He was a foe of refuseniks and dissidents. Not far away was the office of I. R. Shafarevich, a noted number theorist who was an outspoken religious dissident (who later also unfortunately displayed anti-Semitic sentiments).[24]

Not only is mathematics subject to many of the same personal rivalries that one finds in all fields, it is also increasingly relevant to military concerns. Cryptography, ballistic missile control theory, and the design of atomic weapons all depend heavily on mathematics research. Yet despite the obvious intrusion of political and military elements in mathematics, the field is still one in which the person who makes the necessary effort has a good chance of staying aloof from controversial social issues. This possibility was attractive to quite a few Soviet students who had the necessary talents and educational backgrounds for careers in mathematics. It is a field in which there were many talented role models, a distinguished tradition, excellent research institutions, and relatively few obstacles.

ASTRONOMY

Astronomy is a field in which Russia achieved eminence long before the Russian Revolution of 1917. The Pulkovo Observatory near St. Petersburg was a center of outstanding work starting from the date of its establishment in 1839. Its founder, F. G. W. Struve, became famous for his measurements of stellar parallax and his accurate observations of double stars. Struve established not only a tradition of outstanding astronomical research, but also a family line of distinguished astronomers that worked for more than 150 years over four generations; his great-grandson Otto Struve left Russia in 1920 and eventually became a well-known astrophysicist in the United States. The great-grandson long promoted international knowledge of Soviet astronomy by serving as editor of several journals in which the works of Soviet astronomers were often featured.

In 1808, when he was fifteen years old, the German founder of this astronomical dynasty, F. G. W. Struve, was seized near Hamburg by French military recruiters seeking soldiers for Napoleon's armies. Struve escaped by leaping from the second floor of a building and fled to Russia, where he spent the rest of his life.[25] He began his work on double stars at Dorpat University and in 1822 published a catalog of all known double stars, 795 at that time. Only five years later he published another catalog of double stars in which the list was extended to 3,112. After moving to Pulkovo, Struve devised a long-term plan of research to determine the coordinates of all bright stars in the heavens with a high degree of accuracy. The Pulkovo star catalogs, published in 1845, 1865, 1885, 1905, and 1930, were fruits of Struve's vision. They were acclaimed by astronomers throughout the world.

Other important astronomers working before the Revolution included B. Ia. Shveitser and P. K. Shternberg, who investigated anomalous deviations from the perpendicular of plumb weights; F. A. Bredikhin, who developed a theory explaining the shapes of comets; A. A. Belopol'skii, who was a specialist on solar observation and spectral analysis; V. K. Tseraskii and S. N. Blazhko, who discovered and investigated a number of variable stars; and M. A. Koval'skii, who developed a method for studying the rotation of the Galaxy.

Although there was only a small group of professional astronomers in Russia before 1917, governmental and public interest in astronomy was high. The tsarist government correctly saw astronomy as a field where relatively small sums of money could bring considerable international acclaim, and therefore provided the one or two best observatories, such as Pulkovo and Moscow, with the finest equipment available in the world. Good observatories were also established at Odessa (1870), Tashkent (1874), and Kazan' (1901). Professional astronomers were organized in the Russian Astronomical Society, founded in St. Petersburg in 1908, while amateurs established their own astronomical societies in Nizhnyi Novgorod (later Gor'kii) in 1887 and Moscow in 1908.

The work of Soviet astronomers in the 1920s and 1930s did not go much beyond prerevolutionary efforts, and in some respects actually lagged behind the earlier tradition of excellence. This decline is explained primarily by the agony of Soviet astronomy in the thirties, as carefully documented by the American astronomer and historian Robert McCutcheon.[26]

In 1936 and 1937, the height of the "Great Purge" in the USSR, over two dozen of the leading astronomers of the Soviet Union were arrested, and executed or sentenced to long terms in Soviet labor camps. Approximately 20 percent of the astronomers in the Soviet Union disappeared. Most of those sent to camps never returned. Among those seized were B. P. Gerasimovich, director of the Pulkovo Observatory; B. V. Numerov, the director of the Leningrad Astronomical Institute; and A. I. Postoev, director of the Tashkent Observatory. Gerasimovich was shot on November 30, 1937. Numerov died four years later, evidently in one of the special labor camps for scientists. Postoev survived a number of years in the camps and then escaped with the German army in World War II and ended up as a displaced person in the American zone of Germany at the end of the war. Threatened with forced repatriation to the USSR (and almost certain death), Postoev managed to get an invitation from the American astronomer Harlow Shapley to take a position at the Harvard College Observatory. However, the American government denied him a visa on the basis that he might be a security risk. Thus Postoev faced the double tragedy of accusations in the Soviet Union of being a "capitalist wrecker" and in the United States the suspicion of being a Communist spy. Postoev emigrated in 1952 to Brazil, where he worked in an astronomical institute. He died in an automobile accident in 1977.

Another astronomer, N. A. Kozyrev, spent ten years in the camps but survived to relate his story to Alexander Solzhenitsyn, who placed in his *Gulag Archipelago* a graphic description of Kozyrev's fight to retain his astronomical skills while a prisoner.[27] Released in 1948, Kozyrev returned to astronomy and in 1958 attracted attention throughout the world for reporting spectroscopic evidence for an active lunar volcano within the crater Alphonsus.[28] He died in 1983.

Soviet astronomy was heavily damaged on the eve of World War II by the persecution of so many of its leaders. It suffered again in the destruction of observatories during the war, including famed Pulkovo. A number of Soviet astronomers (Iu. N. Fadeev, M. N. Stoilov, V. G. Shaposhnikov, F. F. Rents, V. A. Elistratov, N. V. Tsimmerman) either were killed in the war or died during the blockade of Leningrad. As a result, many important projects, including the work on weak stars, were disrupted. In 1955, however, Pulkovo published a catalog of both bright and weak stars, and in subsequent years a number of other catalogs. In the fifties and sixties another important source of astronomical research was the Biurakan Astrophysical Observatory in Armenia, directed by Academician V. A. Ambartsumian, who had studied at Pulkovo under Belopol'skii.

Soviet astronomy began a period of considerable expansion in the late fifties, spurred, in part, by the International Geophysical Year and the launching by the Soviet Union of the world's first artificial satellite, both in 1957. In 1959 and again in 1965 Soviet astronomers obtained photographs of the back side of the moon with their Luna 3 and Zond 3 space probes, a genuine contribution to our knowledge of the earth's main satellite.

Before the Revolution most of the astronomical telescopes and instruments used in Russia were purchased abroad. The 15-inch refractor telescope that was for long the pride of Pulkovo Observatory was made, for example, by Merz and Mahler. In 1862 the American instrument maker Alvin Clark surpassed the Pulkovo lens with an 18½-inch objective and, a few years later, built a 26-inch lens for Washington Observatory. The tsarist government replied by ordering from the firm of Alvin Clark and Sons a 30-inch refractor, which, when it was installed at Pulkovo in 1884, was again the largest in the world. In 1912 the tsarist government ordered in England an 81-centimeter refractor, an order that was later cancelled and fulfilled instead by Soviet industry. This telescope was eventually destroyed by invading Nazi armies in World War II. In the meantime, much larger instruments were being produced elsewhere, such as the 2.5-meter reflecting telescope at Mt. Wilson in the United States (1917), several 2-meter reflecting telescopes in England manufactured in the thirties, and, finally, the 5-meter reflector placed at Mt. Palomar in the United States in 1948.

Between the wars Soviet astronomers made ambitious plans for large telescopes, but the plans were not fulfilled. The Soviet optical industry was not yet capable of producing some of the instruments. As a result, observational astronomy suffered badly, and Soviet astronomers were forced to turn their attention more to theoretical studies such as astrophysics and cosmology. Between 1920 and 1960 the total inventory of large Soviet telescopes (measured in square meters) increased only insignificantly. After that date, however, rapid growth occurred. One of the early leaders in raising the quality of Soviet optical instrumentation was D. D. Maksutov.

In 1960 the Soviet Union began to build and in 1976 finished the world's largest optical telescope, a 6-meter reflector, at the Special Astrophysical Observatory in Zelenchukskoe in the Northern Caucasus. Although an impressive piece of equipment, it did not perform up to expectations. By the time it was constructed the technology was dated, and the surface of the mirror was not smooth and polished enough to supply the fine resolution that such an extraordinary telescope should provide.

In the 1960s, 1970s, and 1980s, the Soviet Union was one of the handful of countries in the world with research programs across the entire range of astronomy – celestial mechanics; the study of planets, stars, comets, and meteorites; the structure of the galaxy and the universe, cosmology, cosmogony, and astrophysics. In almost all these areas the traditions established in the last years of the tsarist period provided important roots. S. V. Orlov specialized in the research on comets, and B. V. Numerov (before his death in a labor camp) pursued fundamental astrometrical studies. In research on the solar system, V. V. Sharonov, N. N. Pariiskii, O. Iu. Shmidt, and G. A. Tikhov were visible figures. In stellar astronomy and stellar cosmogony some of the leading researchers were S. N. Bliazhko, V. G. Fesenkov, B. V. Kukarkin, P. P. Parenago, L. I. Sedov, and B. A. Vorontsov-Veliaminov. In photographic astronomy S. K. Kostinskii and A. A. Mikhailov (who succeeded Belopol'skii as director of Pulkovo) directed important research programs. In the newer field of radio astronomy, N. D. Papaleksi conducted some of the first Soviet research; in more recent years I. S. Shklovskii continued this work. A. D. Dubiago and M. F. Subbotin were well-known specialists in celestial mechanics. A. Ia. Orlov earned praise for his work as a geodesist.

In observational astronomy the Soviet contribution to world knowledge was not as great as the massive effort should have produced. Among the factors hindering Soviet astronomers were inadequate instruments and computing facilities, weak contact between young Soviet astronomers and young foreign ones, insufficient peer review of projects, and too much distance between research and education in astronomy. The shortages of supplies that plagued all Soviet science were particularly damaging in astronomy, since an astronomer often needs photographic plates and other materials at an exact moment in time. A delay may destroy the whole project.

In theoretical astrophysics, the picture was very different, as has already been mentioned in the section on physics. In astrophysics Russian and Soviet scholars were long prominent in world science. In addition to other names mentioned in this section, leaders in this area were G. A.

Shain, A. L. Zel'manov, I. D. Novikov, N. D. Moiseev, and P. K. Kobushkin. In the 1970s the National Academy of Sciences of the United States surveyed Soviet science and stated in its Kaysen Report that "In theoretical astrophysics the Soviets are universally regarded as being at the very forefront of world efforts. The groups of R. Z. Sagdeev, Ia. B. Zel'dovich, and Iu. S. Shklovskii at the Institute of Space Research, V. L. Ginzburg at the Lebedev Institute, I. M. Khalatnikov at the L. D. Landau Institute of Theoretical Physics, and the Theoretical Planetary Physics Group at the O. Iu. Shmidt Institute of Physics of the Earth are all of such caliber that U.S. astrophysicists would not hesitate to send their best young postdoctoral students to work with these groups." In the seventies and eighties cooperative work between the United States and the Soviet Union was intense, especially in cosmic x-ray sources and relativistic astrophysics.

CHEMISTRY

Despite the fact that the Soviet Union possessed a great army of chemists, many of them talented and well trained, it was not able to achieve in chemistry the position of world eminence that it occupied in mathematics and some areas of physics. Nonetheless, the Soviet Union possessed a fine tradition in chemistry dating back before the Revolution; the great figure here, of course, was Mendeleev (see Chapter 2) but there were other significant chemists in Russia before 1917, such as A. A. Voskresenskii ("the grandfather of Russian chemistry"), N. N. Zinin, A. M. Butlerov, V. V. Markovnikov, N. A. Menshutkin, A. M. Zaitsev, E. E. Vagner, A. E. Favorskii, N. D. Zelinskii, N. S. Kurnakov, L. A. Chugaev, and V. N. Ipat'ev (the latter five worked in Russia both before and after the 1917 Revolution; Ipat'ev emigrated to the West in 1930).[29]

The emergence of a research-oriented community of Russian chemists occurred in the period 1855–65, as documented and analyzed by Nathan Brooks.[30] Individual Russian chemists had, of course, been at work before this time. Lomonosov, as noted in Chapter 1, established the first chemical laboratory in Russia in 1748. However, Lomonosov's interests were so broad that he was not able to conduct much rigorous research. The one young Russian student whom Lomonosov trained as a disciple, V. I. Klement'ev, died at the age of twenty-eight, and thus Lomonosov left no school of Russian chemists after his death. The chemical laboratory established by Lomonosov soon fell into disrepair. University laboratories of chemistry were established at Moscow University and Kazan' University in the early nineteenth century, but until the middle of that century they were used mostly for teaching purposes, not for research. Brooks has shown that around 1860 a shift occurred in the values of the younger Russian chemists.[31] Before that time the loyalties of Russian chemists were primarily local, not international. Career advancement came from serving institutional needs and consulting to the tsarist government, not from advancing knowledge of chemistry. Even though most Russian chemists spent some time studying abroad at such leading centers as Liebig's laboratory at Giessen University, upon their return to Russia they often abandoned the research ethos that reigned among professional European chemists. In the 1860s, however, this situation began to change. Talented young Russian chemists like Mendeleev and Butlerov agitated for the creation of a society of chemists and a Russian-language journal of chemistry; a society and a journal are clear indications of professionalization.

These younger scientists wanted intellectual exchange with and recognition by their colleagues elsewhere in Europe. This desire coincided with a moment in the history of chemistry, symbolized by the first international chemical congress in Karlsruhe in 1860, when chemical terminology was being standardized among all research chemists, and when chemists were increasingly reaching for theoretical explanations of their empirical results.

It was not easy for the older Russian chemists to accept the new professional standards. In his diary for 1861 Mendeleev described the reply he received from Academician C. J. Fritzsche, an older chemist, when Mendeleev encouraged Fritzsche to serve as head of a future chemical society:

> I received a shallow education – not the sort you received. . . . What do you want from me? I do not have the strength to catch up with you. . . . I respect, with my whole soul I respect your views, your direction. Your theoretical frame of mind sometimes is incomprehensible to me. I sympathize with the society you desire. I am wholly ready to assist its realization, but I will not become its head. . . .[32]

In this strikingly honest response one sees that Fritzsche knew that the new research orientation in chemistry, already established in Europe, would soon come to Russia. Even though he was constitutionally incapable of becoming an integral part of that movement, he did not want to stand in its way. Indeed, he promised even to help it take over, to the degree that he could. One suspects that Fritzsche's attitude was more generous than that of the average older Russian chemist. But the new values took over anyway, at least among the well-educated younger chemists, and a professional community of chemists, albeit a small one, was a permanent part of Russian intellectual life from the late 1860s onward.

Probably the most significant Russian chemist of the nineteenth century, besides Mendeleev, was A. M. Butlerov (1828–1886), who taught first at Kazan' University and then at St. Petersburg University. Soviet historians of chemistry gave Butlerov priority over Kekulé and Couper in creating the structural theory of chemistry.[33] Butlerov wrote that "the chemical nature of a compound molecule is determined by the nature of its component parts, by their quantity, and by their chemical structure," and he thought that each chemical compound should have one structure that best explained its properties.[34] Recent work in the West has disputed Butlerov's priority but credited him with the popularization of the structural theory and with significant research on its basis.[35] Relying on his theory Butlerov elucidated the structure of isomers and unsaturated compounds, and he wrote an important book on organic chemistry that was translated into German. He also created a significant school of chemists, scholars who continued to work long into the Soviet period.

A particularly influential member of the Butlerov school, a man who published over five hundred scientific papers and who was a leader both before and after 1917, was N. D. Zelinskii (1861–1953).[36] Zelinskii first brought the concepts of stereochemistry to Russia, he and his students synthesized many new hydrocarbons, and he was active on a great variety of other topics, including petroleum chemistry.

After 1917 the Soviet government spent a great deal of money educating chemists and equipping large centralized institutes of chemical research. In some areas of research, such as studies of the transuranium elements, the level of Soviet research funding was much greater than in the United States.[37] Soviet research in catalysis was quite good, and attracted international attention. In many other areas Soviet chemists had ambitious research programs involving thousands of scientists.[38]

Some distinguished achievements resulted from all this effort, including the winning of a Nobel Prize in chemistry in 1956 by N. N. Semenov. Nonetheless, outstanding accomplishments in chemistry were fewer than the effort warranted. The question naturally arises, therefore, "Why did Soviet chemistry do less well than other prestigious areas of the physical sciences, such as mathematics and physics?" After reviewing the history of Soviet chemistry we will briefly consider this question, which throws light on the influence of social factors on scientific advances.

As in many other areas of Soviet science, schools of researchers formed around a leading scientist were important in the evolution of Soviet chemistry. Several of the most important of these schools were, before the Revolution, V. V. Markovnikov (who, in turn, was a student of Butlerov), who created a new chemical laboratory at Moscow University in 1887, and taught there until his death in 1904;[39] A. M. Zaitsev, also a student of Butlerov, who worked in the chemical laboratory of Kazan' University from 1865 to 1910 and trained perhaps more chemists than any other Russian chemist before the Revolution; L. A. Chugaev, who lived only a few years after the Revolution, but who left behind him a large school of inorganic chemists, including I. I. Cherniaev and A. A. Grinberg;[40] N. D. Zelinskii, who at his death at age ninety-two in 1953 had perhaps the largest school of chemists in the Soviet Union and who was a specialist in organic catalysis; A. E. Arbuzov, a specialist in elemento-organic compounds; A. N. Nesmeianov, also a specialist in elemento-organic compounds and a president of the Academy of Sciences; and N. N. Semenov, a physical chemist internationally known for his theory of chain reactions who was for many years director of the Institute of Chemical Physics of the Academy of Sciences in Moscow.

During the period of rapid industrialization of the Soviet Union in the 1930s and immediately after World War II, Soviet chemists played important roles in developing the chemical industry. Leading scientists in metallurgy were I. P. Bardin, E. V. Britske, G. G. Urazov, A. A. Iakovkin, and A. A. Baikov. In the chemical and petroleum industries, the scientific leaders included L. Ia. Karpov, D. N. Prianishnikov, N. F. Iushkevich, I. Ia. Bashilov, Iu. G. Mamedaliev, S. S. Nametkin, A. D. Petrov, and A. E. Porai-Koshits. The chemist responsible for the early development of synthetic rubber in the Soviet Union was S. V. Lebedev.

When the Soviet Union was catching up with other industrialized nations the chemistry that was required was not, on the whole, on the cutting edge of world knowledge. Once that process was largely completed the necessary chemistry both for winning international laurels and for creating new forms of advanced chemical technology was of a different and more difficult sort. For Soviet chemists to become world leaders new requirements had to be met: close contact with leading chemists in other countries, laboratories equipped with the latest instrumentation, and intimate interaction with the chemical industry. In recent years Soviet chemists attempted to meet these new requirements. In physical chemistry the best laboratories were the Institute of Physical Chemistry in Moscow directed by V. I. Spitsyn and the Institute of Chemical Kinetics and Combustion in Novosibirsk led by Iu. N. Molin. The Institute of Chemical Physics in Moscow directed by N. N. Semenov and the Institute of High Temperatures headed by A. Ia. Sheindlin were also the foci of good work in physical chemistry.

Polymer chemistry in the Soviet Union was centered in the L. Ia. Karpov Physico-Chemical Institute in Moscow and the Institute of High Molecular Compounds in Leningrad. Although the Karpov Institute had a high reputation in the Soviet Union, especially for its novel management system, Soviet work in polymer chemistry continued to lag behind other countries. Part of the reason for this was evidently a policy decision, since the Soviet industrial leaders seemed content to develop plastics after they had seen them originated and tested in the West. Another reason for the lag was the underdevelopment of industrial research facilities located on the site of chemical industries. Most industrial research in the Soviet Union was in centralized industrial ministries.

The Soviet Union possessed several leading areas in chemistry. Catalysis research continued to be strong, as it had been for many years. One of the best centers was the Institute of Chemical Physics in Moscow, where the work of O. V. Krylov was well known; other leading laboratories were the Institute of Elemento-Organic Compounds in Moscow, the Institute of Catalysis in Novosibirsk, and the Zelinskii Institute of Organic Chemistry in Moscow. Yet another strong field of Soviet chemistry was organometallic compounds, a field in which the work of O. A. Reutov had influence abroad.

On the other hand, Soviet research in organic chemistry was not notable. A major problem here, as in physical chemistry, seemed to be the lack of adequate equipment. Organic chemistry has been transformed in recent decades by the development of new research methods based on advanced technology, such as nuclear magnetic resonance, gas-liquid chromatography, mass spectrometry, x-ray crystallography, and computing facilities. The Soviet Union possessed all of this equipment in selected instances, but the instruments never seemed to be in adequate supply and spare parts and maintenance were inadequate. Part of the reason was the absence of a competitive market of instrumentation vendors.

Looking back over the record of Soviet achievement in chemistry, one sees that it was uneven. In those areas where the Soviet Union was strong, it usually had been strong for many years as a result of a scientific school led by one or two scientists. The Soviet strength in catalysis was rooted in the school of N. D. Zelinskii, perhaps the largest school in Soviet chemistry. Similarly, the Soviet strength in organometallic chemistry was rooted in the schools of A. E. Arbuzov and A. N. Nesmeianov.

We have seen that Soviet weaknesses in physical and organic chemistry were at least partially explained by poor equipment and weak ties with foreign research and facilities. Much of the foreign instrumentation went to classified work in laser and solid state chemistry. The fact that Soviet chemistry was largely centralized in massive institutes employing, in some cases, several thousand researchers, meant that strength and weaknesses tended to become ensconced, with little opportunity for change. In countries with less centralized research facilities scientist-entrepreneurs would take the necessary risks in attempting to open up new areas.

A last reason for the relative weakness of Soviet chemistry relative to fields such as mathematics and physics, despite the enormous activity in the field, was that chemistry was much more closely connected to industry than mathematics and physics. The Soviet chemical industry was not very innovative and its links to academic research were weak. As a result, the fruitful interaction between industry and research that chemistry needs was often missing.

THE EARTH AND ATMOSPHERIC SCIENCES

Starting in the eighteenth century with the expeditionary work of the Academy of Sciences, Russia had a deep tradition in the earth sciences. One of Lomonosov's main interests was mineralogy and mining, and he was a strong proponent of the exploration of Siberia's mineral wealth. Participants in the "Great Northern Expedition" of 1733–43 such as I. G. Gmelin, S. P. Krasheninnikov, and G. V. Steller brought back geological and biological specimens that became the center of Russia's first natural history museum, the Kunstkamera of the Academy. This was the initial scientific step toward the description of the mineral wealth of Russia. The visitor to the geological museums of Russia today can still sense the significance of this tradition, one prized by a nation with the richest and most variegated deposits of minerals in the world. No surprise that the Soviet Union would eventually produce the world's largest community of geologists, and, in fact, had more geologists than all the rest of world combined.

Geology in Russia played an important role in the early history of that discipline, but not because of independent theoretical contributions made by Russian geologists. Rather, the Russian Empire provided a rich inventory of geological formations that attracted the attention of West European geologists and provided the raw material for some of the basic descriptive terms of geology. As we saw in Chapters 1 and 2, among the famous European explorers and geologists who probed the terrestrial secrets of the Russian Empire in the eighteenth and nineteenth centuries were Vitus Bering, Peter Simon Pallas, Roderick Murchison, and Alexander von Humboldt. Murchison investigated in 1841 the enormous tract of rock formations in the Russian province of Perm and suggested the term "Permian" to describe this uppermost system of the Palaeozoic era. The term has remained a standard part of geological terminology. Citizens of the Russian Empire, especially Baltic Germans, but increasingly in the nineteenth century ethnic Russians as well, became interested in geology. For many years their main interest was descriptive geology rather than the great theoretical debates waged in Western Europe over such concepts as Neptunism, Vulcanism, Catastrophism, and Uniformitarianism. In 1807 the old Mining College of Peter the Great was reorganized and renamed the Mining Department, and in 1825 the Department began publishing Russia's first journal devoted to geological issues, the *Mining Journal*, and, a few years later, began promoting systematic geological investigations. However, almost all the work continued to be descriptive and empirical, primarily of local interest. No Russian names are ranked with the great geologists of the late eighteenth and early nineteenth centuries, such as James Hutton, Georges Cuvier,

Charles Lyell, Abraham Werner, William Smith, or Adam Sedgwick. Soviet historians of science pointed to early leaders of Russian geology such as V. M. Severgin (1765–1826), N. I. Koksharov (1818–1892), G. E. Shchurovskii (1803–1884), and S. S. Kutorga (1805–1861), who helped found the discipline in Russia, and who accumulated an enormous amount of empirical data.[41] Their theoretical concerns, however, were derivative of Western Europe.

One area in which Russians pioneered was soil science. Indeed, soil science is an example of *reverse* science transfer from Russia to the West, an exception to the general rule. Russian contributions to soil science that were later transferred to Western countries include the soil types *chernozem* (black earth), *krasnozem* (red loam), *podzol* (ash-colored soil), *bielozem* (white soil), *sierozem* (gray soil), and the design of soil maps. The most significant figure here was V. V. Dokuchaev (1846–1903).[42] Dokuchaev taught mineralogy and crystallography at St. Petersburg University, but he made his international reputation in the study of the topsoil of European Russia, particularly the black earth, or *chernozem*. The desire of the Russian government in the late nineteenth century to catalog and populate its vast undeveloped areas provided a stimulus for this effort. Dokuchaev taught that soil is a geobiological formation with an evolutionary history formed not only by the bedrock but also by plants and animals. He often spoke of "genetic soil science" to emphasize the evolutionary nature of soils. He developed a classification of soils and invented terms to fit that classification. Dokuchaev's influence continued long after the Revolution as the Soviet Union remained strong in soil science, a field led for many years by S. S. Neustruev, L. I. Prasolov, and B. B. Polynov.

While soil science in other countries was influenced by Dokuchaev and his pupils, the impact was more in terminology and design than it was in philosophy. Dokuchaev actually promoted what one American student has called a "pan-scientific" view of the importance of soil.[43] Dokuchaev wrote in 1898:

> The greatest and highest charm of natural history, the kernel of natural philosophy [consists in the] existence of an eternal genetic and ever orderly connection between the vegetable, animal and mineral kingdoms on one side, and man, his life and even his spiritual world, on the other. . . . It seems to us [that] soil science, taken in the Russian sense, must be placed in the center of the theory of the interrelations between living beings and inanimate Nature, between man and the rest of the world, i.e. its organic and mineral part.[44]

This viewpoint anticipates the "biosphere" of the Russian geologist Vladimir Vernadskii, who was one of Dokuchaev's students. Few Western intellectuals know that "biosphere," a term now important in ecological thought, was a Russian innovation, and almost nobody has seen that one of its roots was in Dokuchaev's soil science.

Significant improvements in the state of Russian geology came in the last part of the nineteenth century. In 1882 the tsarist government established the Geological Committee in the Mining Department of the Ministry of State Lands and gave it the necessary support to make it a genuine center of geological work. The Geological Committee continued until the revolutions of 1917, and was the seed bed out of which a mature and active Soviet geological research program grew. The Geological Committee published several series (*Proceedings; Russian Geological Library*) that contained much of the best Russian research in geology, paleontology, and the other areas of the earth sciences. The Committee also sponsored investigations of the valuable deposits of coal and iron in the Donets Basin, Krivoi Rog, and the southern Urals. These areas would later be the centers of the Soviet steel industry.

One of the first tasks for the Geological Committee was the compilation and publication of a geological map of European Russia, an assignment completed in 1892. All of Siberia and Central Asia still awaited thorough geological exploration, however, and the major part of this effort was still facing the country at the advent of the Soviet government. A start was made during the construction of the Trans-Siberian Railway at the end of the nineteenth century. Significant work on Siberia, Central Asia, and the Caucasus was performed in these years by I. D. Cherskii, P. A. Kropotkin, A. L. Chekanovskii, P. P. Semenov-Tian-Shanskii, N. A. Severtsov, and I. V. Mushketov. In 1913 the Geological Committee began the compilation of geological maps of Siberia and Central Asia.

Russia was fortunate in having a group of outstanding geologists who continued to work right through the Revolution with relatively little disruption, despite the material hardships of those years. These transitional geologists included A. P. Pavlov, A. P. Karpinskii (first president of the Academy of Sciences in the Soviet period), V. I. Vernadskii, F. Iu. Levinson-Lessing, and A. E. Fersman. Geologists seemed to adjust to the policies of the Soviet government more easily than some other types of scientists, perhaps because their field was further from ideological concerns than biology or physics, and because they were accustomed to the sort of planned and collective expeditionary work favored by the Soviet government over individualized research projects. Geologists played important roles in some of the early planning organs of Soviet science, such as KEPS, the council for the study of the natural resources of Russia, founded in 1915.

As geology in Russia matured, the descriptive concerns of earlier generations were complemented with growing theoretical prowess. In the field of crystallography, E. S. Fedorov (Fyodorov) (1853–1919) brilliantly applied mathematical analysis to the structure of crystals.[45] He helped develop a new classification of crystallographic systems and point group symmetries that became known internationally as the "Fyodorov-Groth nomenclature." Shortly after his death from malnutrition during the food shortages of the Civil War his followers published *The Crystal Kingdom*, a compendium of the work of the researchers gathered around Fedorov.

Another leader in bringing more theoretical considerations to Russian geology, both before and after the Revolution, was V. I. Vernadskii (1863–1945), a distinguished geologist whose ideas are still being explored.[46] As mentioned earlier, Vernadskii was a student of Dokuchaev, whose ideas about the geobiological and evolutionary formation of soils influenced Vernadskii toward looking upon the earth's surface and near atmosphere in a similar fashion. In his early years as a scientist Vernadskii was primarily concerned with mineralogy and crystallography. Around 1910 he became very interested in radioactive minerals. In subsequent years his concerns centered more and more on what he saw as the powerful, perhaps even determining, influence of living organisms on the earth's crust and atmosphere. He believed that the most common gases of the earth's atmosphere – nitrogen, oxygen, and carbon dioxide – are created by living organisms. The distribution and concentration of the various chemical elements in the earth's crust were, in his opinion, frequently the result of life processes. In his last years his writings were often as much philosophical as scientific. He popularized the concepts of the "biosphere" and the "noosphere."

As mentioned earlier, in the 1970s and 1980s Vernadskii became almost a cult figure in the Soviet Union because of his prescience on ecological issues and his emphasis on the interconnection between living beings and the characteristics of the earth and its atmosphere. The many articles and books on Vernadskii published in these years usually downplayed his antipathy to Marxism and his resistance to certain policies of the Communist Party, such as its takeover of the Academy of Sciences at the end of the 1920s.[47]

In the years immediately before and after World War II geology in the Soviet Union became an enormous field of activity. Some geologists played important roles in the exploration of valuable deposits of minerals, coal, and oil and in the establishment of Soviet industry based on these resources. A. D. Arkhangel'skii and I. M. Gubkin were specialists in the geology of petroleum deposits. Gubkin is remembered for his role in developing the "Second Baku," one of the Soviet Union's largest oil fields. P. I. Stepanov specialized in the geology of coal, especially in Ukraine. Mineralogists who rendered valuable service to the Soviet government included A. G. Betekhtin, D. S. Beliankin, and S. S. Smirnov.

One of the major tasks facing the Soviet government was the exploration of remote areas in Asia, Siberia, and the Arctic. Geologists who led this effort included V. A. Obruchev, L. B. Rukhin, D. I. Shcherbakov, and N. N. Zubov. Much of the northern region of Russia and Siberia contains areas of permafrost where the ground never entirely thaws in the summer. As a result of their experience with permafrost, Soviet geologists and engineers like M. I. Sumgin established a new field of permafrost studies that later was important to other northern countries facing construction problems in Arctic regions. The active volcanoes of Kamchatka were yet another feature of the Soviet Union's borderlands that attracted the attention of Soviet geologists; they became the subject of a school of vulcanology founded by A. N. Zavaritskii.

Other Soviet geologists were active across a broad spectrum of activities, such as N. S. Shatskii, who served as a director of the Institute of Geology of the Academy of Sciences; N. I. Andrusov, whose concerns included stratigraphy and paleontology; and A. A. Polkanov, whose many interests were often tied to geochronology.

The Soviet government gave strong support to oceanographic studies and built the world's largest fleet of oceanographic vessels. Their explorations covered all the oceans of the world. Early leaders in this effort were Iu. M. Shokal'skii, K. M. Deriugin, P. P. Shirshov, N. M. Knipovich, and I. I. Mesiatsev.

Geology in the Soviet Union in recent decades was strong on theory and on traditional methods of observation and analysis, resulting in a large quantity of data, but weak in reliable instrumentation, innovative analysis, quality of data, and computer applications. Soviet geology was characterized by a certain conservatism in interpretation. For example, Soviet geologists were very tardy in accepting the revolution in geology caused by the development of plate tectonics. The main obstacle here does not seem to have been ideological or political in any straightforward sense, but, instead, the authority of a few administratively powerful Soviet geologists who had staked their reputations on opposition to plate tectonics.[48]

The strengths and weaknesses of Soviet earth and atmospheric science were linked to the natural features of the country and the bureaucratic characteristics of Soviet science. The richness of the Soviet Union's geological inventory continued to inform the science of geology in many ways, and the increasing mathematization of the study of the earth and its atmosphere resulted in links between traditional strengths in mathematics in the Soviet Union and research in terrestrial and atmospheric phenomena. However, the relative backwardness of that country in instrumentation and computers prevented Soviet geologists from getting the full potential from their voluminous data. Furthermore, the power of institute directors made it difficult for younger geologists with unorthodox ideas (such as plate tectonics) to gain a hearing.

In certain areas, however, such as polar research, climatology, some aspects of space geology, oceanography, and seismology, Soviet research was excellent. In theoretical seismology, for example, the Soviet Union was a world leader for decades, doing important work on the prediction of earthquakes by statistical analysis.[49] One of the founders of seismic studies in the Soviet Union was the geophysicist G. A. Gamburtsev. Another strong area was atmospheric science, including meteorology and climatology. Important figures here in early years were L. S. Berg and N. E. Kochin.[50] Later, mathematical modeling of weather processes became a strong focus, especially at the Siberian branch of the Academy of Sciences, where G. I. Marchuk, subsequently the president of the All-Union Academy, was a major figure. K. A. Kondrat'ev pursued atmospheric research based on data gathered by Soviet rockets. Soviet scientists were also pioneers in certain areas of geophysics, such as the measurement of continental heat flow. A leader in terrestrial thermodynamics was E. A. Liubimova of the Institute of the Physics of the Earth of the Academy of Sciences.

Leading research institutions in the earth and atmospheric sciences were the Institute of the Arctic in Leningrad, and the P. P. Shirshov Institute of Oceanology and the O. Iu. Shmidt Institute of the Physics of the Earth in Moscow. Strong, but less impressive, were the V. I. Vernadskii Institute of Geochemistry and Analytical Chemistry in Moscow. The branch of the Academy of Sciences in Novosibirsk had good research programs in oceanography, climatology, and geophysics. Central Asia and the Caucasus were the locations of much research on earthquakes, where leading institutions were the Institute of Seismological Construction and Seismology in Dushanbe, the Institute of Seismology in Tashkent, and the Institute of Geophysics in Tbilisi. Also strong was the Institute of Geophysics in Kiev. The Far Eastern Branch of the Academy in Vladivostok on the Pacific Ocean developed programs in oceanography, mineralogy, and climatology.

Appendix Chapter B: The biological sciences, medicine, and technology

In the twenty years stretching from 1829 to 1849 five future great biologists of Russia were born, men who would prove that Russian biology was no longer merely descriptive nor limited to imitation of its West European model. They were the physiologist Ivan Sechenov, the embryologist A. O. Kovalevskii, the paleontologist V. O. Kovalevskii, the embryologist I. I. Mechnikov (Metchnikoff), and the physiologist I. P. Pavlov. Although the lives of these men were different in many ways, some common patterns can be seen. All were born in the provinces, and all died in major cities (Moscow, St. Petersburg, Paris). Four of the five came from families of sufficient means to provide for early education at home; the fifth was the son of a priest and had parental encouragement toward education. At the university or postgraduate stage all studied extensively in Western Europe before returning to their homeland. All became internationally known and two of the five won Nobel Prizes. All, at one time or another, were involved in political difficulties, complications that forced one to leave Russia permanently at the age of forty-three and at least one other to consider such a move at a similarly mature age.

Ivan Sechenov (1829–1905) has often been called the "Father of Russian Physiology."[1] His background was typical of many Russian scientists of the last half of the nineteenth century, with parents who were provincial landowners on a modest scale. He graduated from the Military Engineering School in St. Petersburg, worked for a short period as a military engineer, and then returned to studies at Moscow University, where he received an education as a physician. Sechenov did postgraduate research in Germany and France, where he worked with some of the most prominent scientists of his day: Johannes Müller, E. du Bois-Reymond, Karl Ludwig, Hermann Helmholtz, Claude Bernard. Upon his return to Russia he served on the faculties of the St. Petersburg Medico-Surgical Academy, Novorossiisk University in Odessa (where he was a colleague of Mechnikov), St. Petersburg University, and Moscow University.

Sechenov was attracted to neurophysiology and, in particular, to the study of reflexes. In 1863 he published his *Reflexes of the Brain*, a book that became the subject of a great controversy among members of the St. Petersburg intelligentsia, since it was seen as both a scientific and a political treatise. That Sechenov saw the implication is clear from the title that he originally gave the book, but one that was disapproved by the tsarist censor: *An Attempt to Establish the Physiological Basis of Psychological Processes*. Sechenov stated his ambitious thesis boldly in the book, proclaiming that "all acts of conscious or unconscious life are reflexes." Thus, at the very birth of Russian physiology an explicit link to materialism was made, and this tradition has remained a part of Russian and Soviet physiology and psychology to the present day. Not all Russian physiologists and psychologists agreed with Sechenov's philosophy, to be sure. Indeed, the psychologists, many of whom supported idealistic interpretations, did not consider Sechenov a psychologist at all, but a rather naive physiologist working within a narrow framework. Nonetheless, Sechenov's approach earned him notoriety in the politicized atmosphere of late tsarist Russia. In 1866 the book was prohibited for sale by the St. Petersburg censors, and Sechenov was threatened with court action for allegedly undermining public morals. This dramatic history later provided a theme that Soviet philosophers and scientists, following an official philosophy of materialism, would find a fertile one for further development. An irony in the entire story is that, as David Joravsky has shown, Sechenov himself was not a political radical but a rather mild liberal. Being a liberal in tsarist Russia was, however, politically problematic.

On a scientific level Sechenov has a deserved place in the history of physiology. His work in the early 1860s on "central inhibition" showed the effects of thalamic nerve centers on spinal reflexes, while later research pointed to the phenomenon of spontaneous fluctuations of bioelectric potentials in the brain. His concentration on reflexes influenced the later path-breaking researches of the greatest of all Russian physiologists, I. P. Pavlov, as Pavlov himself acknowledged. Sechenov's analysis of sensation as a combination of reception and muscle activity was a new and influential approach. And his work on the chemistry of respiration was the beginning of the investigation in Russia of the physiology of work. Most important of all, however, was the fact that Sechenov successfully declared the independence and vitality of the study of physiology in Russia.

Embryology had a tradition in Russia dating to Karl Ernst von Baer (1792–1876), a Baltic German living in the Russian Empire who spent thirty-two active years in St. Petersburg.[2] Von Baer was one of the great embryologists of history, discovering the mammalian egg and the notochord, describing the rudimentary development of the central nervous system, and writing some of the fundamental early works on animal development. The embryological work was done in Königsberg before he moved to St. Petersburg. After the transfer in 1834 von Baer shifted his attention to anthropology and geography. To many scientists, both in Russia and in Western Europe, von Baer was not representative of Russian science, but instead of the German scientific world of central Europe, despite the fact that he was a loyal citizen of the Russian Empire and actively promoted the research programs of the Academy of Sciences in St. Petersburg in a variety of fields.

Aleksandr Kovalevskii (1840–1901) was the first outstanding embryologist of the Empire who was clearly creating a native Russian tradition in the field.[3] Born into a landowning family in economic decline, Kovalevskii began in engineering, and then shifted to the study of natural science at St. Petersburg University. From 1860 to 1866 Kovalevskii studied in Heidelberg, Tübingen, and Naples, working on comparative embryology. The work that he produced attracted the attention of leading biologists of the day, including Haeckel, von Baer, and Darwin himself. Kovalevskii demonstrated that many organisms, including coelenterates, worms, and echinoderms, develop by invagination from bilaminar sacs. By showing the similarity of development of quite different organisms, Kovalevskii supplied important unifying evidence supporting Darwin's theory of evolution. The work brought him many honors, including membership in the Royal Society and a number of other foreign academies and societies.

Aleksandr Kovalevskii was not nearly as interested in radical politics as his brother Vladimir, a paleontologist, or his sister-in-law Sofiia Kovalevskaia, the eminent mathematician. But even Aleksandr could not escape political complications, much as he wanted to, and in the 1880s he almost emigrated from Russia to escape political interference with his research. In the end he stayed on to become an honored member of the Imperial Academy of Sciences and a professor of St. Petersburg University.

Vladimir Kovalevskii, a founder of evolutionary paleontology, lived a more politically engaged life than his brother.[4] A supporter of the emancipation of women, he made a fictitious marriage to the noted mathematician Sofiia Kovalevskaia so that she might have an opportunity to study abroad. The relationship turned out to be more permanent than either expected, even though troubled by personal and financial problems. Vladimir, like all the leading Russian biologists of his generation, studied abroad, spending the years 1869 to 1874 in Germany, Italy, France, Holland, and Great Britain. The visits were designed to embrace both science and politics, and they succeeded in both of their goals; Vladimir and Sofiia lived in Paris during the Commune of 1870, where they combined sympathy with the radical Communards with devotion to science.

Vladimir Kovalevskii pursued pioneering paleontological research in the evolution of morphological characteristics of mammals and the phylogeny of horses and pigs. He was the first paleontologist to trace the evolution of horses, concentrating on the development of the horse's foot.

Unlike his brother, Vladimir Kovalevskii was never able to secure a proper academic position, although he had a temporary post at Moscow University from 1880 to 1883. His effort to find financial support eventually led him to engage in shady business speculations the failure of which drove him, in 1883, to suicide. His academic career was cut off at the age of forty-one.

I. I. Mechnikov (1845–1916) was one of the greatest of the Russian biologists of the prerevolutionary period.[5] In his early embryological work Mechnikov strove successfully to show that the development of lower animals is similar to that of the higher ones. In the process of studying starfish larva Mechnikov noticed that the mobile cells seemed to be attacking foreign bodies in a way remarkably similar to the action of white blood cells in higher animals and man, despite the fact that the starfish had no vascular system. The continuity of the processes of immunology in animals with vastly different structures became a new illustration for Mechnikov of biological unity. He introduced the term "phagocyte" to the field of immunology, and he became an impassioned defender of his theory of phagocytosis against critics defending a noncellular theory of immunity. In the end a compromise between the schools of thought was achieved, and Mechnikov in 1908 shared the Nobel Prize with Paul Ehrlich, one of the early defenders of the noncellular immunological view.

Because Mechnikov spent the last twenty-eight years of his life at the Pasteur Institute in Paris, he is often thought of as being as much a French as a Russian scientist. Mechnikov developed his main ideas before permanently leaving Russia, however, and throughout the remainder of his life abroad he continued to publish in Russian journals. Much scholarship on Mechnikov has neglected the Russian sources and the Russian aspects of his life. But important as the Russian side of his activities was in forming his intellectual inclinations, Mechnikov was a truly international scientist, having worked in Germany, Italy, and France even before he emigrated from his native land. He was an impressive symbol of the way that leading Russian scientists were emerging on the world stage in the last half of the nineteenth century.

Although Mechnikov believed that science should be kept separate from politics, his life was an illustration of the impossibility of the goal. One of the reasons that Mechnikov left Russia was that student unrest and inadequate government support made research there difficult, especially in provincial institutions such as the University of Odessa, where Mechnikov unsuccessfully tried to lead a bacteriological institute to a level of quality equivalent to the Pasteur Institute in Paris, which he knew so well. Mechnikov was also sensitive to the fact that scientists at the prestigious institutions in Russia, such as Moscow University, failed to recognize the importance of his work, while leaders in Western Europe, including Pasteur, did. Discouraged by the internal strife, political disruptions, economic difficulties, and lack of recognition, Mechnikov in 1888 moved permanently to Paris, where he continued to defend his theory of phagocytosis.

The last member of our quintet of famous Russian biologists of the last decades of the tsarist regime was the best known of all, Ivan Pavlov (1849–1936), a man whose active career continued into the Soviet period.[6] Son of a priest and educated at a religious seminary before turning to the study of natural science at St. Petersburg University, Pavlov never lost his interest in the Orthodox faith. Pavlov always strove to avoid politics, including the dissident liberal version of his predecessor physiologist Ivan Sechenov, to whom Soviet biographers usually link Pavlov ideologically. Distinct though Pavlov may have been from Sechenov politically, he was loyal to the materialistic philosophy of Sechenov, and was convinced that the psychic activity of man could be explained on a physiological basis.[7] His intellectual formation drew on rather different sources: his family background, his admiration for Russian scientists and thinkers such as Sechenov, D. I. Pisarev, I. F. Tsion, and S. P. Botkin, and his work, in 1884–6, in the laboratories of Karl Ludwig in Leipzig and Rudolf Heidenhain in Breslau.

Pavlov's work can be divided into three phases: the physiology of the circulation of the blood (1874–88), the physiology of digestion (1879–97), and higher nervous activity (1902–36). It was the work on conditioned and unconditioned reflexes based primarily on work done in the middle period that brought him the greatest fame and was later further developed into a theory of human behavior. From the standpoint of the history of science the greatest significance of Pavlov derives from his success in bringing psychic activity within the realm of phenomena to be studied and explained by the normal objective methods of natural science.[8] In contrast to the introspective approach of many investigators of mental activity at the turn of the century, Pavlov's method was based on the assumption that psychic phenomena can be understood on the basis of evidence gathered entirely externally to the subject. He was not original in his intention to proceed in this manner, but as a great experimental observer and gifted surgeon he was able brilliantly to combine this methodological assumption with unusual skill in devising and conducting experiments with animals. Whatever limitations in approach later physiological psychologists might find in Pavlov's work, for its time it was truly path-breaking, as was illustrated by his receiving the Nobel Prize in 1904, the first Russian so honored.

In Soviet works on the history of Soviet science, Pavlov was under-standably presented as one of the great heroes, and a vindication of the Marxist materialistic approach to the study of biology and human behavior.[9] In most of these works no indication can be found of tensions between Pavlov and the Soviet government. The situation was considerably more complicated. Pavlov was a member of that generation of Russian scientists whose views were formed long before the revolutions of 1917 and many of whom never accepted Marxist or Soviet philosophical and political principles. Pavlov, for example, was a vociferous critic of the attempt by the Communist Party to take over the Academy of Sciences and to submit it to government control. His main concern, however, was his own research, and when the Soviet government eventually supported that work in an impressive way, making it possible for Pavlov to build an entirely new laboratory, he began to take a much more charitable attitude toward the government. A modus vivendi of sorts emerged that lasted until Pavlov's death in 1936, just as the purges of Soviet intellectuals were entering their most violent stage.

Western observers of Soviet science often have had a rather low opinion of Soviet biology, aware that Lysenko destroyed the field of genetics there in the 1940s and nurtured a form of pseudobiology until the middle 1960s. Even many years later Soviet geneticists had difficulty reaching world levels in their research.

People outside the Soviet Union who know about Lysenkoism are often surprised to hear that in the 1920s a group of Soviet geneticists were world leaders, making breakthroughs that earned them credit for being among the creators of population genetics. But since most of these Soviet geneticists were destroyed or dispersed in the later period of Lysenko, the reconstruction of the historical record has been difficult.

Mark Adams has been the major historian of science resurrecting this period.[10] He pointed out that it was the Soviet geneticist Nikolai Kol'tsov (1872–1940) who brought together in the 1920s at the Institute of Experimental Biology in Moscow a group of researchers who laid the basis for "the new population genetics." The head of the genetics section of the institute was Sergei Chetverikov (1880–1959), a traditional Darwinist specializing in insects, who acquired deep interests in genetics and biometrics. Out of the confluence of these interests the new synthesis of genetic teaching developed. Adams has pointed out that the researchers around Chetverikov made a threefold contribution: They developed a deeper understanding of the influence of genetic and environmental backgrounds on the effects of genes; they bridged the gap between Mendelism and Darwinism; and they founded experimental population genetics by conducting the first genetic analyses of the free-living *Drosophila melanogaster* species of fruit fly.[11]

In order to understand this achievement of Soviet biology, it is necessary to recall that in the early twentieth century many biologists saw a tension, if not a contradiction, between Darwinian evolution and Mendelian genetics. Darwinians envisioned changes in organisms occurring gradually over long periods of time, based on minute variations; Mendelists at first emphasized the extreme stability of the gene, and, then later, incorporating the concept of mutation, depicted stability occasionally interrupted by rather large changes, a different picture from the traditional Darwinian one. Furthermore, the members of the different camps proceeded on the basis of contrasting methods. The followers of traditional Darwinism emphasized descriptive natural history, while the new Mendelians used mathematical approaches. Was there any way all this could be brought together, or was the Darwinian approach destined to be superseded by the Mendelian?

By the second decade of this century, biologists concerned with the problems could be roughly classified into three groups: naturalists working within the late nineteenth-century Darwinian tradition; geneticists studying gene location and mutation, many of them connected with the school of T. H. Morgan at Columbia University; and "biometricians" utilizing highly mathematical methods as developed by Karl Pearson and others. While some hope existed of finding a commonality or "synthesis" of the different approaches, none was apparent.

One of the most important papers pointing a way to such a synthesis was written by Chetverikov in 1926. Chetverikov noted in the opening sentences of his article that Mendelism was greeted with hostility by "outstanding evolutionists" both in Russia and abroad, and then stated his goal as bringing these two approaches together through a clarification of evolution from the standpoint of genetic concepts.[12] Chetverikov then went on to argue that the process of mutation being observed in laboratories also occurred in nature, but that because recessive mutants would be heterozygous, they would not be evident in phenotypical forms. Natural selection would quickly eliminate harmful dominants, but would act more slowly against harmful recessives. Thus, there would be a build-up of hidden recessive mutants in any population.

In the same paper, Chetverikov agreed with T. H. Morgan and others that selection cannot affect genes themselves, but emphasized that genes do not act in isolation from the rest of the genotype:

The very same gene will manifest itself differently, depending on the complex of the other genes in which it finds itself. For it, this complex, this genotype, will be the *genotypic milieu,* within the surroundings of which it will be externally manifested. And as *phenotypically* every character depends for its expression on the surrounding external environment, and is the reaction of the organism to the given external influences, so *genotypically* each character depends for its expression on the structure of the whole genotype, and is a reaction to definite internal influences.

Chetverikov was presenting here extremely sophisticated concepts of population genetics, and he and his students supported these concepts with original experimental work on natural populations.

Another Soviet innovation in genetics in these years was the presentation of the concept of "gene pool." The Russian geneticist A. S. Serebrovskii first formulated the concept in terms of *genofond* ("gene fund"), a word that was brought from the Soviet Union to the United States by Theodosius Dobzhansky, one of the members of the Chetverikov group, who translated it into English as "gene pool." Even today few people know that this term, so common in biological discourse in the world, has a Russian origin. Yet another Soviet researcher, D. D. Romashov, arrived independently at the concept of "genetic drift," developed in the West by Sewall Wright and others.

Some of the young Soviet biologists who took their inspiration from Chetverikov later became world-prominent scientists, including N. V. Timofeev-Resovskii, N. P. Dubinin, and Theodosius Dobzhansky. Unfortunately, their early work in Russia was almost forgotten after the Chetverikov group was destroyed by Stalinism. Chetverikov himself was arrested, sent into exile in 1929, and never returned to his *Drosophila* research. Timofeev-Resovskii emigrated from the Soviet Union to Germany even earlier, in 1925, and returned to his home only many years later.[13] Dobzhansky emigrated to the United States in 1927, where he became a famous geneticist. He frequently acknowledged his debt to the Chetverikov group. Kol'tsov, the director of the institute who had assembled the Chetverikov group, managed through adept maneuvering to maintain his position until 1938, when he was accused of racism and dismissed. Dubinin, a fervent Marxist, was able to continue the research direction of the school longer than the rest, but eventually, he too fell victim to politics at the time of Lysenko's conquest of genetics. In 1948 he abandoned genetics for many years, working as an ornithologist in the Ural Mountains. He was able to make a full return to genetics only in 1965, after Lysenko's overthrow.

The creation of a synthetic theory of evolution, combining Mendelism and Darwinism, was, of course, not solely a Soviet creation, but instead a result of international effort. G. G. Simpson named eighteen founders of the new synthesis, four of whom were Russian: Chetverikov, Timofeev-Resovskii, Dubinin, and Dobzhansky.[14] Until Lysenko made it impossible for them to do so, the Russian geneticists worked in step with geneticists in other countries, helping and being helped. The American geneticist H. J. Muller, for example, in 1922 brought to the Soviet Union over 100 laboratory strains of *Drosophila melanogaster* with known breeding histories, giving a big boost to Chetverikov and other Soviet biologists. In turn, Chetverikov's group made its contribution by establishing experimental genetics of natural populations as a basis for the new synthesis.

Just as there was an early golden period in the history of Soviet genetics, a period now largely forgotten because of the subsequent disaster of Lysenkoism, so also there was a pioneering moment in the history of Soviet conservation theory and community ecology. The historian who has uncovered this fascinating episode, also an American, is Douglas R. Weiner, who points out that "through the early 1930s the Soviet Union was on the cutting edge of conservation theory and practice." Russians were among the pioneers in phytosociology (I. K. Pachoskii, G. F. Morozov, V. N. Sukhachev), the individualistic theory of plant distribution (L. G. Ramenskii), and ecological energetics (V. V. Stanchinskii).[15] Here also, as in genetics, early achievements were blotted out by Stalinist repression. G. A. Kozhevnikov, early exponent of nature preserves in Russia and the Soviet Union, was expelled from his academic positions during the Cultural Revolution, and V. V. Stanchinskii, developer of trophic dynamics, was denounced and arrested. Conservation and ecology suffered reverses in the Stalinist Soviet Union. Remarkably, despite an atmosphere of threats and repression, attempts by scientists and activists to resist the worst excesses of Stalin's "great transformation of nature" and to defend some islands of "free nature" continued throughout his rule and beyond, as Weiner has shown. After Gorbachev came to power in 1985 the full horrors of environmental abuses in the Soviet Union emerged. The wishes of industrialists had triumphed over those of ecologists and biologists. The early achievements remain worthy of study, however, and we should not lose sight of the fact that by maintaining a coherent oppositional movement independent of the Communist Party and the state continuously over the decades, the conservation movement provided not only an inspiration but also a model (and a source of activists) for environmental and citizens' groups generally in the 1980s and 1990s.

Areas of traditional Russian and Soviet strength in biology include ornithology and zoogeography (the tradition extending from G. G. Doppel'mair through P. P. Sushkin, M. A. Menzbir, V. V. Stanchinskii), limnology and hydrobiology (L. S. Berg, G. G. Vinberg, V. S. Ivlev), and descriptive zoology (A. N. Severtsov, S. I. Ognev). Pioneering studies in the role of snow cover on the ecology of animals were conducted by A. N. Formozov, A. A. Nasimovich, and O. I. Semenov-tian-shanskii. Other innovative work in animal ecology was done by D. N. Kashkarov (steppe and desert fauna) and G. F. Gauze ("competitive exclusion"). L. G. Ramenskii was a distinguished plant ecologist as was V. N. Sukachev. A. A. Rode continued Dokuchaev's work on the formation of soils, integrating such neglected factors as soil hydrology and the action of individual species of decomposers on specific classes of decaying organic matter.

As described in Chapter 6 on the Lysenko affair, biology in the Soviet

Union during and after World War II suffered a terrible slump. Not until 1965 was Lysenko overthrown, although geneticists began to try to revive the field before then, working under the protection of prestigious physicists. After Lysenko's demise, Soviet biology continued to rank in world competition behind Soviet physics and mathematics. In recent years the Soviet government made a major effort to catch up. In the middle 1980s and early 1990s a leading center of molecular biology was the Shemiakin Institute of Bio-Organic Chemistry of the Academy of Sciences of the USSR. In order to build up the field, each year the Shemiakin Institute recruited fifty or sixty students from Moscow University to work as interns and eventually to become graduate students. The Shemiakin became the lead organization for an association of research and production facilities aimed at the development of new strains of animals and plants and a range of pharmaceutical products. By the late eighties the Soviet Union was offering products such as interleukin-2 and hepatitis B vaccine for export sale. Despite the progress, however, Soviet biologists did not regain a position of world leadership.

MEDICINE

The history of medicine and public health in Russia and the Soviet Union is a special subject to which inadequate attention has been given by Western scholars. It appears that this history deviates from the common pattern of both the rest of Russian science and the development of medicine in other industrialized countries. Contrary to the common observation that Russian and Soviet science emphasized theory over application, Russian medicine possesses a tradition of great practitioners and community benefactors, of whom the nineteenth-century physicians Nikolai Pirogov (1810–1881) and Sergei Botkin (1832–1889) are outstanding examples. And contrary to the importance of professional autonomy and independent licensing powers by medical organizations in many other countries, medicine in Russia and the former Soviet Union has been characterized by a strong tradition of public service and a relatively weak sense of professional independence. These characteristics have roots in the tsarist period but continued to develop, for somewhat different reasons, in the twentieth century under the Soviet regime. During the period of Communist rule, Soviet physicians never came to represent a profession in the Western sense.

The roots of Russian and Soviet medicine stretch far back into history. The identification of these roots is in part a definitional question, since several areas of the later Soviet Union with ancient medical traditions (Armenia, Azerbaidzhan, Central Asia) became parts of the Russian Empire only in the nineteenth century. The connections to ancient Kiev are, of course, deeper. As early as the tenth century Armenian and Byzantine physicians took care of noble families in Kiev while monks from monasteries provided rudimentary medical care for the rest of the population. With the emergence of Muscovy as a powerful independent principality in the late fifteenth and sixteenth centuries, medical contacts with other countries, including those in Western Europe, began to develop. English, Dutch, and German physicians and apothecaries appeared in Muscovy, at first serving the ruling families, but gradually broadening their practices. By the late seventeenth century in Moscow drugs were being sold to the general population by the Apothecary Bureau (*Aptekarskii prikaz*), a part of the Kremlin bureaucracy. This organization gradually acquired broader medical responsibilities. It provided physicians for the army, and trained practitioners similar to the barber-surgeons of the West, who in Russia were called "treaters" (*lekary*). The Apothecary Bureau underwent several changes in name, and was moved to the new capital of St. Petersburg in the early eighteenth century. There its descendant organizations were known as the Medical Chancery (1721–63) and the Medical Collegium (1763–1803). These state institutions were in charge of medical affairs for the entire Russian Empire until the nineteenth century, when medical education and care began to develop in universities and hospitals.

It is very difficult to say much about the effects of state policies on general health in early modern Russia; as John Alexander, one of the pioneering American historians in this area, remarked, "Our ignorance about these matters is immense."[16] In view of the primitive knowledge of medicine and the scarcity of physicians (only 2,000 for a population of 40 million people in 1800) state policies probably had little effect. During the great plagues that periodically swept the Empire the state authorities were nearly helpless. The best preventive measure seemed to be flight from the afflicted areas, a recourse more available to the upper classes than the lower ones.[17] We do not possess reliable statistics on longevity or infant mortality for much of the tsarist period, but in the widespread conditions of poverty, alcoholism, and filth, the reality was surely grim. Before one concludes, however, that old Russia was uniquely backward in public health, one should recall that much of the world before 1800 faced similar conditions. And even in the most prosperous countries of Western Europe physicians were powerless in the face of most diseases.

In the beginning of the nineteenth century, medical education in Russia began to come alive. By 1814 Russia possessed six institutions preparing physicians, the Medical-Surgical Academy in St. Petersburg, and the medical faculties of the universities of Moscow, Vilna, Kazan', Dorpat, and Khar'kov. Between 1803 and 1840 the number of physicians more than tripled to 6,879. Then the growth rate slowed, but the total number of physicians reached almost 12,000 by 1870.[18]

Few Russian medical doctors acquired the social prestige and authority of their most privileged colleagues in Western Europe or North America. Throughout the nineteenth century most Russian physicians were servants to the state and most of them came from nonprivileged social ranks. Medicine was not a favored educational choice among noblemen, who often saw it as rather unpleasant manual work. For commoners, however, medicine was considerably more attractive, since it provided social mobility and, on the basis of "The Table of Ranks," even the possibility of achieving noble status. In the early nineteenth century sons of clergy and exseminarians were particularly common among medical students; later in the century the progeny of low-ranking civil servants, shopkeepers, soldiers, impoverished nobility, and even peasants increasingly entered the field.

Russia was far ahead of Western Europe in providing medical education for women, even though the absolute numbers were not large. In 1882, when there were only ten women doctors in England and seven in France, 227 women doctors had been educated in Russia.[19] The motivations of women to become physicians were a part of the upsurge of reformist and radical ideas among the Russian intelligentsia in the last half of the nineteenth century. The tsarist government vacillated between yielding to and resisting the pressure for educational reforms, including women's education. During times of retrenchment, such as the years immediately after the assassination of Alexander II in 1881, women were sharply restricted in their educational opportunities. Even in the most liberal periods the women medical students were the objects of much prejudice, and studied in special "medical courses for women," rather than the regular medical schools. Nonetheless, compared to other countries, women in Russia were pioneers in medicine, as they were in many other areas of science.[20]

Nikolai Ivanovich Pirogov (1810–1881) was one of Russia's most outstanding physicians of the nineteenth century. He came from a family of peasant background, but his father was a minor official. Educated originally at Moscow and Dorpat universities, he spent two years doing medical research at the University of Berlin, where he became thoroughly familiar with the latest trends in West European medicine. He proved to be a gifted surgeon, medical administrator, and scholar, and in 1841 was appointed to a chair at the St. Petersburg Medical and Surgical Academy. He became an authority in anesthetics, developed innovative techniques in osteoplastic surgery, invented new surgical instruments, and published significant works on cholera and clinical surgery. He also served as a military surgeon in battle in the Caucasus in the 1840s and in the Crimean War in the 1850s.

Pirogov's fame in Russian society at large, rather than merely among medical specialists, arose mainly from his activity in promoting educational reform. As curator of the Odessa and Kiev school districts he defended humanistic disciplines, greater democracy in education, and the expansion of the educational system. He was one of the initiators of the project for establishing Odessa University. He was sharply critical of Russia's corrupt bureaucracy and called for a more egalitarian educational system, including better opportunities for women and Jews. His frankness in criticizing tsarist officials created powerful enemies, who successfuly worked to achieve Pirogov's dismissal in 1866 from all educational posts. From that date until his death he again concentrated on questions of medicine and public health.

In 1881, the fiftieth jubilee of Pirogov's professional career and also the last year of his life, Pirogov's medical colleagues honored him by establishing a medical society dedicated to Pirogov's lifelong goals of advancing public health and medical education. It was the first medical society in Russia. After Pirogov's death the new society was renamed "The Society of Russian Physicians in Memory of N. I. Pirogov." It would remain active until the Russian Revolution. In the late nineteenth century it became deeply involved in *"zemstva* medicine," to be discussed below. By the early twentieth century the Pirogov Society became a center of political opposition to the tsarist regime.

One of the striking characteristics of Russian medicine in the last half of the nineteenth century was the strength of the idea of public service. As Nancy Frieden has written, "Russian physicians adopted a distinctive social position, an overriding interest in public health and welfare. . . . Working primarily in the public sector, their daily tasks affected the general welfare and their concerns centered on public needs."[21] The origin of this ethos can be found in the *zemstvo* reforms of the 1860s, and its most powerful institutional exponent became the Pirogov Society.

The *zemstva* were local organs of self-government, primarily in rural areas, which Alexander II and his advisors introduced in 1864 as a part of the Great Reforms. Although elections of members of these assemblies were weighted so that the nobility would have controlling influence, the majority of the population, including the peasants, was able to participate. Throughout the last decades of the nineteenth century the imperial government successfully resisted the attempts by some leaders of the movement to create a central organization. On the local level, however, the zemstva accomplished much good work, especially in the areas of primary education, public health, and agriculture.

In the late nineteenth century zemstva medicine became an innovative free rural health program that has a special place in the history of public health. All zemstva physicians were not the self-sacrificing and idealistic servitors of the public good that much of the literature describes, but many certainly were. The zemstva program established a strong tradition in Russia, lasting long into the Soviet period, that physicians were predominantly state employees who emphasized preventive medicine and community service. Their first task was not the curing of disease by means of medical science (often still inadequate to the task), but the prevention and control of disease through sanitation and education.

In the very last years of the tsarist period the ethos of the Pirogov Society and of zemstva medicine began to erode somewhat, both for scientific reasons and for political ones. Scientifically, enough progress was being made in the actual control of disease to undermine the older populist view that what really counted was the prevention, not the curing of disease. Research-oriented physicians in the cities were beginning to criticize the often poorly qualified village doctors who were at the center of zemstva ideology. At the same time, it became increasingly clear that the distrust of the central government so embedded in the zemstva movement was hindering the solution of enormous health problems. Some assistance from the central government was absolutely necessary. Thus, medicine was in flux in Russia at the time of the Revolution. The old emphasis on preventive medicine, public service, and decentralized organization was still strong, but new approaches were beginning to appear.[22]

The history of medicine and public health during the Soviet period of history is a fascinating subject crying out for careful, empirical research. The Soviet Union was the first country in the world to announce a system of free medical care for all its citizens. During the first decades of its existence the Soviet government made remarkable progress in expanding medical education and care. Before the Revolution, in 1913, the Russian Empire possessed about 207,000 hospital beds and 23,000 physicians; by 1940 the Soviet Union had almost 800,000 hospital beds and about 150,000 physicians. By the late 1960s the Soviet Union possessed more physicians per capita than any other country in the world, with the possible exception of Israel. Similar impressive strides were made in basic health indicators, with general mortality declining from 30.2 per 1,000 in 1913 to 18.3 in 1940.

Western analysts of Soviet health have found that even in the early years of the Soviet government, progress was not nearly as uniform and equitable as Soviet officials claimed. Christopher Davis has shown that almost from the beginning of the Soviet period health care was distributed unequally among different socioeconomic groups, geographic regions, and political factions.[23] A more complete picture of the history of Soviet public health awaits much more research. No doubt that picture will contain a mixture of genuine achievements and serious shortfalls, and will, therefore, look much more like the picture of public health in most other nations than the early admirers of the Soviet system thought would be the case.

Although much that goes on in Soviet medicine is similar to medicine worldwide, Soviet medical authorities developed several distinctive approaches. They made early promises that attracted attention throughout the world but later failed to be fulfilled. Writing in 1933, one Soviet authority announced that the Soviet Union was on a path "about which not one capitalist country can dream: the real and genuine prevention of disease."[24] Soviet public health leaders particularly stressed an environmentalist theory of disease, a belief that social conditions are major factors in the origin of disease. Even as late as 1974, the author of the article "Medicine" in the best-known Soviet encyclopedia wrote,

> Soviet medical science believes that the actual source of disease must be sought in the unfavorable effect exerted by environmental factors – physiological, biological, and social. Moreover, it believes that the effect of the diverse causes of diseases depends on working and living conditions, the character of socioeconomic relations, and the condition of the body itself, which reacts actively, and not passively, to external influences.[25]

Soviet writers often termed this environmentalist approach to disease "Marxist-Leninist," but it should be apparent that in its stress on community medicine and sanitation it shared many of the ideals of the pre-revolutionary traditions of *zemstva* medicine and the Pirogov Society. It is, of course, not accidental that both nineteenth-century "Pirogovtsy" and early Soviet Marxists shared this reformist, environmentalist approach to medicine, since both were committed to creating a more just and healthy society, and both saw social action as more important and more feasible than advanced medical science in achieving that goal.

Such a view of disease naturally leads to great emphasis on preventive medicine and social hygiene. This approach was one of the explanations for the creation in the Soviet Union of a vast system of sanitoria and rest homes. It also explained the importance given by many Soviet doctors to long-term care. Early Soviet leaders in the development of theories of preventive disease and health-resort medicine were N. A. Semashko and Z. P. Solov'ev.[26]

Many early Western observers praised the emerging Soviet health system. The distinguished American historian of medicine Henry Sigerist wrote in 1937, "I have come to the conclusion that what is being done in the Soviet Union today is the beginning of a new period in the history of medicine. All that has been achieved so far in five thousand years of medical history, represents but a first epoch: the period of curative medicine. Now a new era, the period of preventive medicine, has begun in the Soviet Union."[27] Even Mark Field, a more critical analyst of Soviet medicine than Sigerist, wrote in 1967, "Soviet socialized medicine has been one of the more impressive and positive achievements of the Soviet regime, and has probably met with the approval of the great majority of the population."[28] And Field stressed that the attractiveness of Soviet medicine was particularly great in underdeveloped countries. A former minister of health of India told Field, "We simply cannot afford a medical system of the American type; for our needs, and with our resources, the Soviet model is infinitely more relevant."[29]

Scholars have recently examined the record of Soviet medicine and public health more critically. Susan Gross Solomon has shown that social hygiene, with its emphasis on socioeconomic factors as causes of disease, was welcomed in Soviet Russia so long as its radical critique was directed against the tsarist legacy, but already by the late twenties it had become apparent that the criticism could be equally well leveled against Soviet social policies. When Stalin launched a strenuous industrialization program near the end of the decade, ignoring public health in the process, he came to regard the social hygienists as dangerous opponents, and moved against them.[30] Thus much of the early promise of Soviet social hygiene was not fulfilled.

Over seventy years after the Russian Revolution the record of Soviet medicine looks much different than it did to its early admirers. The most dramatic achievements occurred in the first forty or so years; in recent decades not only did the progress slow but a serious deterioration occurred. The national crude death rate reached a low point in 1964 of 7.1 per 1,000 and then began a rise to 10.1 by 1979. Something similar happened to life expectancy and infant mortality rates. Life expectancy for men at birth declined from 66 to 62 years, while for women it leveled off at 74 years. Most ominous was the infant mortality rate, an indicator that is often cited by public health experts as a sensitive barometer of the general health of a society, and even of its quality of life. The infant mortality rate in the Soviet Union reached a low of 22.9 per 1,000 live births in 1971 and then increased to an estimated 31.1 in 1976, a 36 percent increase in only five years.[31] The situation was even worse than these statistics indicated, because they are based on Soviet methods of accounting and definitions not generally accepted elsewhere. Western experts who have adjusted the Soviet statistics to the internationally accepted methodology of the United Nations have estimated that the Soviet infant mortality rate never fell below 26.2, and that by 1976 was 35.6. This latter infant mortality rate was aproximately three times that of the United States in 1976. And the United States, saddled with its own serious social problems, was far from being the world leader in reducing these rates.

After the advent in 1985 of Mikhail Gorbachev to the leadership of the Soviet Union, Soviet physicians and public health leaders used the newly acquired freedom of press to criticize sharply their health system. In 1988 the minister of public health of the USSR, Evgenii Chazov, referred to the pre-Gorbachev period in the following way:

We took pride in our system of health care for the people. But we kept quiet about the fact that in the level of infant mortality we ranked 50th in the world, after Mauritius and Barbados. We were proud that we had more physicians and more hospitals than any other country in the world, but we kept quiet about the fact that in average life expectancy we ranked 32nd in the world. . . . In terms of the level of allocations from the gross national product for the health care of the people, we are somewhere around 75th out of 126 countries.[32]

In the late 1980s and early 1990s newspapers were filled with a litany of complaints about the health system, including descriptions of shortages of simple medical supplies, unsanitary conditions in hospitals, corruption of physicians, and a system of privilege that left only poor and tardy care to people without political connections.

TECHNOLOGY

Many contemporary Western observers of the former Soviet Union know that technological backwardness was a major problem for that nation. These same observers often assume that the history of the development of technology in that country was a fairly recent one, at least compared to the leading Western nations. They even sometimes assume that in the nineteenth century Russia was nearly totally undeveloped, and that only in the twentieth century did a closing of the technological gap between the Soviet Union and Western nations begin.

A closer look at the history of technology in Russia and the Soviet Union reveals a much more complicated story. Instead of a picture of a steady closing of a gap, starting in the twentieth century, we see a jagged curve stretching back for several centuries, with individual high points of excellence achieved long before 1917, only to be lost again in succeeding years. And this repetitive story of momentary achievement followed by obsolescence continues in the twentieth century. The problem of technology in the history of Russia and the Soviet Union does not seem to be primarily one of the transfer of technology from other nations, something that has been going on there for centuries, but of sustaining the application and production of technology in the economy and constantly improving that technology. Technology is not something that, once obtained, develops on its own. It requires a culture and an economy that are receptive and stimulating. Creating such a culture and economy has been Russia's real problem in the fostering of technology.

Evidence of isolated early achievements in technology is abundant in Russian history. As early as the sixteenth century the casting technology of the Moscow Cannon Yard astonished Western visitors. Here were cast the largest church bell ever made and hundreds of heavy cannons for the Russian armies. Some of the bells and cannons were highly decorated. Originally tutored by Western foundrymen, the Muscovites developed their own procedures, which they kept secret from foreign visitors.[33]

In 1632 the Dutchman Andrei Vinius established near Tula south of Moscow a factory for the manufacture of armaments that has had a continuous history to the present day. Peter the Great gave Russian technology, especially that which could be used for military or naval purposes, a great boost in the early eighteenth century. Not only did he import foreign technicians but he also sent Russian mechanics abroad for education. One of them, Andrei Nartov, became a master machinist who developed lathes, mint presses, guns, and canal locks.

Westerners are often surprised to hear that at the end of the eighteenth century Russia was the largest exporter of iron in the world, producing in its Urals factories about one-third of the world supply of iron.[34] The single largest purchaser of Russian iron was England. In 1766 a Russian inventor, I. I. Polzunov, developed a 32-horsepower steam engine to pump water out of mines. Soviet historians made much of Polzunov's engine, which preceded that of James Watt by several years and was evidently an improvement over the earlier Newcomen engine. However, Polzunov's engine frequently broke down and was soon forgotten; its importance was not as a practical achievement but as an illustration of early Russian interest in steam technology.

A Russian father-and-son team, the Cherepanovs, produced a steam locomotive that could pull a 60-ton load in 1835. As was the case with Polzunov's engine, the Cherepanov locomotive was not pursued further. Russian railways based on foreign technology developed rather early, however; the first steam railway in Russia was opened to the public in 1837, the same year as the first one in Austria and only five years after the first steam railway in France.[35] The two largest cities in Russia, Moscow and St. Petersburg, were connected by railway a year before New York and Chicago were similarly connected.[36] American engineers helped establish near St. Petersburg in the 1840s a large and impressive locomotive factory. In 1847 the Philadelphia engineer Joseph Harrison, Jr., wrote that the factory, employing primarily Russian workmen, could build locomotives faster than "any establishment in any country that we have any knowledge of."[37] For a few years foreign observers agreed that the tsarist empire possessed the largest and best equipped railways in the world. Yet Russia expanded its railways much more slowly than the leading Western powers and soon fell behind again in railway engineering. By 1855 Russia possessed only 653 miles of railway, compared to 17,398 in the United States and 8,054 in England.[38]

An illustrative example of how Russian technology often advanced in spurts, aided by foreign assistance, only to fall behind later can be found in the Tula arms factories. As mentioned, they were established by a Dutchman in the seventeenth century; at first they employed the most modern methods. Already by the time of Peter the Great in the early eighteenth century they were lagging behind West European technology. Peter ordered their modernization and especially the greater utilization of water power. He brought in Swedish, Danish, and Prussian gunsmiths to teach Russian apprentices. After Peter's death, the policy of obtaining foreign assistance continued. Catherine the Great took an interest in the Tula factories in the last third of the eighteenth century and sent Russian gunsmiths to England to improve their skills. During the Napoleonic Wars the Tula factories were major suppliers of guns of different calibers to the Russian armies. Nonetheless, after the successful defense of Russia, the gunsmiths of Tula were once again forced to turn to the West to modernize their methods. In 1817 a master English gunsmith, John Jones, was brought with his family to Tula to manufacture gun locks by means of dies instead of the previous manual forging.[39] He also introduced the use of drop-hammers and tried to produce interchangeable parts for his guns. Jones so impressed his employers that in 1826 one of the tsar's inspectors reported that the Tula factory "had been improved to such an extent that not a single weapon factory in the world can be compared with it." In that same year Tsar Nicholas I visited the factory to test the new guns. He chose thirteen guns from a pile of several hundred, ordered them disassembled, the parts mixed up, and thirteen guns then reassembled. According to the government report, all thirteen guns worked perfectly, proving that interchangeability of parts had been achieved. The tsar was elated.

The tsarist claim for true interchangeability of gun parts in 1826 was probably false. Historians of technology who have examined parts of surviving guns from Tula of this period conclude that they were not truly interchangeable.[40] Just as Eli Whitney in the United States misled inspectors by selecting a few specimens out of many for examination, the Tula gunsmiths probably misled the tsar by having ready for him several hundred exceptionally well-machined guns.[41]

The tsar's conviction in 1826 that the Tula gun factories were as good as any in the world may have lulled him into neglecting their further advancement. The supervisors of the Tula works rewarded most fulsomely the master craftsmen who produced richly ornamented "presentation arms" for gifts to the ruling family and top officers, not ordinary machinists for producing standard but modern weapons for the infantry. By the time of the Crimean War in 1853–6 Russia's weapons were once again of poorer quality than those of its enemies. The ordinary soldiers in the tsarist armies in that conflict were armed with smooth-bore guns, some even of the flint-lock type, while the Allies had rifles with cylindro-ogival "Minie" bullets.

The underlying cause of Russia's difficulty in keeping up with its competitors in armaments was socioeconomic. In England and France the Industrial Revolution was underway. In those countries weapons were the evolving products of a constantly improving and enlarging industrial base. In Russia weapons were being produced without the help of such a base, and consequently the effort advanced in lurches whenever exposure of backwardness forced importation once again of advanced Western technology.

The low level of education of Russian workers was also a major barrier to technical advancement. Merritt Roe Smith has demonstrated that in the United States one of the reasons that the Springfield armory gradually gained ascendancy over the Harper's Ferry armory during the nineteenth century was that Massachusetts workers were better educated and more sophisticated than Virginia ones at that time.[42] Yet the Virginia workers were incomparably better educated than any in Russia in the nineteenth century, where no effective system of public education yet existed.

In the development of electricity a similar pattern emerged of an early promising start followed by a slump. When the streets and public gardens of Paris and London were first electrified in the late 1870s and 1880s, the method of illumination was the arc light, patented in Paris in 1876 by the Russian inventor Pavel Iablochkov. At the World Exposition in Paris in 1878 Iablochkov's lights were exhibited to great acclaim. The new street lamps were popularly referred to as "Russian lights."[43] Impressed by his success in Western Europe, Iablochkov returned to Russia and attempted to manufacture and sell his lights there. Without an adequate market at home and unable to innovate rapidly enough to keep up with foreign firms, Iablochkov's company failed miserably.[44] The major cities of the Russian Empire were eventually electrified by foreigners. Electrification of the countryside proceeded very slowly, and was a major task of the Soviet government even decades after the 1917 Revolution.

In the last decades of the tsarist empire further examples of the alternation of periods of technological achievement followed by retardation continued to appear. For example, in 1900 Russia was the world's largest producer of crude oil. In 1912 a plant in Moscow began producing a domestic automobile in small numbers. Igor Sikorsky even designed and built a four-engine airplane in 1913; more than fifty of these large planes were used as bombers in World War I. But in backward tsarist Russia there was no possibility of a mass market for automobiles, and little chance for development of an aeronautical industry, outside of military needs. Petroleum was used in Russia primarily to produce kerosene for illumination and heat, not for gasoline. When popular demand for automobiles in other countries, particularly the United States, caused a dramatic jump in the need for gasoline, the center of production of petroleum products moved elsewhere.

The pattern of cycles of innovation and stagnation in technology continued into the Soviet period. When the Soviet Union rapidly industrialized in the late 1920s and the 1930s, it adopted the latest technology from abroad. At various moments the Soviet government granted concessions to foreign concerns, formed joint stock companies, hired foreign consultants, or simply illegally copied Western technology. In the period 1929–45 about 175 technical assistance agreements were arranged between the Soviet Union and Western companies; the latter included the best and largest firms in the world: Ford, International Harvester, Krupp, Pennsylvania Railroad, Pratt and Whitney, Siemens, Standard Oil, Union Oil Products, Babcock and Wilcox, Bucyrus Erie, Caterpillar Tractor, Dupont, Metropolitan-Vickers, and many others.[45]

Some Western observers believed that by modernizing after other industrial nations the Soviet Union would benefit from its ability to pick and choose the latest technology. Professor Alexander Gerschenkron wrote a series of famous articles in which he spoke of "the advantages of backwardness," noticing that the Soviet steel and automobile industries were being constructed on the very latest Western models.[46] The great steel plant of Magnitogorsk was consciously modeled on the United States Steel plant at Gary, Indiana, where the leading Soviet metallurgical engineer, Ivan Bardin, had once worked. Gary at that time represented the last word in steel-making. The large auto plant in the city of Gor'kii was built by engineers and workers from the Ford Motor Company, and was modeled on the River Rouge plant in Detroit, considered to be the most modern complete auto plant in the world.

Once Soviet factory directors achieved independence in production, they tried to shed the foreign tutelage. Soviet leaders believed that, after industrial parity had been achieved, what they saw as the inherent superiority of a socialist economy would then lead to further qualitative improvements and technological originality. Time demonstrated, however, that they overestimated the economic efficiency of socialism and they underestimated the deadening effects on technology of their social and political backwardness.

Once basic industrialization had been attained by the early 1960s, the debilitating effects of a state-owned and centralized economy and the absence of a sustaining environment showed up ever more clearly. Although the Soviet Union became the world's largest producer of many basic industrial products, including oil, steel, cement, and machine tools, it offered the world market very little in terms of original or qualitatively superior products. And even quantitatively it began in the 1970s to slip in relative terms. After becoming the second-largest economy in the world, surpassed only by the United States, the rate of industrial growth of the Soviet Union sagged dramatically. This failure of the Soviet economy to continue to grow and innovate was a major cause of the economic and political reforms initiated by Mikhail Gorbachev in the last half of the 1980s. Those reforms included a turn once again to Western companies for technological assistance by means of joint ventures and other agreements.

After the coming to power of Gorbachev in the mideighties, engineers from USX, the descendant of U. S. Steel, and the Ford Motor Company were asked to come to the USSR and to examine the possibility of modernizing the Magnitogorsk and Gor'kii plants that the American companies had originally helped build. In both instances the American engineers found that surprisingly little had changed since the original construction of the Soviet plants. Magnitogorsk had become the rust belt of the Soviet Union, a monument to the inefficient production of steel. By the 1980s American steel plants were not doing so well either, but it just so happened that Gary, the old model for Magnitogorsk, was one of the few American steel plants to be thoroughly modernized in the intervening years. At Gor'kii, the Ford engineers found that some of the original equipment of the 1930s had been used as late as the 1970s, and that the basic layout of the plant and its management principles were those of the River Rouge plant in the thirties. Several Ford engineers referred to the Gor'kii plant as a museum to the art of automobile manufacturing. The Ford Motor Company eventually declined the invitation to form a joint venture for the modernization of Gor'kii, partly because of the formidable nature of the task, but even more, it seems, because of their worries about being able to extract their profits.

As another example of the fits and starts of Soviet technology, one might notice that in the 1950s the Soviet Union was at a world level in the development of computers, only to fall behind in later years. In 1950 the Soviet computer designer S. A. Lebedev produced the MESM, the first electronic, stored-program, digital computer in continental Europe (only the U.S. and British efforts were earlier).[47] Furthermore, the MESM was developed totally independently of Western efforts. As was usually the case with early computers in other countries, only one MESM was produced, but in the 1960s the Soviet Union manufactured about 250 fairly successful second-generation computers, the BESM-6. Although the BESM-6 was a good computer, already it was less independent and less distinguished, relative to its time, than the MESM. In subsequent years, the Soviet Union had more and more difficulty maintaining the pace of competition. While it continued to develop some indigenous computer designs (and the former Soviet Union still does today), eventually it gave up the attempt to develop an independent series of manufactured computers, shifting instead to IBM standards (as did several other countries involved in the early development of computers).

A similar pattern can be seen in the Soviet nuclear power program. The USSR was the first country in the world to produce a usable nuclear power plant, and rapidly expanded the use of nuclear energy. Soviet engineers failed, however, to keep up with the advances in instrumentation, reliability, and safety procedures developed in other countries. The disaster at Chernobyl in 1986 was a shock both to the Soviet public and the Soviet nuclear power industry. In the late 1980s and early 1990s many planned nuclear power plants were canceled. Western experts who visited Soviet nuclear power plants were alarmed by the potential for accidents.

SPACE TECHNOLOGY

An area of technology where the Soviet Union made spectacular gains in the last half of the twentieth century was space exploration. The USSR was the first country in the world to launch an artificial satellite, first to launch a person into space, and first to perform a "space walk" in which a cosmonaut left his space vehicle. And although the United States was the first to send astronauts to the moon, the Soviet Union continued in the seventies and eighties a very active and impressive space program.

The Soviet space program illustrates both the strengths and the weaknesses of Soviet technology. It was a centralized effort to which the government assigned first priority in obtaining talent and necessary materials. In its early phases it was heavily dependent on military technology, being based on rockets developed for intercontinental ballistic missiles. These are the sorts of activities in which the Soviet system traditionally did well in contrast to development of technology dependent on the decentralized civilian and consumer economies.

Russia possessed one of the great pioneers in the conceptualization of space travel in the figure of Konstantin Tsiolkovskii (1857–1935), an autodidact who gained recognition for his work only late in life. Tsiolkovskii elaborated a theory of multistage rockets and he also proposed using clusters of rockets in order to achieve great speeds. He explored the mathematical relationship between the velocity of a rocket at any one moment, the velocity of the gas particles expelled from the nozzle of the rocket, the mass of the rocket, and the mass of the expended fuel. In 1897 he constructed the first wind tunnel in Russia, and he conducted a number of experiments with models of airfoils. However, his main achievement was the advancement of the idea of space travel and the derivation of basic principles rather than the actual designing and building of working rockets.

When one looks closely at the early history of the Soviet space program one is filled with admiration for its early leader, Sergei Korolev, the "Chief Designer" whose name was kept secret until his death in 1966.[48] Korolev performed miracles in fulfilling the demands of the government under incredibly difficult conditions. Arrested in 1937 and thrown into one of Stalin's labor camps, Korolev worked on rocket technology for many years in a special prison laboratory, or *sharashka*, of the type described so vividly in Alexander Solzhenitsyn's novel *The First Circle*. After Stalin's death in 1953 Korolev was rehabilitated and drawn into work on military missiles. Ordered to develop a rocket of sufficient power and range to reach the United States, Korolev was confronted with the problem that the large rocket engines necessary for this task produced temperatures in the walls of the nozzles greater than any Soviet alloys could withstand. Special heat-resistant alloys were used in large American rocket engines such as the Atlas and the Saturn that were unavailable to Korolev. Consequently, he adopted a dramatically different approach. He clustered smaller rocket engines in pods of four or five. The rocket that sent up the world's first satellite in 1957 had a four-chamber cluster. The rocket that sent into orbit the first man, Iurii Gagarin, was a giant "cluster of clusters" rocket with a total of twenty engines. Making all these engines and their associated fuel pumps and systems work simultaneously was an engineering accomplishment of the first magnitude. It was not the most direct and efficient solution of the problem, but it worked.

No sooner would Korolev fulfill one of Khrushchev's demands for a space spectacular than the Soviet leader would present him with another. The most extreme of these requests for presentation technology was undoubtedly one in 1963 that the Soviet Union launch three men in one space vehicle before the United States succeeded in launching two in a single capsule. To carry out that order Korolev had to choose cosmonauts of small stature, ask them to drop the precaution of wearing bulky space suits, and pack them in a small sphere so tightly that they were arranged around each other like pretzels. But the effort succeeded, and on time.[49]

After Khrushchev's overthrow in 1964 and after the United States surpassed the Soviet Union with its successful landing of astronauts on the moon in 1969 the Soviet space program became less hectic. Soviet authorities maintained that they had never been engaged in a race with the United States to put a man on the moon, but in 1990 a group of American aeronautical engineers from the Massachusetts Institute of Technology was shown an old Soviet lunar lander intended for that purpose and were told that the Soviet Union only abandoned the race to the moon when it was clear the Americans would be first. In the seventies and eighties the Soviet space program made steady progress at a time when the American program was uneven. By the late eighties and early nineties, however, the Soviet space program encountered increasing criticism from a now-vocal public that saw it as a drain on resources needed in the domestic economy. The crisis and demise of the USSR in

1991 left one of its astronauts stranded in space, not to come down until 1992, after the country had disappeared. The space program of the former Soviet Union seemed to be similarly hanging in air, without adequate funding or public support.

Although the explanations for the cycles of advance and retardation are somewhat different for each different technology, social and economic barriers, rather than technical ineptitude, are the common factors. It is even worth noticing that a number of the retarding characteristics of the society of the Tula Armory in the nineteenth century continued to exist quite recently, and, in fact some still do. Workers in Tula in the nineteenth century were tied to their place of residence and were subject to strict regulation, just as workers were in Stalin's Soviet Union. (In fact, the system of *propiska,* or enforced residence permits, still existed in the Soviet Union in 1991.) Both in Tula and in the Soviet Union of the period described here there was no free market, and no system of competitive bidding by private contractors. In both tsarist Russia and the Soviet Union social hierarchies reigned in the towns and the workplaces. Under both the tsars and Communist rule it was more important to please the political authorities by announcing spectacular and showy achievements (what might be called "presentation technology") than it was to be efficient or cost-effective. And living standards in both societies depended more on rank and access to influential people and institutional services than on achievements or salaries. These conditions created in both tsarist Russia and the Soviet Union a society where the inertial forces were enormous. A graphic description of these inertial forces was given in fictional form many years ago by Vladimir Dudintsev in his novel *Not by Bread Alone,* in which a lone inventor of a new means of producing steel pipes vainly fights the Soviet bureaucracy. Dudintsev was silenced by the censors of Brezhnev's Soviet Union, but under Gorbachev he emerged with a new novel describing resistance to scientific and technological innovation.

Despite the disappointing overall record in technological creativity, Soviet engineers and industrial designers still made a number of distinguished accomplishments, even in the early years.[50] For example, the Soviet Union was a pioneer in the production of synthetic rubber. The chemist I. I. Ostromislenskii actually began work on synthetic rubber before the Revolution, and this line of research was successfully continued after 1917 by B. V. Buizov and S. V. Lebedev. By 1929 the Soviet Union was producing sodium-butadiene synthetic rubber in significant quantities. Although Germany, England, and the United States were beginning to produce synthetic rubber of different types in the same years, Soviet scientists and engineers developed their own approaches. Another early Soviet innovation was the "Ramzin 'once-through' steam boiler," a highly economical high-pressure boiler, for which its inventor, Leonid Ramzin, won a Stalin Prize in 1943. Ramzin's development of this boiler has a special poignancy, as he had been brought to trial in 1930 and accused (see Chapter 3) of being one of the leaders of the Industrial Party, a group allegedly committing industrial sabotage. Many of Ramzin's colleagues suffered imprisonment and death, but Ramzin managed to survive and continue to serve as an industrial engineer. Long after Ramzin's death the Soviet government admitted that the charges had been false.

The Soviet Union was also a leader in the development of the turbodrill for use in oil well drilling. In the 1950s and 1960s the Soviet Union relied on this indigenous technology while the United States and other countries used a different type of rotary drill. The Soviet innovation was at first a success, but in later years it proved uneconomical as other types of drills were developed and as deeper and deeper oil wells were needed. Robert Campbell has shown that Soviet continued reliance on the turbodrill was in part based on preference for engineering criteria over economic ones.[51]

One of the strongest areas of Soviet technology, particularly after World War II, was metallurgy. Here is one of the few fields where Soviet innovations competed successfully on international markets. By the late 1960s Soviet industries had sold licenses to Western companies for use of Soviet welding techniques and for the production of liquid cores and molds. The Paton Institute of the Ukrainian Academy of Sciences was world famous as a center of metallurgical research and application. In the mideighties, Soviet metallurgists sold licenses to U.S. companies for electromagnetic casting, large-diameter gas-pipe welding, and ion-gun hardening of industrial cutting tools.[52]

Soviet claims for technological excellence in other fields are not as well documented as those already mentioned in synthetic rubber, thermal power, oil drilling, and metallurgy, but several of them deserve attention. They include Peter Kapitsa's invention of an inexpensive method of producing liquid oxygen, several medical techniques, a method for the production of acetic acid, a motorless agricultural combine, and a host of weapons innovations, such as machine guns and tank armor. Even as long ago as the 1930s Soviet engineers proved skilled at taking standard technologies and scaling them up for the achievement of world records that brought publicity and prestige. As Kendall Bailes has shown, Stalin placed great emphasis on the feats of his pilots, whom he referred to as his "falcons," and by 1938 was claiming sixty-two world records, including the longest, highest, and fastest flights.[53] (The emphasis on aviation records is another example of presentation technology.) In the postwar period, the Soviet Union constructed the world's first atomic-powered icebreaker and the world's largest nuclear power stations.

Notes

CHAPTER 1. RUSSIAN SCIENCE BEFORE 1800

1 Alexander Vucinich saw antirationalism as an essential characteristic of the Russian religious tradition, writing that "the antirationalism of Russian Orthodoxy was intrinsically incompatible with science as a mode of inquiry." Alexander Vucinich, *Science in Russian Culture: A History to 1860* (Stanford, Calif.: Stanford University Press, 1963), p. 387. Even Alexander Schmemann, an admirer of Eastern Orthodoxy, observed that "Russian psychology was from the first marked by . . . ritualism and . . . a somewhat hypertrophical, narrowly liturgical piety." In the Muscovite period, he continued, there developed "a simple fear of books and knowledge." Alexander Schmemann, *The Historical Road of Eastern Orthodoxy,* translated by Lydia W. Kesich (New York: Holt, Rinehart and Winston, 1960), pp. 300, 319. Sir J. A. R. Marriott denied that Russia was a part of European culture, and, again, put strong emphasis on religion: "Russia is not, and has never been, a member of the European family. . . . Even Poland, thanks to its adherence to the western form of Christianity, had some affinities with Europe. Russia during long centuries had none." Sir J. A. R. Marriott, *Anglo-Russian Relations, 1689–1943* (London: Methuen, 1944), p. 1.

2 By the early 1980s, the Soviet Union had 10 to 30 percent more scientists and engineers than the United States, depending on the definition of degrees and fields. For a study of the growth of Soviet research personnel, see Louvan E. Nolting and Murray Feshbach, "R and D Employment in the USSR," *Science* (February 1, 1980): 493–503.

3 According to the Soviet historian of science V. P. Zubov, Kiev at the end of the tenth century was the largest city in Europe. *Istoriia estestvoznaniia v Rossii,* vol. I, part 1 (Moscow, 1957), p. 10.

4 Ihor Sevcenko, "Remarks on the Diffusion of Byzantine Scientific and Pseudo-Scientific Literature among the Orthodox Slavs," *Slavonic and East European Review* 599, no. 3 (July 1981): 321–2.

5 Ibid., p. 378

6 *The Russian Primary Chronicle: Laurentian Text,* translated and edited by Samuel Hazzard Cross and Olgerd P. Sherbowitz-Wetzor (Cambridge, Mass.: The Mediaeval Academy of America, 1953), p. 137.

7 The *Izbornik* of 1076 opens with praise of reading books. *Izbornik 1076 goda* (Moscow: Nauka, 1965), pp. 151–4.

8 G. P. Fedotov, *The Russian Religious Mind: Kievan Christianity, the 10th to the 13th Centuries* (New York: Harper Torchbook, 1960), p. 378.

9 Thomas à Kempis, *The Imitation of Christ*, translated by E. M. Blaiklock (London: Hodder and Stoughton, 1979), p. 24.

10 Roger Hunt, "Shere Pleasure," *Surrey County Magazine* 22 (June 1991): 22.

11 The terms "equipment" and "needs" are taken both from Soviet and Western scholarship; see Sevcenko, "Diffusion of Byzantine Scientific and Pseudo-Scientific Literature among the Orthodox Slavs."

12 Schmemann, *Historical Road of Eastern Orthodoxy*, p. 301.

13 For a sympathetic treatment of Old Russian culture, see Georges Florovsky, "The Problem of Old Russian Culture," *Slavic Review* 21 (March 1962): 1–15.

14 Ia. S. Lur'e, *Ideologicheskaia bor'ba v russkoi publitsistike kontsa XV-nachala XVI veka* (Moscow-Leningrad, 1960); J. L. I. Fennell, *Ivan the Great of Moscow* (London: Macmillan, 1961).

15 Jack V. Haney, *From Italy to Muscovy: The Life and Works of Maxim the Greek* (Munich: W. Fink, 1973); Dimitri Obolensky, "Maximos the Greek," in his *Six Byzantine Portraits* (Oxford: Clarendon Press, 1988); pp. 201–19.

16 *Le Opere di Galileo Galilei*, vol. XI (Florence: G. Barbera, 1901), pp. 68–9.

17 I am indebted to Marc Raeff for insights on Peter's reign, both from his published works and from personal conversations. See, for example, Marc Raeff, *Origins of the Russian Intelligentsia: The Eighteenth Century Nobility* (New York: Harcourt, Brace and World, 1966); Raeff, ed., *Peter the Great Changes Russia* (Lexington, Mass.: D. C. Heath, 1972).

18 G. D. Komkov, B. V. Levshin, and L. K. Semenov, *Akademiia nauk SSSR: kratkii istoricheskii ocherk* (Moscow, 1974), p. 16. Considerable debate exists among historians on whether the meeting actually occurred. See Valentin Boss, "Did Peter the Great Meet Newton?" in his *Newton and Russia: The Early Influence, 1698–1796* (Cambridge, Mass.: Harvard University Press, 1972), pp. 9–18.

19 Boss, ibid., p. 3 and passim.

20 On the early Academy, see K. V. Ostrovitianov (ed.), *Istoriia akademii nauk SSSR*, vol. I (Moscow, 1958); A. S. Lappo-Danilevskii, *Petr velikii, osnovatel' imperatorskoi akademii nauk v St. Peterburge* (St. Petersburg, 1870); M. I. Sukhomlinov (ed.), *Materialy dlia istorii imperatorskoi akademii nauk*, vol. II (St. Petersburg, 1885); A. Kunik, *Sbornik materialov dlia istorii imperatorskoi akademii nauk v XVIII veke* (St. Petersburg, 1865); G. D. Komkov, B. V. Levshin, and L. K. Semenov, *Akademiia nauk SSSR: kratkii istoricheskii ocherk*, vol. I (Moscow, 1974); Alexander Lipski, "The Foundation of the Russian Academy of Sciences," *Isis* 34 (Dec. 1953): 349–54; and Ludmilla Burgess, "The Russification of the St. Petersburg Academy of Sciences and Arts in the Eighteenth Century," unpublished master's degree thesis, history and philosophy of science, University of Melbourne, 1974.

21 For an analysis of the percentage of the members of the Academy who were Russians and Germans in the eighteenth and nineteenth centuries, see A. S. Lappo-Danilevsky, "The Development of Science and Learning in Russia," in J. D. Duff (ed.), *Russian Realities and Problems* (Cambridge: Cambridge University Press, 1917), pp. 173–4. For those interested in an ethnic and

genealogical analysis of the members of the Academy in the last half of the nineteenth century, a valuable source is T. K. Lepin, Ia. Ia. Lus, and Iu. A. Filipchenko, "Deistvitel'nye chleny akademii nauk za poslednie 80 let (1846–1924)," *Izvestiia buro po evgenike* 3 (1925): 7–49.

22 On Lomonosov, see B. N. Menshutkin, *Russia's Lomonosov, Chemist, Courtier, Physicist, Poet* (Princeton, N.J.: Princeton University Press, 1952); G. E. Pavlova and A. S. Fedorov, *Mikhail Vasilievich Lomonosov: His Life and Work*, translated by A. Aksenov (Moscow, 1984); and B. M. Kedrov, "Lomonosov," *Dictionary of Scientific Biography*, vol. III (1973), pp. 467–72. Russian works include B. G. Kuznetsov, *Tvorcheskii put' Lomonosova* (Moscow, 1961) and M. I. Radovskii, *Lomonosov i Peterburgskaia akademiia nauk* (Moscow, 1961). His collected works are available in *Polnoe sobranie sochinenii*, 10 vols. (Moscow-Leningrad, 1950–9).

23 See Henry M. Leicester, *Mikhail Vasil'evich Lomonosov on the Corpuscular Theory* (Cambridge, Mass.: Harvard University Press, 1970).

24 For an example of such claims in English, see B. B. Kudryatsev, *The Life and Work of Mikhail Vasilyevich Lomonosov* (Moscow, 1954), esp. pp. 41, 58, 64, and passim.

25 Kedrov, "Lomonosov."

26 Quoted in Leicester, *Lomonosov on the Corpuscular Theory*, p. 111, from Lomonosov's "Meditations on the Cause of Heat and Cold."

27 For example, see A. A. Morozov, *Mikhail Vasil'evich Lomonosov, 1711–1765* (Moscow, 1950); p. 339.

28 Quoted in Leicester, *Lomonosov on Corpuscular Theory*, p. 25.

29 See the discussion in ibid., pp. 25–6.

30 Ibid., p. 46. See Ia. G. Dorfman, *Lavuaz'e* (Moscow-Leningrad, 1948); p. 183.

31 See V. I. Grekov, *Ocherki iz istorii russkikh geograficheskikh issledovanii v 1725–1765 gg* (Moscow, 1960); L. S. Berg, *Ocherki po istorii russkikh geograficheskikh otkrytii* (Moscow, 1946); F. A. Golder, *Bering's Voyages*, vols. I and II (New York: American Geographical Society, 1922, 1925); N. A. Figurovskii (ed.), *Istoriia estestvoznaniia v Rossii*, vol. I (Moscow, 1957); V. Gnucheva, *Geograficheskii departament akademii nauk XVIII veka* (Moscow, 1958); V. Gnucheva, *Materialy dlia istorii ekspeditsii akademii nauk v XVIII i XX vekakh* (Moscow, 1940).

32 Raymond H. Fisher, *Bering's Voyages: Whither and Why* (Seattle: University of Washington Press, 1977).

33 Walter McDougall, *The Heavens and the Earth: A Political History of the Space Age* (New York: Basic Books, 1985), pp. 118–22.

34 R. W. Home, "The Scientific Education of Catherine the Great," *Melbourne Slavonic Studies* 11 (1976): 18–22; also see Valentin Boss, *Newton and Russia: The Early Influence, 1698–1796* (Cambridge, Mass.: Harvard University Press, 1972).

35 Vucinich, *Science in Russian Culture*, vol. I, p. 78.

36 *Leonhardi Euleri Opera omnia* (Berlin, 1911), vol. 13, ch. 2, p. 182.

37 Vucinich, *Science in Russian Culture*, vol. I, p. 156.

38 Quoted in B. H. Sumner, *Peter the Great and the Emergence of Russia* (London: The English Universities Press, 1950), pp. 208–9. I have changed Sumner's word "stooks" to "shocks of grain" to render it more understandable.

CHAPTER 2. SCIENCE IN NINETEENTH-CENTURY RUSSIA

1 Allen McConnell maintains persuasively that the contrast between the "idealistic and liberal" periods of Alexander I's youth and the "reactionary and mystical" period of his last years has been exaggerated by historians. According to McConnell, Alexander was a paternalistic tsar who even in his most liberal dreams hoped to grant the nation a constitution himself and did not wish to consult his citizens about the process. See Allen McConnell, *Tsar Alexander I, Paternalist Tsar* (New York: Crowell, 1970).

2 Alexander Vucinich, *Science in Russian Culture: A History to 1860*, vol. I (Stanford, Calif.: Stanford University Press, 1963), p. 235.

3 /Cynthia H. Whittaker, *The Origins of Modern Russian Education: An Intellectual Biography of Count Sergei Uvarov, 1786–1855* (DeKalb: Northern Illinois Press, 1984).

4 Ibid., p. 188.

5 See W. Bruce Lincoln, *In the Vanguard of Reform: Russia's Enlightened Bureaucrats, 1825–1861* (DeKalb: Northern Illinois University Press, 1982).

6 See Ann Hibner Koblitz, *A Convergence of Lives: Sofia Kovalevskaia, Scientist, Writer, Revolutionary* (Boston: Birkhäuser, 1983).

7 Quoted in Michael T. Florinsky, *Russia: A History and an Interpretation*, vol. II (New York: Macmillan, 1953), p. 1035.

8 I am especially indebted to Gregory Crowe for help on Lobachevskii. Crowe has written the best work on the genesis of Lobachevskii's non-Euclidean geometry. Gregory Crowe, "The Life and Work of Nikolai Ivanovich Lobachevsky: A Study of the Factors Leading to the Discovery and Acceptance of the First Non-Euclidean Geometry," unpublished senior thesis, Harvard University, Cambridge, Mass., 1986.

9 William Kingdon Clifford, *Lectures and Essays* (London: Macmillan, 1879); pp. 297–8. The historian of mathematics E. T. Bell also used the phrase "Copernicus of geometry." E. T. Bell, *Men of Mathematics* (New York: Simon and Schuster, 1937), p. 294.

10 See V. F. Kagan, *Lobachevskii* (Moscow-Leningrad, 1948), pp. 20–5.

11 The Aksakov family chronicle describes this gymnasium in considerable detail; S. T. Aksakov entered the gymnasium before Lobachevskii, in 1800. See S. T. Aksakov, *Chronicles of a Russian Family*, translated by M. C. Beverley (New York: E. P. Dutton, 1924). B. A. Rosenfeld stated that Lobachevskii and his two brothers received public scholarships to the gymnasium. B. A. Rosenfeld, "Lobachevsky, Nikolai Ivanovich," *Dictionary of Scientific Biography*, vol. VIII (1973), p. 428. I am also grateful to Dr. Charles Duffy, Massachusetts Maritime Academy, who worked in the Lobachevskii archives in Kazan', for information on the family and financial situation of Lobachevskii.

12 The prejudice was not, evidently, absolute. One of the teachers in the Kazan' gymnasium in 1801 was a thoroughly Russified Tatar who had been educated at Moscow University. He taught Russian grammar and mathematics. See Aksakov, *Chronicles of a Russian Family*, pp. 353–4.

13 A defense of the Gauss-Bartels-Lobachevskii "link thesis" is Morris Kline, *Mathematical Thought from Ancient to Modern Times* (New York: Oxford Univer-

sity Press, 1972), pp. 877–9. While this thesis is, I believe, mistaken, it is not absurd or completely groundless.

14   V. F. Kagan, *Lobachevskii* (Moscow-Leningrad, 1948); B. L. Laptev, "Teoriia parallel'nykh priamykh v rannikh rabotakh Lobachevskogo," *Istoriko-matematicheskie issledovaniia* 4 (1951); G. L. Luntz, "O rabotakh N. I. Lobachevskogo po matematicheskomu analizu," *Istoriko-matematicheskie issledovaniia* 2 (1949); A. P. Norden, ed., *Ob osnovaniiakh geometrii* (Moscow, 1956); B. A. Rosenfeld, "Interpretatsii geometrii Lobachevskogo," *Istoriko-matematicheskie issledovaniia* 9 (1956). The best general article on Lobachevskii in English, with a good bibliography, is B. A. Rosenfeld, "Lobachevsky," *Dictionary of Scientific Biography*, vol. VIII, (1973), pp. 428–35. Also, see Crowe, V. F. Kagan, N. *Lobachevsky and His Contribution to Science* (Moscow: Foreign Languages Publishing House, 1957), and Alexander Vucinich, "Nikolai Ivanovich Lobachevskii: The Man behind the First Non-Euclidean Geometry, *Isis* 53 (1962): 465–81. A valuable new source is B. V. Fedorenko (ed.), *Novye materialy k biografii N. I. Lobachevskogo* (Leningrad: Nauka, 1988).

15   Kenneth May wrote, "Although the friendships of Gauss with Bartels and W. Bolyai suggest the contrary, careful study of the plentiful documentary evidence has established that Gauss did not inspire the two founders of non-Euclidean geometry. Indeed, he played at best a neutral, and on balance a negative, role, since his silence was considered as agreement with the public ridicule and neglect that continued for several decades and were only gradually overcome, partly by the revelation, beginning in the 1860s, that the prince of mathematicians had been an underground non-Euclidean." Kenneth O. May, "Gauss, Carl Friedrich," *Dictionary of Scientific Biography*, vol. V (1972), p. 302.

16   Norman Daniels, *Thomas Reid's Inquiry: The Geometry of Visibles and the Case for Realism* (Stanford: Stanford University Press, 1989).

17   V. Kagan, N. *Lobachevsky and His Contribution to Science* (Moscow: Foreign Languages Publishing House, 1957), p. 29.

18   On the split between Uvarov and Magnitskii over the fate of Kazan', see Whittaker, *Origins of Modern Russian Education*, pp. 74–6.

19   Kagan, *Lobachevsky*, 1957; Alexander Vucinich, "Nikolai Ivanovich Lobachevskii: The Man behind the First Non-Euclidean Geometry," *Isis* 53 (1962): 495; and N. P. Zagoskin, *Istoriia imperatorskogo kazanskogo universiteta za pervyia sto let ego sushchestvovaniia*, Vols. I–IV (Kazan', 1902–4).

20   Ferdinand Schweikart in the period 1807–19 developed a form of hyperbolic geometry that he called "astral geometry." His approach was limited and his work is best described as a part of the prehistory of non-Euclidean geometry. See Werner Burau, "Schweikart, Ferdinand Karl," *Dictionary of Scientific Biography*, vol. XIII (1975), p. 255.

21   See the introduction to Lobachevskii's *Geometrical Researches on the Theory of Parallels*, translated by G. B. Halsted (LaSalle, Ill.: Open Court Publishers, 1914). Despite many statements to the contrary, it seems that Gauss never carried out the triangle experiment either; see Arthur Miller, "The Myth of Gauss' Experiment on the Euclidean Nature of Physical Space," *Isis* 63 (1972): 345–8.

22 Ia. Depman, "Novoe o N. I. Lobachevskom (K voprosu o retsenzii v 'Syne otechestva')," *Trudy instituta istorii estestvoznaniia* 2 (1948): p. 561; also E. P. Faidel' and K. I. Shafranovskii, "Pechat v Rossii o trudakh N. I. Lobachevskogo (1834–1856)," *Vestnik akademii nauk SSSR* 3 (1944).

23 N. I. Lobachevskii, "Gëomëtrie imaginaire," *Journal für die reine und angewandte Mathematik* 17 (1837); *Geometrische Untersuchungen zur Theorie der Parallellinien* (Berlin: Fincke, 1840).

24 For a discussion of Bolyai, with reference to Lobachevskii, see V. F. Kagan, "Stroenie neevklidovoi geometrii u Lobachevskogo, Gaussa i Bol'ai," *Trudy instituta istorii estestvoznaniia* 2 (1948): 323–89, esp. p. 329.

25 G. W. Dunnington, *Carl Friedrich Gauss, Titan of Science: A Study of His Life and Work* (New York: Exposition Press, 1955), p. 214.

26 P. Stäckel, W. and J. Bolyai, *Geometrische Untersuchungen*, vol. I (Leipzig: B. G. Teubner, 1913), p. 76.

27 In particular, B. M. Kedrov, *Den' odnogo velikogo otkrytiia* (Moscow, 1958); and his *Filosofskii analiz pervykh trudov D. I. Mendeleeva o periodicheskom zakone (1869–1871)* (Moscow, 1959).

28 N. A. Figurovskii, *Dmitrii Ivanovich Mendeleev* (Moscow, 1961), p. 27.

29 Ruth Amende Roosa, "The Association of Industry and Trade, 1906–1914; An Examination of the Economic Views of Organized Industrialists in Pre-Revolutionary Russia," unpublished Ph.D. dissertation, Columbia University, New York, 1968.

30 J. W. van Spronsen, *The Periodic System of Chemical Elements; a History of the First Hundred Years* (Amsterdam: Elsevier, 1969).

31 *Sto let periodicheskogo zakona khimicheskikh elementov* (Moscow, 1969); see, in particular, N. A. Figurovskii's "Sistematizatsiia khimicheskikh elementov do otkrytiia periodicheskogo zakona D. I. Mendeleevym," pp. 15–41.

32 For example, van Spronsen, *Periodic System of Chemical Elements*, p. 132.

33 B. M. Kedrov, *Den' odnogo velikogo otkrytiia* (Moscow, 1958).

34 Bernadette Bensaude-Vincent, "Mendeleev's Periodic System of Chemical Elements," *British Journal of the History of Science* 19 (1986); 3–17. We should be careful, however, not to assume that Mendeleev's views at the time of his creation of the periodic table in 1869 were identical to those he expressed in revised and later editions of his *Principles of Chemistry*, to which most Western literature refers, because only the later editions were translated into Western languages. The evolution of Mendeleev's views after 1869 needs to be studied more thoroughly.

CHAPTER 3. RUSSIAN INTELLECTUALS AND DARWINISM

1 James Allen Rogers, "Charles Darwin and Russian Scientists," *Russian Review*, 19 (1960): 382.

2 Ecclesiastical criticism of Darwin did not really emerge until thirty years after the publication of the *Origin of Species*, and even then the objection was not usually to evolution but to Darwin's description of it. See George Kline's excellent article, "Darwinism and the Russian Orthodox Church," in Ernest

J. Simmons (ed.), *Continuity and Change in Russian and Soviet Thought* (Cambridge, Mass.: Harvard University Press, 1955); pp. 307–28.

3 Quoted by N. Umov, "Po povodu sbornika" in M. M. Kovalevskii et al., *Pamiati Darvina* (Moscow, 1910), p. 1. Also referred to by Alexander Vucinich, *Darwin in Russian Thought* (Berkeley: University of California Press, 1988); p. 16.

4 Charles Darwin, *On the Origin of Species* (London: John Murray, 1859), p. 43.

5 Ibid., p. 236.

6 D. Pisarev, "Progress v mire zhivotnykh i rastenii," *Polnoe sobranie sochineniia*, vol. III (St. Petersburg, 1894), col. 453.

7 Ibid., col. 452.

8 For an interesting discussion of Zaitsev's views, see G. Berliner, "Varfolomei Zaitsev, publitsist shestidesiatykh godov," in V. P. Koz'min (ed.), *V. A. Zaitsev: Izbrannye sochineniia v dvukh tomakh*, vol. I (Moscow, 1934), pp. 15–48.

9 Zaitsev, *Izbrannye sochineniia v dvukh tomakh*, vol. I, p. 230.

10 Ibid., p. 232.

11 Ibid., p. 228.

12 Ibid., p. 229.

13 Ibid., pp. 229–30.

14 Darwin, *Origin of Species*, p. 134.

15 N. D. Nozhin, "Po povodu statei 'Russkago slova' o nevol'nichestve," *Iskra* 8 (1865), p. 115.

16 Ibid.

17 Darwin, *Origin of Species*, p. 75.

18 N. D. Nozhin, "Nasha nauka i uchenye," *Knizhnyi Vestnik* (April 15, 1866): 175.

19 Quoted in William F. Woehrlin, *Chernyshevskii: The Man and the Journalist* (Cambridge, Mass: Harvard University Press, 1971), p. 135.

20 Chernyshevskii, "Proiskhozhdenie teorii blagotvornosti bor'by za zhizn'," *Polnoe sobranie sochinenii*, vol. 10 (Moscow, 1951), p. 758.

21 Ibid., p. 770.

22 Quoted in Alexander Vucinich, *Darwin in Russian Thought* (Berkeley: University of California Press, 1988). p. 150.

23 See, in particular, Daniel Todes, *Darwin without Malthus: The Struggle for Existence in Russian Evolutionary Thought* (Oxford: Oxford University Press, 1989); Vucinich, *Darwin in Russian Thought*.

24 Todes, *Darwin without Mathus*, pp. 159–65.

25 Ibid., p. 163.

26 Ibid., p. 92.

27 Barry G. Gale, "Darwin and the Concept of Struggle for Existence: A Study in the Extrascientific Origins of Scientific Ideas," *Isis* 63, no. 218 (September 1972): 321–44. Also, see Edward Manier, *The Young Darwin and His Cultural Circle* (Dordrecht, Holland: D. Reidel, 1978), passim, esp. p. 200.

28 P. A. Kropotkin, *Mutual Aid: A Factor of Evolution* (Boston: Extending Horizon Books, 1955), p. 5.

29 Ibid., p. 128.

30 L. J. Blacher, "Kovalevsky, Vladimir Onufievich," *Dictionary of Scientific Biography*, vol. VII (1973), p. 480.
31 Vucinich, *Darwin in Russian Thought*, pp. 118–46.
32 Todes, *Darwin without Malthus*, p. 41.
33 Vucinich, *Darwin in Russian Thought*, p. 125.
34 Ibid., p. 135.
35 V. V. Rozanov, *Priroda i istoriia* (St. Petersburg, 1903), pp. 25–37.
36 Ibid., p. 35.
37 Quoted in Vucinich, *Darwin in Russian Thought*, p. 243.

CHAPTER 4. THE RUSSIAN REVOLUTION AND THE SCIENTIFIC COMMUNITY

1 Clarence Crane Brinton, *The Anatomy of Revolution* (New York: Prentice-Hall, 1952), p. 219.
2 Ibid., p. 224.
3 Sheila Fitzpatrick used a similar periodization in *The Russian Revolution* (New York: Oxford University Press, 1982).
4 Andrei Amalrik, "On Détente," *New York Times*, October 22, 1975.
5 But Amalrik observed that the mainspring was still unwinding in China.
6 Dr. Nathan Brooks is currently working on a history of the Russian Physical and Chemical Society and the formation of a community of professional chemists in Russia. See his "The Formation of the Russian Chemical Community (1800–1917)," unpublished Ph.D. dissertation, Columbia University, New York, 1988.
7 For a defense of the strength of the tradition of rationalist and naturalist thought in pre-1917 Russia, see Alexander Vucinich, *Science in Russian Culture: A History to 1860*, and his *Science in Russian Culture, 1860–1917* (Stanford, Calif.: Stanford University Press, 1963, 1970).
8 For discussions of the Ledentsov Society, a philanthropic society, and other embryonic groups supporting science in the early years of the twentieth century, see M. S. Bastrakova, "Organizatsionnye tendentsii russkoi nauki v nachale XX v.," in *Organizatsiia nauchnoi deiatel'nosti* (Moscow, 1968), pp. 150–86, and Vucinich, *Science in Russian Culture, 1860–1917*.
9 For general works on the Academy of Sciences see Petr Pekarskii, *Istoriia imperatorskoi akademii nauk*, 2 vols. (St. Petersburg, 1870, 1873); K. V. Ostrovitianov et al. (eds.), *Istoriia akademii nauk SSSR*, 2 vols. (Moscow, 1958, 1964); G. A. Kniazev and A. V. Kol'tsov, *Kratkii ocherk istorii akademii nauk SSSR* (Moscow-Leningrad, 1957); Alexander Vucinich, *The Soviet Academy of Sciences* (Stanford, Calif.: Stanford University Press, 1956); Loren R. Graham, *The Soviet Academy of Sciences and the Communist Party, 1927–1932* (Princeton, N.J.: Princeton University Press, 1967); and G. D. Komkov, B. V. Levshin, and L. K. Semenov, *Akademiia nauk SSSR: kratkii istoricheskii ocherk* (Moscow, 1974).
10 Examples of two scientists who participated in Bolshevik political activity in 1917 are the astronomer P. K. Shternberg and the engineer L. Ia. Karpov. Other scientists who soon indicated a favorable attitude toward the new

regime were K. A. Timiriazev, K. E. Tsiolkovskii, I. V. Michurin, A. N. Bakh and I. M. Gubkin. See I. S. Smirnov, *Lenin i sovetskaia kul'tura* (Moscow, 1960). The assessment of political attitudes is, of course, a difficult enterprise; among Soviet historians there was considerable disagreement about the portion of the scientific intelligentsia that was sympathetic to the Revolution. See, for example, G. A. Kniazev and A. V. Kol'tsov, *Kratkii ocherk istorii akademii nauk SSSR* (Moscow-Leningrad, 1957), and the criticism of the opinion expressed there in Smirnov, *Lenin i sovetskaia kul'tura*, p. 235.

11 *Obshchestvennyi vrach* 9–10 (1917): 79–80, as cited in S. A. Fediukin, *Velikii oktiabr' i intelligentsiia: iz istorii vovlecheniia staroi intelligentsii v stroitel'stvo sotsializma* (Moscow, 1972), p. 43.

12 For a bitter description of early Soviet terror, see Pitirim A. Sorokin, *Leaves from a Russian Diary – and Thirty Years After* (Boston: Beacon Press, 1950; New York: Kraus Reprint Co., 1970). Similar accounts are in Sergey Melgounov (sic), *The Red Terror in Russia* (London: J. M. Dent and Sons, 1926); P. Miliukov, *Rossiia na perelome* (Paris, 1924); Alexander Solzhenitsyn also portrays the early repression of intellectuals in *The Gulag Archepelago*, 2 vols. (New York: Harper & Row, 1973, 1975). For a Soviet account of the shooting of nine hundred people, including the well-known geodesist Davidov, see Fediukin, *Velikii oktiabr' i intelligentssia*, pp. 95–6.

13 A balanced and interesting description of early attitudes of the technical intelligentsia toward the Soviet government is Kendall Bailes, *Technology and Society under Lenin and Stalin* (Princeton, N.J.: Princeton University Press, 1978). In addition, see the work by Fediukin cited above, and also Fediukin's *Sovetskaia vlast' i burzhuaznye spetsialisty* (Moscow, 1965).

14 Vladimir N. Ipatieff (American spelling for Ipat'ev), *The Life of a Chemist* (Stanford, Calif.: Stanford University Press, 1946), p. 260.

15 See P. N. Pospelov (ed.), *Lenin i akademiia nauk: sbornik dokumentov* (Moscow, 1969).

16 S. Belomortsev, "Bol'shevizatsiia akademii nauk," *Posev*, 46 (November 18, 1951), p. 11.

17 "Izvestiia rossiiskoi akademii nauk," Seriia VI, no. 14 (1918): 1395, as cited in Fediukin, *Velikii oktiabr'i intelligentsiia*, p. 147.

18 M. S. Bastrakova, *Stanovlenie sovetskoi sistemy organizatsii nauki (1917–1922)* (Moscow, 1973), pp. 97–9.

19 Ibid., p. 98.

20 S. F. Ol'denburg, "Vladimir Andreevich Steklov," *Nauchnyi rabotnik* 5–6 (1926), p. 3.

21 M. S. Bastrakova, *Stanovlenie sovetskoi sistemy organizatsii nauki (1917–1922)* (Moscow, 1973), pp. 99–100.

22 A valuable collection of documents concerning science in the early years is *Organizatsiia nauki v pervye gody sovetskoi vlasti (1917–1925)* (Leningrad, 1968). Included in it is a photostatic copy of Lenin's noted "Draft Plan of Scientific-Technological Works," written, it is thought, in April 1918. The importance of the plan was not its immediate effects (although addressed to the Academy, it was evidently never transmitted), but the evidence it presents for assuming that Lenin thought the Academy should be a controlling center of Soviet science. See A. V. Kol'tsov, *Lenin i stanovlenie akademii nauk kak tsentra*

sovetskoi nauki (Leningrad, 1969), p. 76; and E. N. Gorodetskii, "K istorii Leninskogo plana nauchno-technicheskikh rabot," in *Iz istorii revoliutsionnoi i gosudarstvennoi deiatel'nosti V. I. Lenina* (Moscow, 1960), pp. 191–232. Also, see *Akademiia nauk SSSR – shtab sovetskoi nauki* (Moscow, 1968).

23   See Joel Shapiro, "A History of the Communist Academy, 1918–1936," unpublished Ph.D. dissertation, Columbia University, New York, 1976.

24   Iu. Larin (M. A. Lur'e) maintained that Lenin and he advanced grand plans for electrification, canal-building, and construction of industry in 1918 not so much because they believed that this could be done right away, but in order to attract the "summit" of the technical intelligentsia. The Bolshevik leaders wanted to show, said Larin, that, far from being madmen and oppressors, they were planning technical projects of a type about which prerevolutionary engineers could only dream. Iu. Larin, *Intelligentsiia i sovety: khoziaistvo, burzhuaziia, revolutsiia, gosapparat* (Moscow: Gosizdat, 1924), pp. 11–12.

25   For a survey of the activity of KEPS, see B. A. Lindener, *Raboty rossiiskoi akademii nauk v oblasti issledovaniia prirodnykh bogatstv Rossii; obzor deiatel'nosti KEPS za 1915–1921 gg.* (Petrograd, 1922); also, Bastrakova, *Stanovlenie . . . ,* passim.

26   Two examples of such literature are Ipatieff, *The Life of a Chemist,* and I. P. Bardin, *Zhizn' inzhenera* (Moscow, 1938).

27   In his memoirs Ipat'ev gives much evidence of his unaltered social and political views after the Revolution and describes his unsuccessful effort to retain his private farm, on which he practiced the methods of scientific agriculture. He was little interested in politics, much preferring to give his efforts to technical problems. His social outlook remained remarkably stable through many years of the work with the Soviet regime, followed by emigration to the United States.

28   In an interesting analysis of the hostility of Russian workers toward the intelligentsia, Larin maintained that many of them, as former peasants, merely transferred their dislike of the landowners to the factory engineers. Thus, even the family members of supervisory engineers were regarded as "members of the ruling class." Iu. Larin, *Intelligenty i sovety: khoziaistvo, burzhuaziia, revoliutsiia, gosapparat* (Moscow, 1924).

29   The issues were raised very pointedly by A. Shliapnikov, leader of the Workers' Opposition. In a 1919 article he said that when the working masses fully realize the extent to which they are being pushed aside by the specialists, their dissatisfaction will grow in a very dangerous way. A. Shliapnikov, "O spetsialistakh," *Pravda,* March 27, 1919, p. 1. For general discussions of the Workers' Opposition, see Leonard Shapiro, *The Communist Party of the Soviet Union* (New York: Random House; 1960), and Robert Daniels, *Conscience of the Revolution* (Cambridge, Mass.: Harvard University Press, 1960).

30   Alexandra Kollontai, "The Roots of the Workers' Opposition," Solidarity Pamphlet (London, 1968), p. 6.

31   Much has been written on the proletarian culture movement in literature and art, but little on its role in science. This omission is regrettable, since one of the architects of the movement was A. A. Bogdanov, a physician and political activist very interested in science and dedicated to its furtherance. See Loren R. Graham, "Aleksandr Aleksandrovich Bogdanov," *Dictionary of*

Scientific Biography (Supplementary vol., 1977). For different views on the proletarian culture movement, see Edward J. Brown, *The Proletarian Episode in Russian Literature, 1928–1932* (New York: Columbia University Press, 1953); M. Kim, *Kommunisticheskaia partiia-organizator kul'turnoi revoliutsii v SSSR* (Moscow, 1955); Anatoli Lunacharskii, *Stati o sovetskoi literature* (Moscow, 1958), pp. 176–7; Herman Ermolaev, *Soviet Literary Theories, 1917–1934* (Berkeley: University of California Press, 1963); L. Trotsky, *Literature and Revolution* (New York: International Publishers, 1925); Lynn Mally, *Culture of the Future: The Proletkult Movement in Revolutionary Russia* (Berkeley: University of California Press, 1989).

32 V. Pletnev, "Na ideologicheskom fronte," *Pravda*, September 27, 1922.

33 The Soviet historian Fediukin maintained that in the worst period of civil war and famine the qualified technical specialists received salaries "more than five or six times exceeding the rate of pay of the people's commissars and the head of the government – V. I. Lenin." *Velikii oktiabr' i intelligentsiia*, p. 99.

34 N. Krupskaia, "Proletarskaia ideologiia i proletkul't," *Pravda*, October 8, 1922; I. I. Skvortsov-Stepanov, "Chto takoe spets i kak ego delaiut?" *Pravda*, October 28, 1922; Ia. Iakovlev, "O proletarskoi kul'ture' i proletkul'te," *Pravda*, October 24, 1922.

35 A photostatic copy of the article with Lenin's marginalia may be found in *Voprosy kul'tury pri diktature proletariata* (Moscow-Leningrad, 1925).

36 A typical statement by Lenin was "We must take all of culture which capitalism left to us and build socialism out of it. We must take all science, technology, all knowledge, art. Without them we cannot construct the life of a communist society. And science, technology and art are in the hands and heads of the specialists." Quoted in Fediukin, *Velikii oktiabr' i intelligentsiia*, p. 71. For Lenin's warning about the Academy, see "Iz vospominanii A. V. Lunacharskogo o pozitsii V. I. Lenine po voprosu o reforme akademii nauk v 1919 g.," in P. N. Pospelov (ed.), *Lenin i akademiia nauk* (Moscow, 1969), pp. 62–3. For an interesting early work on Lenin's attitudes toward the bourgeois specialists, see S. Girinis, *Lenin o spetsakh* (Moscow, 1924).

37 Krupskaia, "Proletarskaia ideogiia i prolekul't."

38 Iakovlev, "O 'proletarskoi kul'ture' i proletkul'te."

39 For interesting information on the changes in mood involved in the opening of foreign trade and the role of technology in reconstruction, see the work on the life of the prominent engineer and political activist: Lubov Krassin, *Leonid Krassin: His Life and Work* (London, 1929).

40 Fediukin, *Velikii oktiabr' i intelligentsiia*, p. 256.

41 *Smena vekh* (Prague: Politika, 1921).

42 Soviet literature and memoirs abound with examples. One of the most illuminating, relating to a somewhat later period, comes from the memoirs of N. P. Dubinin, a defender of Mendelian genetics soon to be defeated by Lysenko:

Once in 1935 S. G. Levit and I came out of the House of Scholars. In the vestibule we met T. D. Lysenko and I. I. Prezent in identical yellowish fur-trimmed jackets of some kind or another, and wearing rough, slouched caps. They looked like either fishermen or workers from some northern port. Seeing them, Levit laughed.

"Trofim Denisovich," Levit said, turning to Lysenko, "Why do you dress like that, so people can recognize you from a distance?"

"No," Lysenko steamed, "This is our own form of dress, one that is special for us. And you, there, Comrade Levit, that hat on your head is a true identification sign, we can see clearly what class you prefer!"

Within a year or two, Levit, who was actually a sincere Marxist, despite his dress, was arrested, subsequently to die in incarceration. N. P. Dubinin, *Vechnoe dvizhenie* (Moscow, 1973), p. 160.

43 Lenin himself criticized Spengler in "K desiatiletnemu iubileiu 'Pravda,' " *Pravda*, May 5, 1922. For the interest of Soviet scholars in Spengler, see Fediukin, *Velikii oktiabr' i intelligentsiia*, p. 250.

44 Criticism of Bergson was frequent among Russian radicals even before the Revolution, and it continued through the Soviet period. In the twenties, however, interest in Bergson grew among some intellectuals. See G. V. Plekhanov, "H. Bergson, tvorcheskaia evoliutsiia," *Izbrannye filosofskie proizvedeniia*, vol. III (Moscow, 1957); N. O. Losskii, *Intuitivnaia filosofiia Bergsona* (Petrograd, 1922); F. Challaye, "Bergson vu par les Soviets," *Preuves*, 4, no. 44 (1954): 62–3.

45 For an early, fairly favorable reaction to Freud, see B. Bykhovskii, "O metodologicheskikh osnovaniiakh psikhoanaliticheskogo ucheniia Freida," *Pod znamenem marksizma* (November-December 1923); for another view, see I. Sapir, "Freidizm, sotsiologiia, psikhologiia," *Pod znamenem marksizma* (July–August 1929).

46 An example of a defense of eugenics written in Soviet Russia is Iu. A. Filipchenko, *Chto takoe evgenika?* (Petrograd, 1921). Soviet Marxists were at first uncertain what attitude to adopt toward eugenics, but eventually they moved strongly against it. A Marxist critique of eugenics, near the beginning of the wave of criticism that caused the movement to disappear in Russia, is V. Slepkov, "Nasledstvennost' i otbor u cheloveka," *Pod znamenem marksizma* 4 (1925): 102–22.

47 Alexander Erlich, *The Soviet Industrialization Debate, 1924–1928* (Cambridge, Mass.: Harvard University Press, 1960).

48 The geology division of the Academy of Sciences was frequently criticized for studying "science for science's sake," and, after the reorganization of the Academy at the end of the twenties, geology research (including KEPS) was reorganized on much more utilitarian grounds. See Graham, *Soviet Academy*, esp. pp. 38–43 and 164–7.

49 Lenin's *Materialism and Empirio-Criticism* played an important role here; there were also numerous articles on the subject in the twenties in the journals *Pod znamenem marksizma* and *Vestnik kommunisticheskoi akademii*. For an example in English, see A. M. Deborin, "Lenin and the Crisis of Contemporary Physics," *Otchet o deiatel'nosti akademii nauk za 1929 g.*, Vol. I (Leningrad, 1930), Appendix.

50 See "Gigiena," *Bol'shaia sovetskaia entsiklopediia*, vol. XVI (Moscow, 1929), cols. 609–34, esp. col. 614.

51 The charge of ideological deviation among old engineers was a common theme in the twenties, peaking in the Shakhty trial in 1928 and the accusation that there existed an "Industrial Party" that aimed to restore capitalism. The episode is discussed in most standard histories of the Soviet Union. See also Bailes, *Technology and Society under Lenin and Stalin*, and *Protsess 'Prompartii' 25 noiabria – 7 dekabria 1930 g.: stenogramma sudebnogo protsessa* (Moscow, 1931).

52 Examples of how far some Soviet scientists went down the path of the German race hygiene and race anthropology movement can be seen in J. Philiptschenko (Iu. Filipchenko), "Die russische rassenhygienische Literatur, 1921–1925," *Archiv fur Rassen- und Gesellschaftsbiologie*, 17, no. 3 (1925): 346–8; N. K. Koltzoff (Kol'tsov), "Die rassenhygienische Bewegung in Russland," *Archiv fur Rassen und Gesellschaftsbiologie* 17, no. 1 (1925): 96–9. One Soviet physical anthropologist even concluded that, on the average, the brains of peasants and workers weigh less than those of members of the intelligentsia! For a favorable review of this work in the major German race hygiene journal, see *Archiv fur Rassen- und Gesellschaftsbiologie* 20, no. 2 (1927): 193–8. For a German emigré socialist critique of the race hygiene movement see M. Levin, "Stimmen aus dem deutschen Urwalde," *Unter den Banner des Marxismus*, vol. 2, (1928). Soviet criticism of racist concepts and many uses of physical anthropology became common after the late twenties.

53 For a discussion of the "Great Break" with special reference to the academic world, see David Joravsky, *Soviet Marxism and Natural Science, 1917–1932* (New York: Columbia University Press, 1961).

54 Sheila Fitzpatrick, "Cultural Revolution as Class War," in her *Cultural Revolution in Russia, 1928–1932* (Bloomington: Indiana University Press, 1978), pp. 8–40; also, see her "Cultural Revolution in Russia, 1928–1932," *Journal of Contemporary History* 9, no. 1 (1974).

55 An incomplete Soviet account of the reconstruction campaign in the Academy of Sciences is in V. D. Esakov, *Sovetskaia nauka v gody pervoi piatiletki: osnovnye napravleniia gosudarstvennogo rukovodstva naukoi* (Moscow, 1971).

56 Loren R. Graham, *The Soviet Academy of Sciences and the Communist Party, 1927–1932* (Princeton, N.J.: Princeton University Press, 1967).

57 The four academicians were S. F. Platonov, N. P. Likhachev, M. K. Liubavskii, and E. V. Tarle, the latter of whom returned after two years to become one of the USSR's best-known scholars, winning the Stalin Prize three times. All were historians.

58 There are a number of instances when the Academy of Sciences is known to have refused to obey the Communist Party's will, and no doubt more exist. Thus in December 1936 a few of the members refused to vote in favor of expelling Academicians Ipat'ev and Chichibabin as a result of their emigration at a time when unanimous votes on such issues were expected. After World War II, Nikita Khrushchev was not able during his tenure as Party leader to force the election of a scientist he wished to become a member. In the 1970s a number of academicians refused to join in the condemnation of their fellow member Andrei Sakharov. Zhores A. Medvedev, "New President of the Soviet Academy," *Nature* 258 (December 18, 1975): 566.

59 The intractability of the Academy of Sciences as a problem for the Marxists is revealed graphically in Lunacharskii's query:

> Just what could we demand of the Academy? That it suddenly, all in a big crowd, transform itself into a Communist gathering, that it suddenly cross itself in a Marxist fashion, put its hand on *Capital*, swearing that it is a genuine Bolshevik? . . . Everyone knows that a genuine conversion of this sort could not be.

A. V. Lunacharskii, "K 200-letiiu vsesoiuznoi akademii nauk," *Novyi mir* (1925): 110. For a general discussion of Lunacharskii's role in defending the place of science and culture after the Revolution, see Sheila Fitzpatrick, *The Commissariat of Enlightenment: Soviet Organization of Education and the Arts under Lunacharsky* (Cambridge: Cambridge University Press, 1970).

60 An interesting study of Marat and his scientific work is Norman Mandelbaum, "Jean-Paul Marat: The Rebel as Savant (1743–1788): A Case Study in Careers and Ideas at the End of the Enlightenment," unpublished Ph.D. dissertation, Department of History, Columbia University, New York, 1977.

61 N. K. Piksanov, *Gor'kii i nauka* (Moscow, 1948); S. F. Ol'denburg, "Maksim Gor'kii i uchenye," *Gor'kii i nauka* (Moscow, 1965); A. B. Derman, *Akademicheskii intsident* (Simferopol, 1923; University Microfilm Reprint, Ann Arbor, Mich., 1961).

62 See Joel Shapiro, "A History of the Communist Academy, 1918–1936," Ph.D. dissertation, Department of History, Columbia University, New York, 1976; Fitzpatrick, *Commissariat of Enlightenment*, p. 83; and M. N. Pokrovskii, *Izbrannye proizvedeniia*, Vol. 4, edited by M. N. Tikhomirov, V. M. Khvostov, L. G. Beskrovnii, O. D. Sokolov (Moscow-Leningrad, 1967), pp. 457–61.

63 Roger Hahn, *The Anatomy of a Scientific Institution: The Paris Academy of Sciences, 1666–1803* (Berkeley: University of California Press, 1971), pp. 135–6.

64 It is true that in the early years after the Russian Revolution the idea of a free association was popular, but it soon yielded place to more centralized plans. See, for example, Fitzpatrick, *Commissariat of Enlightenment*, p. 75.

65 For discussions of the problem of coordinating Soviet science, see G. I. Fed'kin, *Pravovye voprosy organizatsii nauchnoi raboty v SSSR* (Moscow, 1958); and G. A. Lakhtin, *Organizatsiia sovetskoi nauki: istoriia i sovremennost'* (Moscow: Nauka, 1990).

66 See David Joravsky, *The Lysenko Affair* (Cambridge, Mass.: Harvard University Press, 1970); Zhores A. Medvedev, *The Rise and Fall of T. D. Lysenko* (New York: Columbia University Press, 1969); and Loren R. Graham, *Science, Philosophy and Human Behavior in the Soviet Union* (New York: Columbia University Press, 1987).

CHAPTER 5. THE ROLE OF DIALECTICAL MATERIALISM

1 Margaret Jacob, "The Church and the Formulation of the Newtonian World View," *Journal of European Studies* 1 (1971): 128–48.

2 Charles C. Gillispie, "The *Encyclopedie* and the Jacobin Philosophy of Science: A Study in Ideas and Consequences," in Marshall Clagett (ed)., *Critical*

Problems in the History of Science: Proceedings (Madison: University of Wisconsin Press, 1959), pp. 255–308.

3 Quoted in Sidney W. Fox (ed.), *The Origins of Prebiological Systems and of Their Molecular Matrices* (New York: Academic Press, 1965), pp. 53–5.

4 These nine points are taken from my *Science, Philosophy and Human Behavior in the Soviet Union* (New York: Columbia University Press, 1987), pp. 62–3.

5 David Joravsky, *Soviet Marxism and Natural Science* (New York: Columbia University Press, 1961).

6 Quoted in James V. Wertsch, "L. S. Vygotsky's 'New' Theory of Mind," *American Scholar* (Winter 1988): 81.

7 Jerome S. Bruner, "Introduction," in L. S. Vygotsky, *Thought and Language*, edited and translated by Eugenia Hanfmann and Gertrude Vakar (Cambridge, Mass.: MIT Press, 1962), pp. vi, x.

8 Wertsch, "Vygotsky's 'New' Theory of Mind," p. 87.

9 The translators commented, "Although our more compact rendition would be called an abridged version of the original, we feel that the condensation has increased clarity and readability without any loss of thought content or factual information." Eugenia Hanfmann and Gertrude Vakar, translators' Preface to Vygotsky, *Thought and Language*, p. xii.

10 Lev Vygotsky, *Thought and Language*, translated and edited by Alex Kozulin (Cambridge Mass.: MIT Press, 1986).

11 Wertsch, "Vygotsky's 'New' Theory of Mind," p. 83.

12 L. S. Vygotsky, *Izbrannye psikhologicheskie issledovaniia* (Moscow, 1956), pp. 91–2; also see p. 105.

13 Vygotsky, *Thought and Language* (edited and translated by Hanfmann and Vakar), pp. 18–20.

14 Ibid., pp. 135–6.

15 Ibid., p. 49.

16 Ibid., p. 43.

17 Ibid., p. 51.

18 Katerina Clark and Michael Holquist, *Mikhail Bakhtin* (Cambridge, Mass.: Harvard University Press, 1984), pp. 229–30.

19 Joseph Stalin, *Marxism and Linguistics* (New York: International Publishers, 1951), p. 36.

20 Quoted by O. L. Zangwill, "Psychology: Current Approaches," in *The State of Soviet Science* (Cambridge, Mass.: MIT Press, 1965), p. 122.

21 A. R. Luria, *Higher Cortical Functions in Man* (New York: Basic Books, 1966), p. 540.

22 Michael Cole, Vera John-Steiner, Sylvia Scribner, and Ellen Souberman (eds.), *Mind in Society: The Development of Higher Psychological Processes – L. S. Vygotsky* (Cambridge, Mass.: Harvard University Press, 1978); Michael Cole and Sylvia Scribner, *The Psychology of Literacy* (Cambridge, Mass.: Harvard University Press, 1981); Michael Cole and Sheila Cole (eds.), *The Making of Mind: A Personal Account of Soviet Psychology* (Cambridge, Mass.: Harvard University Press, 1979).

23 Wertsch, "Vygotsky's 'New' Theory of Mind," p. 89.

24 C. H. Waddington, "That's Life," *New York Review of Books*, February 29, 1968, p. 19.

25 Notes taken from interview, Moscow, August 1971.

26 See *Aleksandr Ivanovich Oparin (Materialy k biobibliografii uchenykh SSSR, Seriia biokhimii, vypusk 3)* (Moscow and Leningrad, 1949), p. 5.

27 When I accused Oparin in the 1971 interview of being a supporter of Lysenko, he replied, "It is easy for you, an American, to make such accusations. If you had been here in those years, would you have had the courage to speak out and be imprisoned in Siberia?" Of course, he had a point, but he ignored the fact that some of his colleagues did speak out and some even survived (interview, Moscow, August 1971). For evidence of Oparin's resistance to Lysenko's followers, see A. I. Ignatov, "Mezhdunarodnyi simpozium po proiskhozhdeniiu zhizni na zemle," *Voprosy filosofii* 1 (1958): 154; and A. P. Skabichevskii, "Problema vozniknoveniia zhizni na zemli i teoriia akad. A. I. Oparina," *Voprosy filosofii* 2 (1953): 150–5.

28 A. I. Oparin in J. D. Bernal, *The Origin of Life* (London: Weidenfeld and Nicolson, 1967), p. 203.

29 Oparin in Bernal, ibid., p. 228.

30 For a more detailed discussion of Oparin and his work, see Graham, *Science, Philosophy, and Human Behavior in the Soviet Union*.

31 A. I. Oparin, *Life, Its Nature, Origin and Development* (New York: Academic Press, 1964).

32 Ibid., p. 4.

33 Ibid., p. 5.

34 Ibid., p. 6.

35 Ibid., p. 9.

36 For one of his strongest defenses of Lysenko, see his *Znachenie trudov tovarishcha I. V. Stalina po voprosam iazikoznaniia dlia razvitiia sovetskoi biologicheskoi nauki* (Moscow, 1951), pp. 10–15.

37 Zhores A. Medvedev, *The Rise and Fall of T. D. Lysenko* (New York: Columbia University Press, 1969), pp. 137–8.

38 Ibid., p. 214.

39 I am spelling the name as "Fock" rather than "Fok" (which would be consistent with the transliteration system used elsewhere in the book) because of Fock's personal insistence that his name be spelled in non-Russian sources as it was spelled by his Dutch ancestors who came to Russia at the time of Peter the Great.

40 Manuscript copy, "The Nature of the Physical World," Trinity College Library, Cambridge, pp. 309–11.

41 N. Bohr, "The Quantum Postulate and the Recent Development of Atomic Theory," *Nature* 121 (Suppl. April 14, 1928): 580, 584.

42 Paul Forman, "Weimar Culture, Causality, and Quantum Theory, 1918–1927: Adaptation by German Physicists and Mathematicians to a Hostile Intellectual Environment," *Historical Studies in the Physical Sciences* 3 (1971): 1–115.

43 V. A. Fock, "Nil's Bor v moei zhizni," *Nauka i chelovechestvo 1963*, vol. II, (Moscow, 1963), pp. 518–19.

44 N. R. Hanson, "Five Cautions for the Copenhagen Interpretation's Critics," *Philosophy of Science* (October 1959): 327.

45 Quoted in M. E. Omelyanovsky (sic), *Dialectics in Modern Physics* (Moscow,

1979), p. 144. The following section is taken from my "The Soviet Reaction to Bohr's Quantum Mechanics," in Herman Feshbach et al. (eds.), *Niels Bohr, Physics and the World: Proceedings of the Niels Bohr Centennial Symposium* (New York: Harwood Academic, 1988), pp. 305–17, esp. 311–13.

46 V. A. Fock, "Nil's Bor v moei zhizni," pp. 518–19.

47 V. A. Fock, "Ob interpretatsii kvantovoi mekhaniki," in P. N. Fedoseev et al. (eds.), *Filosofskie problemy sovremennogo estestvoznaniia* (Moscow, 1959), p. 235, and V. A. Fock, "Zamechaniia k stat'e Bora o ego diskussiiakh s Einshteinom," *Uspekhi fizicheskikh nauk* 66, no. 4 (December 1958): 602.

48 Omelyanovsky, *Dialectics in Modern Physics*, p. 50.

49 N. Bohr, "Can Quantum-Mechanical Description of Physical Reality be Considered Complete?" *Physical Review* 48 (October 15, 1935): 699.

50 Ibid., p. 697.

51 Niels Bohr, "Quantum Physics and Philosophy: Causality and Complementarity," in his *Essays 1958–1962 on Atomic Physics and Human Knowledge* (New York: Interscience Publishers, 1963), p. 3.

52 Ibid., p. 6.

53 See, esp., Omelyanovsky, *Dialectics in Modern Physics*, pp. 47, 50, 54, 57, 58, and 311–13.

54 Aage Bohr, "Preface," in N. Bohr, *Essays 1958–1962 on Atomic Physics and Human Knowledge*, p. vi.

55 Graham, *Science, Philosophy, and Human Behavior in the Soviet Union*, and Graham, *Between Science and Values*.

56 *Atti del convegno sulla relativita generale: problemi del'energia e onde gravitazionali* (Florence, 1965), pp. 1–12.

57 Graham, *Science, Philosophy and Human Behavior in the Soviet Union*.

58 Ibid., pp. 386–91.

59 This definition of mathematics occurs in all three editions of Kolmogorov's article. See *Bol'shaia Sovetskaia Entsiklopediia* (hereafter *BSE*), vol. 38 (Moscow, 1938), col. 359; *BSE*, vol. 26 (Moscow, 1954), p. 464; *BSE*, vol. 15 (Moscow, 1974), p. 467.

60 This statement was added to the last two editions: *BSE*, vol. 26 (Moscow, 1954), p. 464; *BSE*, vol. 15 (Moscow, 1974), p. 467. Kolmogorov also praised dialectical materialism at official occasions, such as a 1949 celebration at Moscow University of Stalin's seventieth birthday. A. N. Kolmogorov, "Matematika Stalinskoi epokhi," Arkhiv MGU, fond 2, op. 4, ed. khr. 3, esp. pp. 11–12. I am grateful to Gregory Crowe, History of Science Department, Harvard University for providing this source to me.

61 F. P. Ramsey, "Mathematics, Foundations of," *Encyclopedia Britannica*, vol. 15 (1941), p. 83.

62 Ibid.

63 A. N. Whitehead, "Mathematics, Nature of," *Encyclopedia Britannica*, vol. 15 (1941), pp. 87, 88.

64 Stephen W. Hawking, *A Brief History of Time: From the Big Bang to Black Holes* (New York: Bantam Books, 1988), pp. 48–50; 104–5; 130–2. For a discussion of the connection between the big bang theory and Western cultural and religious attitudes, see Daniel J. Kevles, "The Final Secret of the Universe?" *New York Review of Books*, May 16, 1991, pp. 27–32.

CHAPTER 6. STALINIST IDEOLOGY AND THE LYSENKO AFFAIR

1 For example, Valerii Soifer, *Vlast' i nauka: istoriia razgroma genetiki v SSSR* (Tenafly, N. J.: Hermitage, 1989); M. D. Akhundov and L. B. Bazhenov, "Fenomen ideologizirovannoi nauki," unpublished manuscript, 1990; A. A. Pechenkin, "The 1949–1951 Anti-Resonance Campaign," unpublished manuscript, 1990; D. A. Aleksandrov and N. L. Krementsov, "Opyt putevoditelia po neizvedannoi zemle: predvaritel'nyi ocherk sotsial'noi istorii sovetskoi nauki (1917–1950-e gody)," *Voprosy istorii estestvoznanii i tekhniki* 4 (1989): 67–87; Sergei Diachenko, "Podvig," *Ogonek* 47 (1987): 10–12; Aleksei Adzhubei, "Tri pis'ma," *Ogonek* 47 (1987), p. 13; Valerii Soifer, "Gor'kii plod," *Ogonek* 1 and 2 (1988); Valery N. Soyfer (sic), "New Light on the Lysenko Era," *Nature* 339 (June 8, 1989): 415–19; "Takim on byl: beseda s presidentom VASKHNIL A. A. Nikonovym," *Izvestiia*, November 24, 1987, p. 3; Yevgeniya Albats (sic), "Genius and the Villains," *Moscow News* 46 (November 22–29, 1987): 10; "Podvig uchenogo: k 100-letiiu so dnia rozhdeniia Nikolaia Ivanovicha Vavilova," *Nedelia* 46 (November 16–22, 1987); 6–7; "Velikii podvig rytsaria nauki," *Pravda*, November 25, 1987; Raisa Berg, *Acquired Traits: Memoirs of a Geneticist from the Soviet Union*, translated by David Lowe (New York: Viking, 1988); A. Takhtadzhian, "Kontinenty Vavilova," *Literaturnaia gazeta*, November 25, 1987, p. 12; "The Right to a Good Name," *Moscow News* 5 (1988): 2; "Scientists Restore Truth to Soviet History," *Moscow News* 51 (December 27, 1987 – January 3, 1988); Vladimir Venzher, "I did my best to fight against such stereotypes," *Moscow News* 40 (1987): 9; and "Ia byl by schastliv otdat' sebia polnost'iu moei rodine," *Sovetskaia rossiia*, October 4, 1987, p. 4. M. G. Iaroshevskii (ed.), *Repressirovannaia nauka* (Leningrad: Nauka, 1991).

2 For refutations of the view that Lysenko's popularity was rooted in human inheritance see Zhores A. Medvedev, *The Rise and Fall of T. D. Lysenko*, translated by I. Michael Lerner (New York: Columbia University Press, 1969); David Joravsky, *The Lysenko Affair* (Cambridge, Mass.: Harvard University Press, 1970); Loren R. Graham, *Science, Philosophy and Human Behavior in the Soviet Union* (New York: Columbia University Press, 1987).

3 The account here follows, in part, my *Science, Philosophy, and Human Behavior in the Soviet Union*.

4 Julian Huxley, *Heredity East and West: Lysenko and World Science* (London: Schuman, 1949), p. 17.

5 See the discussion of this case in P. S. Hudson and R. H. Richens, *The New Genetics in the Soviet Union* (Cambridge: Cambridge University Press, 1946).

6 Soifer, *Vlast' i nauka: Istoriia razgroma genetiki v SSSR*, p. 121.

7 T. D. Lysenko, "Iarovizatsiia – eto milliony pudov dobavochnogo urozhaia," *Izvestiia*, February 15, 1935, p. 4.

8 T. D. Lysenko, *Agrobiology* (Moscow, 1954).

9 Diachenko, "Podvig," p. 11; also, Joravsky, *Lysenko Affair*, p. 119.

10 Quoted in Graham, *Science and Philosophy in the Soviet Union* (New York: Alfred Knopf, 1972), p. 216.

11 Diachenko, "Podvig," p. 12.

12 Joravsky, *Lysenko Affair*, pp. 112–30, esp. 116.

13 A touching account of his death is in Diachenko, "Podvig."

14 Yevgeniia Albats (sic), "Genius and the Villains," *Moscow News* 46 (November 22–29, 1987): 10.
15 Yuri Trofimovich Lysenko, "Letter to the Editor," *Moscow News* 50 (December 13, 1987): 2.
16 D. Pyasetsky (sic), "Letter to the Editor," *Moscow News* 5 (1988): 2.
17 I. I. Shmal'gauzen, "Predstavleniia o tselom v sovremennoi biologii," *Voprosy filosofii* 2 (1947): pp. 177–83.
18 Soifer, *Vlast' i nauka*, pp. 369–70.
19 See Graham, "Lysenko and Zhdanov," *Science and Philosophy in the Soviet Union*, pp. 443–50.
20 Soifer, *Vlast' i nauka*, p. 392.
21 Ibid., p. 401.
22 K. O. Rossiianov, "Stalin kak redaktor Lysenko," paper given at Second Conference on the Social History of Soviet Science, Moscow, 1990 (received by personal communication).
23 Graham, *Science, Philosophy, and Human Behavior in the Soviet Union*, pp. 140–2. Also, see M. L. Golubovskii (ed.), *A. A. Liubishchev: V zashchitu nauki, stat'i i pis'ma, 1953–1972* (Leningrad: Nauka, 1991).

CHAPTER 7. SOVIET ATTITUDES TOWARD THE SOCIAL AND HISTORICAL STUDY OF SCIENCE

1 The history of medicine was studied in nineteenth-century Germany, and a chair in the history of science was established at the Collége de France at the end of the nineteenth century, but the Institute of the History of Science and Technology of the Academy of Sciences of the USSR, established in 1932 as described below, was the first institution or department devoted to the entire field of the history of science and technology.
2 S. R. Mikulinskii, "Entsiklopedist XX veka," in his edited volume *V. I. Vernadskii: Trudy po vseobshchei istorii nauki* (Moscow: Nauka, 1988), pp. 22–3.
3 See the works cited by Mikulinskii, ibid., p. 23. They include works on Lomonosov, Kant, Goethe, general surveys of the history of knowledge, and biographical works on a number of Russian scientists of the nineteenth and twentieth centuries.
4 V. I. Vernadskii, "Mysli o sovremennom znachenii istorii znanii," *Trudy komissii po istorii znanii*, Izdatel'stvo akademii nauk SSSR (Leningrad, 1927), and other issues in this series. Although he died in 1886 Butlerov was a figure in the debate over Marxism and resonance chemistry in the 1950s and was described by some Soviet historians of science as greater than Kekulé in the development of structural chemistry. See Loren R. Graham, *Science, Philosophy and Human Behavior in the Soviet Union* (New York: Columbia University Press, 1987), pp. 300–11.
5 Ibid., p. 6.
6 Ibid., p. 2.
7 Vernadskii, "Khimicheskii sostav zhivogo veshchestva v sviazi s khimiei zemnoi kory," in his *Biogeokhimicheski ocherki 1922–1932* (Moscow, 1940).
8 Kendall Bailes, *Science and Russian Culture in an Age of Revolutions: V. I.*

Vernadsky and His Scientific School, 1863–1945 (Bloomington: Indiana University Press, 1989).

9   Vernadskii, "Mysli o sovremennom," p. 9.

10 See Loren R. Graham, *The Soviet Academy of Sciences and the Communist Party, 1927–1932* (Princeton, N.J.: Princeton University Press, 1967), pp. 99–102, 131–138.

11 The standard source on Bukharin is Stephen F. Cohen, *Bukharin and the Bolshevik Revolution: A Political Biography, 1888–1938* (New York: Alfred A. Knopf, 1973).

12 V. D. Esakov, "N. I. Bukharin i akademiia nauk," *Priroda* 9 (1988): 94.

13 See, in particular, the Introduction and ch. 1–3 and 5 of his *Historical Materialism* (Ann Arbor: University of Michigan Press, 1969), esp. pp. 9–83 and 104–29.

14 N. Bukharin, "Theory and Practice from the Standpoint of Dialectical Materialism," in Bukharin et al. (eds.), *Science at the Cross Roads* (London: Frank Cass, 1971): 11–33.

15 Quoted in Loren Graham, "Rehabilitation of Nikolai Bukharin," *History of Science Society Newsletter* 18, no. 1 (January 1989): 3.

16 Bukharin, "Theory and Practice from the Standpoint of Dialectical Materialism," pp. 12–13.

17 See Loren R. Graham, "Bukharin and the Planning of Soviet Science," *Russian Review* (April 1964): 135–48.

18 V. D. Esakov, "N. I. Bukharin i akademiia nauk," p. 96.

19 S. R. Mikulinskii, "V. I. Vernadskii kak istorik nauki," in his edited *V. I. Vernadskii: Trudy po vseobshchei istorii nauki* (Moscow: Nauka, 1988), p. 26.

20 *Izvestiia*, May 10, 1988.

21 A helpful overview of Soviet attitudes toward the history of science is Alexander Vucinich, "Soviet Marxism and the History of Science," *Russian Review* 41, no. 2 (1982): 123–43.

22 Esakov, "N. I. Bukharin i akademiia nauk," p. 96.

23 Arnold Thackray, "History of Science," in Paul T. Durbin (ed.), *A Guide to the Culture of Science, Technology, and Medicine* (New York: Free Press, 1980), pp. 14–15.

24 *Isis*, 72, no. 263 (September 1981).

25 W. F. Bynum, E. J. Browne, and Roy Porter (eds.), *Dictionary of the History of Science* (Princeton, N. J.: Princeton University Press, 1981), pp. 145–6.

26 Wolf Schafer (ed.), *Boris Hessen Revisited* (tentative title) (Cambridge, Mass.: MIT Press, forthcoming).

27 Bukharin et al. (eds.), *Science at the Cross Roads*, pp. 182–3.

28 Ibid., p. 211.

29 J. D. Bernal, *The Social Function of Science* (London: George Routledge Sons, London, 1939), p. 406.

30 Hyman Levy, *Modern Science* (London: H. Hamilton, 1939), p. 97.

31 Robert K. Merton, *Science, Technology and Society in Seventeenth-Century England* (New York: Howard Fertig, 1970), pp. 142–3, 163, 185–7, 201, 206.

32 Stephen Toulmin, "From Form to Function: Philosophy and History of Science in the 1950s and Now," *Daedalus* 106 (Summer 1977): 150.

33 See my "The Socio-Political Roots of Boris Hessen: Soviet Marxism and the

History of Science," *Social Studies of Science* (November 1985), part of which is reproduced here.

34 David Joravsky, *Soviet Marxism and Natural Science, 1917–1932*, esp. 233–249. See also Loren Graham, *Between Science and Values* (New York: Columbia University Press, 1981), pp. 88–98.

35 Arthur Stanley Eddington noted in his Gifford Lectures of 1927, published in 1928 as *The Nature of the Physical World*, that he was "combatting" materialism on the basis of the new physics: see Graham, *Between Science and Values*, n. 14, p. 80. L. Houllevigue remarked that "The atom dematerialises, matter disappears," in his *L'Evolution des Sciences* (Paris: A. Colin, 1914), pp. 87–8.

36 Boris Hessen and V. P. Egorshin, "Ob otnoshenii tov. Timiriazeva k sovremennoi nauke," *Pod znamenem marksizma* 2–3, (1927): n. 7, 192.

37 Ibid., p. 193.

38 Boris Hessen, "Predislovie k stat'iam A. Einshteina i Dzh. Dzh. Tomsona," *Pod znamenem marksizma* 4 (1927): n. 7, p. 158.

39 See Alexander Vucinich, "Soviet Physicists and Philosophers in the 1930s: Dynamics of a Conflict," *Isis* 71, no. 257 (June 1980): 236–50; and Joravsky, *Soviet Marxism and Natural Science*.

40 *Raznoglasii na filosofskom fronte* (Moscow-Leningrad, 1931), p. 240.

41 Ibid., p. 72.

42 Ibid., p. 71.

43 Ibid.

44 Ibid., p. 234.

45 Ibid., p. 71.

46 Ibid., p. 279.

47 See M. Mitin, "Ocherednye zadachi raboty na filosofskom fronte v sviazi s itogami diskussii," *Pod znamenem marksizma* 3 (1931): 14; and V. E. L'vov, "Nauka i zhizn': Al'bert Einshtein v soiuze s religiei," *Novyi Mir* 10 (1931): 195. The latter article appeared after the London conference, but L'vov's opinion of Einstein's article was well known earlier in the year.

48 E. Kol'man, "Vreditel'stvo v nauke," *Bol'shevik* 2 (1931): 75–6.

49 E. Kol'man, "Boevye voprosy estestvoznaniia i tekhniki v rekonstruktivnyi period," *Pod znamenem marksizma* 3 (1931): 56–78.

50 Ibid., p. 57.

51 Ibid., p. 77.

52 Interview of Loren Graham with E. Kol'man, Moscow, August 22, 1971; Letter, E. Kol'man to Loren Graham, April 22, 1977; and E. Kolman, *My ne dolzhny byli tak zhit* (New York: Chalidze Publications, 1982).

53 B. Hessen, "The Social and Economic Roots of Newton's *Principia*," in *Science at the Crossroads*, 2d. ed. (London: Frank Cass, 1971), p. 171.

54 Ibid., pp. 190–1.

55 Ibid., pp. 182–3.

56 Helpful works on the development of science studies in the Soviet Union are Linda L. Lubrano, *Soviet Sociology of Science* (Columbus: American Association for the Advancement of Slavic Studies, 1976), and her "Soviet Science Specialists: Professional Roles and Policy Involvement," in Richard Remnek (ed.), *Scientists and Policymaking in the Soviet Union* (New York: Praeger, 1977);

Yakov Rabkin, "Naukovedenie: The Study of Scientific Research in the Soviet Union," *Minerva* 14, no. 1 (1976): 61–78, and his "Science Studies as an
Area of Scientific Exchange," in J. R. Thomas and U. Kruse-Vaucienne
(eds.), *Soviet Science and Technology: Domestic and Foreign Perspectives* (Washington, D.C.: National Science Foundation, 1977), pp. 69–82; and G. M.
Dobrov, "Science Policy and Assessment in the Soviet Union," *International
Social Sciences Journal* 25, no. 3 (1973): 305–25.

57 Quoted by Rabkin, "Science Studies as an Area of Scientific Exchange,"
from I. Borichevskii, "Naukovedenie kak tochnaia nauka," *Vestnik znaniia* 12
(1926): 786.

58 S. F. Ol'denburg (ed.), *Nauka i nauchnye rabotniki SSSR*, 4 vols. (Leningrad,
1926–1934); O. Iu. Shmidt and B. Ia. Smushkevich, *Nauchnye kadry i
nauchnoissledovatel'skie uchrezhdeniia SSSR* (Moscow, 1930); L. V. Sergeevich,
"Zadacha sobiraniia nauki," *Nauchnyi Rabotnik* (September 1926): 31–4; and
Borichevskii, "Naukovedenie kak tochnaia nauka."

59 Loren R. Graham, *The Soviet Academy of Sciences and the Communist Party,
1927–1932* (Princeton, N.J.: Princeton University Press, 1967), pp. 49–55.

60 J. D. Bernal, *The Social Function of Science* (London: Routledge, 1946); Derek J.
de Solla Price, *Little Science, Big Science* (New York: Columbia University
Press, 1963).

61 See Rabkin, "Naukovedenie" and "Science Studies as an Area of Scientific
Exchange."

62 See Mark Adams, "The Founding of Population Genetics: Contributions of
the Chetverikov School, 1924–1934," *Journal of the History of Biology* 1 (1968):
23–39; "Towards a Synthesis: Population Concepts in Russian Evolutionary
Thought, 1925–1935," 1 (1970): 107–29; and "From Gene Fund to Gene Pool:
On the Evolution of Evolutionary Language," *Studies in the History of Biology*
3 (1979): 241–85.

63 Maxim Karpinsky, "Scientists Restore Truth to Soviet History," *Moscow News*
(December 20, 1987).

64 D. A. Aleksandrov and N. L. Krementsov, "Opyt putevoditeliia po neizvedannoi zemle: predvaritel'nyi ocherk sotsial'noi istorii sovetskoi nauki
(1917–1950-e gody)," *Voprosy istorii estestvoznaniia i tekhniki* 4 (1989): 67–80.
Also see M. D. Akhundov and L. B. Bazhenov, "U istokov ideologizirovannoi nauki," *Priroda* 2 (1989): 90–9.

CHAPTER 8. KNOWLEDGE AND POWER IN RUSSIAN AND
SOVIET SOCIETY

1 Kendall E. Bailes, *Science and Russian Culture in an Age of Revolutions: V. I.
Vernadsky and His Scientific School, 1863–1945* (Bloomington: Indiana University Press, 1990).

2 Cited in Jeremy R. Azrael, *Managerial Power and Soviet Politics* (Cambridge,
Mass.: Harvard University Press, 1966), p. 53.

3 Azrael, *Managerial Power and Soviet Politics;* Kendall E. Bailes, *Technology and
Society under Lenin and Stalin* (Princeton, N.J.: Princeton University Press,
1978).

4 Edwin T. Layton, *The Revolt of the Engineers: Social Responsibility and the American Engineering Profession* (Cleveland: Press of Case Western Reserve University, 1971).

5 I am grateful to Kendall Bailes for his analysis of technocracy; see Bailes, *Technology and Society under Lenin and Stalin: Origins of the Soviet Technical Intelligentsia, 1917–1941* (Princeton, N.J.: Princeton University Press, 1978), esp. pp. 95–121.

6 Bailes and Azrael notice his significance but do not provide many details; see Bailes, *Technology and Society under Lenin and Stalin* and Azrael. See also my *The Ghost of the Executed Engineer*, (Cambridge: Harvard University Press, 1993).

7 P. Pal'chinskii, editorial, *Poverkhnost' i nedra*, 17, no. 1 (January 1926): 2–3.

8 P. Pal'chinskii, "Gornaia ekonomika," *Poverkhnost' i nedra* 18, no. 2 (February 1926): 12.

9 P. Pal'chinskii, "Ekonomicheskaia geologiia," *Poverkhnost' i nedra* 20, no. 4 (April 1926): 7.

10 Bailes, *Technology and Society under Lenin and Stalin*, pp. 117–18.

11 See Bailes, *Technology and Society under Lenin and Stalin*.

12 Ibid., p. 104.

13 Ibid., p. 108.

14 John Lowenhardt, *The Soviet Politburo*, translated by Dymphna Clark (New York: St. Martin's Press, 1982), p. 60.

15 I am grateful to Thomas P. M. Barnett for his research in his "Post-Stalinist Trends in the Soviet Politburo: The Development of Technocracy?" unpublished Government Department paper, Harvard University, Cambridge, Mass., January 29, 1987.

16 Ibid.

17 The next four paragraphs of the text rely heavily on my introductory chapter in Loren Graham (ed.), *Science and the Soviet Social Order* (Cambridge, Mass.: Harvard University Press, 1990).

18 See Robert Darst, Jr., "Environmentalism in the USSR: The Opposition to the River Diversion Projects," *Soviet Economy* 4 (July–September 1988): 223–52.

19 In Soviet newspaper and journal articles after Gorbachev, the psychological element was clearly recognized. A family that rents its own land and cares for it separately was called the "master of the farm" (*khoziain na ferme*). See, for example, "Khoziain na ferme," *Pravda*, September 3, 1988, p. 1.

20 In the section on Sakharov, I rely heavily on my article, "Knowledge and Power," *The Sciences* 20, no. 8 (October 1980): 14–32.

21 Interview by Loren Graham of Peter Kapitsa, Moscow, spring 1982.

22 A. Sakharov, *Sakharov Speaks* (London: Collins and Harvill Press, 1974), p. 32.

23 A. Sakharov, *Memoirs*, translated by Richard Lourie (New York: Alfred A. Knopf, 1990), p. 217.

CHAPTER 9. THE ORGANIZATIONAL FEATURES OF SOVIET SCIENCE

1 Loren R. Graham, "The Formation of Soviet Research Institutions: A Combination of Revolutionary Innovation and International Borrowing," in Don

K. Rowney and G. Edward Orchard (eds.), *Russian and Slavic History* (Columbus: Slavica, 1977), pp. 49–75.

2   Ibid.

3   S. F. Ol'denburg, "Vpechatleniia o nauchnoi zhizni v Germanii, Frantsii i Anglii," *Nauchnyi rabotnik* (February 1927): 89.

4   A. F. Ioffe, "Vpechatleniia ot poezdi po amerikanskim laboratoriiam," *Nauchnyi rabotnik* (April 1926): 59–65.

5   S. F. Ol'denburg, "Britanskaia konferentsiia i nauchnye issledovaniia," *Nauchnyi rabotnik* (February 1927): 93–97.

6   Iu. A. Filipchenko, "Iz vpechatlenii zagranichnoi poezdki," *Nauchnyi rabotnik* (January 1925): 150–9; and A. E. Fersman, "Nedelia sovetskikh uchenykh v Berline i ee mezhdunarodnoe znachenie," *Nauchnyi rabotnik* (September 1927): 76–83.

7   "Dokumente aus der Gründungszeit der Kaiser Wilhelm-Gesellschaft," in *25 Jahre Kaiser Wilhelm-Gesellschaft zur Förderung der Wissenschaften*, I (Berlin, 1936); Friederich Schmidt-Ott, *Erlebtes und Erstrebtes* (Wiesbaden, 1952); F. Glum, "Zehn Jahre Kaiser-Wilhelm-Gesellschaft zur Förderung der Wissenschaften," *Die Naturwisenschaften* (May 6, 1921): 293–300.

8   Adolf Harnack, "Denkschrift," in *25 Jahre Kaiser Wilhelm-Gesellschaft*, pp. 25–6, 39.

9   "Khronika," *Nauchnyi rabotnik* (March 1929): 66.

10 The helplessness of the Communist Party before the Academy during the early twenties is illustrated in the query of A. V. Lunacharskii, minister of education: "Just what could we demand of the Academy? That it suddenly, all in a big crowd, transform itself into a Communist gathering, that it suddenly cross itself in a Marxist fashion, put its hand on *Capital*, swearing that it is a genuine Bolshevik? . . . Everyone knows that a genuine conversion of this sort could not be." Lunacharskii, "K 200 letiiu vsesoiuznoi akademii nauk," *Novyi mir* no. 10 (1925): 109.

11 Loren Graham, *The Soviet Academy of Sciences and the Communist Party, 1927–1932* (Princeton, N.J.: Princeton University Press, 1967), pp. 141–47.

12 P. S. Osadchii wrote in 1928: "We must not ignore the fact that the concentration of scientific work in the new institutes, separated from the universities, has been disadvantageous to the latter, reducing them to purely pedagogical organizations with a low level of scholarly work." Osadchii, "Nauka v planovoi rabote sotsialisticheskogo stroitel'stva," *Nauchnoe slovo*, no. 1, (1928): 17–18.

13 "Khronika," *Nauchnyi rabotnik* (March 1929): 67.

14 See, for example, N. Finkel', "Kapitalizm i issledovatel'skaia rabota," *Molodoi bol'shevik* 14–15 (1931): 22–30.

15 See G. A. Lakhtin, *Organizatsiia sovetskoi nauki: istoriia i sovremennost'* (Moscow: Nauka, 1990), pp. 62–3.

16 *Nauchnye dostizheniia v promyshlennosti i raboty nauchno-tekhnicheskogo otdela VSNKh SSSR* (Moscow, 1925), pp. 41–2.

17 V. Zhamin, "Intensifikatsii nauki," *Ekonomicheskie nauki*, no. 4 (1985): 34.

18 Bruce Parrott, *Politics and Technology in the Soviet Union* (Cambridge, Mass.: MIT Press, 1983).

19 D. A. Senior, "The Organization of Scientific Research," *Survey* (July 1964): 21.

20 A. N. Nesmeianov, "Zagliadyvaia v budushchee nauki," *Pravda*, December 31, 1955, p. 2.

21 The following section parallels my article, "Reorganization of the U.S.S.R. Academy of Sciences," in Peter H. Juviler and Henry W. Morton, *Soviet Policy-Making: Studies of Communism in Transition* (New York: Frederick A. Praeger, 1967), pp. 133–62.

22 I. P. Bardin, "Most mezhdu teoriei i praktikoi," *Izvestiia*, August 28, 1959.

23 *Pravda*, July 2, 1959.

24 Thane Gustafson, "Why Doesn't Soviet Science Do Better Than It Does?" in Linda L. Lubrano and Susan Gross Solomon, *The Social Context of Soviet Science* (Boulder, Colo.: Westview Press, 1980), p. 31.

25 *Dostizheniia i perspektivy*, no. 4 (1977): 30–6.

26 "Kliuchevaia rol' nauki," *Pravda*, October 17, 1986, p. 3.

27 See Harley D. Balzer, *Soviet Science on the Edge of Reform* (Boulder, Colo.: Westview Press, 1989).

28 G. I. Marchuk, "Perestroika nauchnoi deiatel'nosti akademicheskikh uchrezhdenii v svete reshenii XXVII s˝ezda KPSS," *Vestnik akademii nauk SSSR*, no. 1 (1987): 5.

29 "Zakon ob individual'noi trudovoi deiatel'nosti," *Pravda*, November 21, 1986, pp. 1, 3.

30 Sharon L. Leiter, "Small Is Beautiful: The New Soviet Scientific-Technical Cooperatives," unpublished paper, Center for Russian and East European Studies, University of Virginia, Charlottesville, p. 23.

31 Ibid., p. 17.

32 The MNTKs were established December 12, 1985; see *Pravda*, December 13, 1985; Gorbachev discussed them at the 27th Party Congress.

33 Marchuk, "Perestroika nauchnoi . . .," p. 13.

34 One of the best sources for this debate is the weekly newspaper *Poisk* (The Search), devoted to issues of science and education.

35 Interview of Maksim Frank-Kamenetskii by Loren Graham, Columbus, October 26, 1990.

36 Vera Tolz, "The First Elections to the Russian Academy of Sciences," *Radio Free Europe/Radio Liberty Research Report* (February 14, 1992): 48–51; and Aleksei E. Levin, "Changes in Russian Science Administration and Policy," *Radio Free Europe/Radio Liberty Research Report* (February 14, 1992): 52–56.

37 See Shafarevich's "Russofobiia" in *Nash sovremennik* (Nos. 6 and 11), 1989.

38 They were listed in "Ot organizatsionnogo komiteta Rossiiskoi akademii nauk," *Poisk* (December, 6–12, 1991): 3.

39 "Rezoliutsiia konferentsii nauchnykh rabotnikov akademii nauk," *Poisk* (December 13–19, 1991): 3.

40 "Materialy rabochikh grupp i avtorskie razrabotki chlenov orgkomiteta k konferentsii nauchnykh rabotnikov AN SSSR," Moscow, 1991. (Materials distributed to participants of the conference.)

41 Interview of Velikhov by Loren Graham, Moscow, October 3, 1991.

42 "Address of the President of the Academy of Sciences of the USSR to the International Congress of Scientists, December 10, 1991, Moscow." (Materials distributed to participants of the conference.)

286   *Notes to pp. 195–212*

43   Steven Dickman, "Soviet Science: A Struggle for Survival," *Science* 254 (December 20, 1991): 1716.
44   The Royal Society, "Academies of Sciences in the Constituent Republics of the Former Soviet Union: A Current Appraisal," 1992, para. 48. Manuscript in my possession.
45   David P. Hamilton, "Piecemeal Rescue for Soviet Science," *Science* 255 (March 27, 1992): 1632–1634.

APPENDIX CHAPTER A. THE PHYSICAL AND MATHEMATICAL
SCIENCES

1   A. Einstein, "Bemerkung zu der Arbeit von A. Friedmann, 'Über die Krümmung des Raumes'," *Zeitschrift für Physik* no. 11 (1911); and "Notiz zu der Arbeit von A. Friedmann 'Über die Krümmung des Raumes'," *Zeitschrift für Physik* 16 (1923).
2   V. Fock, "Konfigurationsraum und Zweite Quantelung," *Zeitschrift für Physik* 75 (1932), p. 622; "Verallgemeinerung und Losung der Diracschen statistischen Gleichung," *Zeitschrift für Physik* 49 (1928): 339; L. Landau and R. Peierls, "Extension of uncertainty principle to relativistic quantum theory," *Zeitschrift für Physik* 69 (1931): 56; I. E. Tamm, "Exchange forces between neutrons and protons and Fermi theory," *Nature* 133 (1934): 981; "Interactions of neutrons and protons," *Nature* 134 (1934): 1010.
3   J. Frenkel (sic), *Lehrbuch der Elektrodynamik*, vols. I–II (Berlin: J. Springer Verlag, 1926, 1928).
4   *Review of US–USSR Interacademy Exchanges and Relations* (hereafter the "Kaysen Report") (Washington, D.C.: National Academy of Sciences, 1977), p. 102.
5   Paul Josephson, "The Ioffe Physico-Technical Institute and the Birth of Soviet Physics," unpublished Ph.D. dissertation, Massachusetts Institute of Technology, Cambridge, Mass., 1986.
6   M. S. Sominskii, *Abram Fedorovich Ioffe* (Moscow-Leningrad, 1964), pp. 337–94.
7   Aleksei B. Kozhevnikov, "The Roots of Soviet Physics in the 1930s: The Rise and Tragedy of Kharkov Physico-Technical Institute," paper presented to the 18th International Congress of History of Science, Hamburg, August 5, 1989.
8   Ibid., p. 4.
9   F. Janouch, "Lev D. Landau: His Life and Work," colloquium given at CERN, Geneva, 1979.
10   Mark Kuchment, "Active Technology Transfer and the Development of Soviet Microelectronics," in Charles Perry and Robert Pfaltzgraff, Jr., *Selling the Rope to Hang Capitalism?* (Washington, D.C.: Pergamon-Brassey, 1987), pp. 60–9.
11   Kaysen Report, p. 104.
12   Kapitsa maintained that he was not a pacifist and did not, in principle, oppose atomic weapons. It was Beria's disrespect for scientists that irritated Kapitsa most of all. See "Peter Kapitsa: The Scientist Who Talked Back to Stalin," *Bulletin of Atomic Scientists* (April 1990): 26–33.

13   See the discussion of Fock's critique in Loren R. Graham, *Science, Philosophy and Human Behavior in the Soviet Union* (New York: Columbia University Press, 1987), pp. 367–75.

14   See, for example, his *Memoirs* (New York: Alfred A. Knopf, 1990).

15   Kaysen Report.

16   A. M. Liapunov, *Pafnuty L'vovich Chebyshev* (Khar'kov, 1895). Also, A. P. Iushkevich, "P. L. Chebyshev i Peterburgskaia Matematicheskaia Shkola," in his *Istoriia matematiki v Rossii do 1917 goda* (Moscow, 1968), pp. 332–42; and N. S. Ermolaeva, *Peterburgskie matematiki i teoriia analiticheskikh funktsii,* Preprint No. 27 (1989), Institute of the History of Science and Technology, Academy of Sciences of the USSR, Moscow, 1988. I am grateful to Gregory Crowe, History of Science Department, Harvard University, for drawing the latter reference to my attention.

17   See Esther R. Phillips, "Nicolai Nicolaivich Luzin and the Moscow School of the Theory of Functions," *Historia Mathematica* 5 (1978): 275–305.

18   Ibid., p. 288.

19   Roger Cooke, Department of Mathematics, University of Vermont, is currently doing research on the Moscow school of analysts.

20   See the discussion of Aleksandrov in Loren Graham, *Science, Philosophy, and Human Behavior in the Soviet Union* (New York: Columbia University Press, 1987).

21   See Aleksey E. Levin, "Anatomy of a Public Campaign: 'Academician Luzin's Case' in Soviet Political History," *Slavic Review* (Spring 1990): 90–108; and Allen Shields, "Years Ago: Luzin and Egorov," *The Mathematical Intelligencer* 9, no. 4 (1987): 24–7. Also, A. P. Iushkevich, "Delo akademika N. N. Luzina," *Vestnik Akademii Nauk SSSR* 4 (1989): 102–13.

22   For a discussion of Shmidt, see Graham, pp. 386–91.

23   A. Ia. Khinchin, "O vospitatel'nom effekte urokov matematiki," *Matematicheskoe prosveshchenie* 6 (1961).

24   I. R. Shafarevich, "Russofobiia," *Nash sovremennik* (Nos. 6 and 11), 1989.

25   Z. K. Novokshanova (Sokolovskaia), *Vasilii Iakovlevich Struve* (Moscow, 1964), p. 13.

26   Robert A. McCutcheon, "The 1936–1937 Purge of Soviet Astronomers," *Slavic Review* (Spring 1991): 100–17; also see his "The Purge of Soviet Astronomy: 1936–37, with a Discussion of Its Background and Aftermath," unpublished master's thesis, Georgetown University, Washington, D.C., 1985.

27   Alexander Solzhenitsyn, *The Gulag Archipelago,* vol. I, translated by T. P. Whitney (New York, Harper & Row, 1973), p. 484.

28   "Eruption of a Volcano on Moon Reported by Russian Scientist," *New York Times,* November 13, 1958, pp. 1, 12.

29   Iu. I. Solov'ev, *Istoriia khimii v Rossii: Nauchnye tsentry i osnovnye napravleniia issledovanii* (Moscow, 1985). Biographies of all the chemists mentioned in the previous sentence in the text except Menshutkin and Ipat'ev can be found in I. V. Kuznetsov (ed.), *Liudi russkoi nauki* (Moscow, 1961). Biographies of Menshutkin and Ipat'ev (Ipatiev) can be found in *Dictionary of Scientific Biography,* vols. IX and VII (1974 and 1973).

30   Nathan M. Brooks, "The Formation of a Community of Chemists in Russia:

1700–1870," unpublished Ph.D. dissertation in history, Columbia University, 1988.

31 Ibid.

32 Ibid., p. VIII–17.

33 G. V. Bykov, *Aleksandr Mikhailovich Butlerov* (Moscow, 1961); Bykov, "The Origin of the Theory of Chemical Structure," *Journal of Chemical Education* 39 (1962): 220–4; and Bykov, *Istoriia klassicheskoi teorii khimicheskogo stroeniia* (Moscow, 1960).

34 G. V. Bykov, "Butlerov, Aleksandr Mikhailovich," *Dictionary of Scientific Biography*, vol. II (1970), p. 622.

35 A. J. Rocke, "Kekulé, Butlerov, and the Historiography of the Theory of Chemical Structure," *British Journal of the History of Science* 14 (1981): 27–57.

36 S. S. Nametkin, "Nikolai Dmitrievich Zelinskii (biograficheskii ocherk)," *Uchenye zapiski Moskovskogo universiteta* 3 (1934): 21–6.

37 Kaysen Report, p. 111.

38 Surveys of work in Soviet chemistry include N. M. Zhavoronkov (ed.), *Sovetskaia nauka i tekhnika za 50 let: Razvitie obshchei, neorganicheskoi i analiticheskoi khimii v SSSR* (Moscow, 1967); V. V. Korshak, *Sovetskaia nauka i tekhnika za 50 let: Razvitie organicheskoi khimii v SSSR* (Moscow, 1967), and Ia. I. Gerasimov, *Sovetskaia nauka i tekhnika za 50 let: Razvitie fizicheskoi khimii v SSSR* (Moscow, 1967).

39 A. F. Plate, G. V. Bykov, and M. S. Eventova, *Vladimir Vasil'evich Markovnikov: ocherk zhizni i deiatel'nosti, 1837–1904* Moscow, 1962).

40 O. E. Zviagintsev, Iu. I. Solov'ev, and P. I. Starosel'skii, *Lev Aleksandrovich Chugaev* (Moscow, 1965).

41 See S. R. Mikulinskii and A. P. Iushevich (eds.), *Razvitie estestvoznaniia v Rossii* (Moscow, 1977); and D. P. Grigor'ev and I. I. Shafranovskii, *Vydaiushchiesia russkie mineralogi* (Moscow-Leningrad, 1949).

42 G. V. Kir'ianov, *Vasilii Vasil'evich Dokuchaev, 1846–1903* (Moscow, 1966); V. A. Esakov and A. I. Solov'ev, *Russkie geograficheskie issledovaniia evropeiskoi rossii i urala v XIX–nachale XX v.* (Moscow, 1964); pp. 76–90; V. A. Esakov, "Dokuchaev, Vasily Vasilievich," *Dictionary of Scientific Biography*, vol. IV (1971), pp. 143–6.

43 David I. Spanagel, "Russian Soil Science," unpublished paper, History of Science Department, Harvard University, Cambridge, Mass., June 7, 1989.

44 K. D. Glinka, "Dokuchaiev's Ideas in the Development of Pedology and Cognate Sciences," *Russian Pedological Investigations*, vol. 1 (Leningrad: Academy of Sciences, 1927), pp. 16–17.

45 I. I. Shafranovskii, *Evgraf Stepanovich Fedorov* (Moscow-Leningrad, 1963); A. Meniailov, "Fyodorov (or Fedorov), Evgraf Stepanovich," *Dictionary of Scientific Biography*, vol. V (1972), pp. 210–14.

46 Kendall Bailes, *Science and Russian Culture in an Age of Revolutions: V. I. Vernadsky and His Scientific School, 1863–1945* (Bloomington: Indiana University Press, 1990); I. A. Fedoseyev, "Vernadsky, Vladimir Ivanovich," *Dictionary of Scientific Biography*, vol. XIII (1976), pp. 616–20.

47 For examples of Soviet work on Vernadskii, see N. P. Shcherbak, *Vladimir Ivanovich Vernadskii* (Kiev, 1988); K. M. Sytnik, *V. I. Vernadskii: zhizn' i deiatel'nost' na Ukraine* (Kiev, 1988); S. Mikulinskii, *V. I. Vernadskii kak*

issledovatel' istorii i teorii razvitiia nauki (Moscow, 1989); R. K. Balandin, *Vernadskii: zhizn', mysl', bessmertie* (Moscow, 1988). For Vernadskii's opposition to the Soviet takeover of the Academy, see Loren R. Graham, *The Soviet Academy of Sciences and the Communist Party, 1927–1932* (Princeton N.J.: Princeton University Press, 1967), pp. 131–8.

48 Robert M. Wood, "Geology vs. Dogma: The Russian Rift," *New Scientist* (June 12, 1980): 234–7.

49 *Review of US–USSR Interacademy Exchanges and Relations* (Washington, D.C.: Academy of Sciences, 1977), p. 107.

50 E. Pavlovskii, *Pamiati akademika L. S. Berga: sbornik rabot po geografii i biologii* (Moscow-Leningrad, 1955); and P. I. Polubarinova-Kochina, *Zhizn' i deiatel'nost' N. E. Kochina* (Leningrad, 1950).

APPENDIX CHAPTER B. THE BIOLOGICAL SCIENCES,
MEDICINE, AND TECHNOLOGY

1 Daniel Todes, "From Radicalism to Scientific Convention: Biological Psychology in Russia from Sechenov to Pavlov," unpublished Ph.D. dissertation, University of Pennsylvania, Philadelphia, 1981; and his "Biological Psychology and the Tsarist Censor: The Dilemma of Scientific Development," *Bulletin of the History of Medicine* 58 (1984): 529–44; David Joravsky, *Russian Psychology: A Critical History* (Oxford: Basil Blackwell, 1989); M. N. Shaternikov, "The Life of I. M. Sechenov," in I. M. Sechenov, *Selected Works* (Moscow-Leningrad, 1935), pp. vii–xxxvi; and I. M. Sechenov, *Avtobiograficheskie zapiski* (Moscow, 1945).

2 L. Ia. Bliakher, "Karl Maksimovich Ber," in I. V. Kuznetsov (ed.), *Liudi russkoi nauki: biologiia* (Moscow, 1962); Timothy Lenoir, "Kant, von Baer und das kausal-historische Denken in der Biologie," *Berichte zur Wissenshaftsgeschichte*, no. 8 (1985): 99–114; S. R. Mikulinskii, "Vzgliady K. M. Bera na evoliutsii v dodarvinksii period," *Annaly biologii*, no. 1 (1959): 287–362.

3 L. Ia. Bliakher, "Aleksandr Onufrievich Kovalevskii," in I. V. Kuznetsov (ed.), pp. 157–72; K. N. Davydov, "A. O. Kovalevskii kak chelovek i kak uchenyi (vospominaniia uchenika)," *Trudy instituta estestvoznaniia i tekhniki*, 31 (1960): 326–63.

4 A. A. Borisiak, *V. O. Kovalevskii: ego zhizn' i nauchnye trudy*, (Leningrad, 1928); L. Sh. Davitashvili, *V. O. Kovalevskii* (Moscow, 1951).

5 Semyon Zalkind, *Ilya Mechnikov: His Life and Work* (Moscow, 1959); Olga Metchnikoff, *Life of Elie Metchnikoff* (Boston: Houghton Mifflin, 1921); also see the discussion of Mechnikov in Alexander Vucinich, *Darwin in Russian Thought* (Berkeley: University of California, 1988); and Daniel Todes, *Darwin without Malthus: The Struggle for Existence in Russian Evolutionary Thought* (Oxford: Oxford University Press, 1989).

6 B. P. Babkin, *Pavlov: A Biography* (Chicago: University of Chicago Press, 1949); E. A. Asratian, *Ivan Petrovich Pavlov* (Moscow, 1981); also see Joravsky, *Russian Psychology*. Daniel Todes of Johns Hopkins is writing a biography of Pavlov.

7 Joravsky, *Russian Psychology*, pp. 52–83.

8   Loren R. Graham, *Science, Philosophy, and Human Behavior in the Soviet Union* (New York: Columbia University Press, 1981), pp. 158–62.

9   An example, although not quite as hagiographic as some, is Asratian's *Pavlov.*

10 Mark Adams, "The Founding of Population Genetics: Contributions of the Chetverikov School, 1924–1934," *Journal of the History of Biology* 1, no. 1 (1968): 23–39; "Towards a Synthesis: Population Concepts in Russian Evolutionary Thought, 1925–1935," *Journal of the History of Biology* 3, no. 1 (1970): 107–29; "From Gene Fund to Gene Pool: On the Evolution of Evolutionary Language," *Studies in the History of Biology* 3 (1979): 241–85; "Sergei Chetverikov, the Kol'tsov Institute, and the Evolutionary Synthesis," in Ernst Mayr and William Provine (eds.), *The Evolutionary Synthesis: Perspectives on the Unification of Biology* (Cambridge, Mass.: Harvard University Press, 1980), pp. 242–78.

11 Adams, "Founding of Population Genetics."

12 S. S. Chetverikov, "O nekotorykh momentakh evoliutsionnogo protsessa s tochki zreniia sovremennoi genetiki," *Zhurnal eksperimental'noi biologii* 2 (1926): 3–54. An English translation is by Malina Barker, edited by I. M. Lerner, in *Proceedings of the American Philosophical Society* 105 (1961): 167–95.

13 Timofeev-Resovskii much later became the subject of a biographic novel that attracted much attention in Gorbachev's Soviet Union: D. A. Granin, *Zubr* (Leningrad, 1987). In 1991 he was the subject of several films, such as "Okhota na zubra," and "Riadom s zubrom," which were the talk of Moscow.

14 G. G. Simpson, *The Meaning of Evolution* (New Haven, Conn.: Yale University Press, 1949), p. 278, as cited in Adams, "Founding of Population Genetics," p. 23.

15 Douglas R. Weiner, *Models of Nature: Ecology, Conservation, and Cultural Revolution in Soviet Russia* (Bloomington: Indiana University Press, 1988).

16 John T. Alexander, "Medical Chancery," *The Modern Encyclopedia of Russian and Soviet History*, vol. 21 (Gulf Breeze, Fla.: Academic International Press, 1981), p. 173.

17 John T. Alexander, *Bubonic Plague in Early Modern Russia: Public Health Urban Disaster* (Baltimore: Johns Hopkins University Press, 1980); Roderick E. McGrew, *Russia and the Cholera, 1823–1832* (Madison: University of Wisconsin Press, 1965).

18 Nancy Mandelker Frieden, *Russian Physicians in an Era of Reform and Revolution, 1856–1905* (Princeton, N.J.: Princeton University Press, 1981), p. 28.

19 Christine Johnson, "Medical Courses for Women," *The Modern Encyclopedia of Russian and Soviet History*, vol. 21 (Gulf Breeze, Fla.: Academic International Press, 1961), p. 174; also see Jeanette E. Tuve, *The First Russian Women Physicians* (Newtonville, Mass.: Oriental Research Partners, 1984).

20 See Ann Hibner Koblitz, "Science, Women, and the Russian Intelligentsia: The Generation of the 1860s," *Isis* 79 (1988): 208–26.

21 Frieden, *Russian Physicians, 1856–1905*, p. 17.

22 The best source on these trends is Susan Gross Solomon and John F. Hutchinson (eds.), *Health and Society in Revolutionary Russia* (Bloomington: Indiana University Press, 1990).

23 Christopher Davis, "Economic Problems of the Soviet Health Service: 1917–1930," *Soviet Studies* 35, no. 3 (1983): 343–61.

24 M. Landis, "Zdravookhranenie," *Bol'shaia sovetskaia entsiklopediia*, vol. 26 (Moscow, 1933), p. 531.

25 "Medicine," *Great Soviet Encyclopedia* (New York: Macmillan, 1977), p. 649 (translation of 3d ed., Moscow, 1974).

26 N. A. Semashko, *Okhrana zdorov"ia rabochikh i rabotnits, krest"ian i krest"ianok* (Moscow, 1927); Z. P. Solov'ev, *Stroitel'stvo sovetskogo zdravookhraneniia* (Moscow, 1932).

27 Henry E. Sigerist, *Socialized Medicine in the Soviet Union* (New York: W. W. Norton, 1937), p. 308.

28 Mark G. Field, *Soviet Socialized Medicine: An Introduction* (New York: The Free Press, 1967), p. 202.

29 Ibid., p. 204.

30 See Solomon's forthcoming *Caring for the Body Politic*.

31 Christopher Davis and Murray Feshbach, *Rising Infant Mortality in the U.S.S.R. in the 1970s*, U.S. Department of Commerce, Series P-95, No. 74, Washington, D. C., 1980.

32 E. I. Chazov, "Speech," *Pravda*, June 30, 1988, p. 4, translated in *Current Digest of the Soviet Press* 40, no. 27 (1988): 9.

33 Edward V. Williams, *The Bells of Russia: History and Technology* (Princeton, N.J.: Princeton University Press, 1985), p. 52.

34 William L. Blackwell, *The Beginnings of Russian Industrialization, 1800–1860* (Princeton, N.J.: Princeton University Press, 1968), p. 19 and passim.

35 Richard M. Haywood, *The Beginnings of Railway Development in Russia and the Reign of Nicholas I, 1835–1842* (Durham, N.C.: Duke University Press, 1969), p. xvii.

36 English-language sources on the history of Russian railways include Haywood, *Beginnings of Railway Development in Russia*; and J. N. Westwood, *A History of Russian Railways* (London: George Allen and Unwin, 1964).

37 Quoted in Merritt Roe Smith, "Becoming Engineers," unpublished manuscript of August 31, 1987, p. 32.

38 Haywood, *Beginnings of Railway Development in Russia*, p. 242.

39 See Iosif Gamel, *Description of the Tula Weapon Factory in Regard to Historical and Technical Aspects* (New Delhi: Amerind, 1988).

40 Edwin A. Battison, "Introduction," in Gamel, *Tula Weapon Factory*, pp. xxi–xxv.

41 Merritt Roe Smith, "Eli Whitney and the American System of Manufacturing," in Carroll Pursell, Jr., *Technology in America: A History of Individuals and Ideas* (Cambridge, Mass.: MIT Press, 1981), pp. 45–61.

42 Merritt Roe Smith, *Harper's Ferry Armory and the New Technology: The Challenge of Change* (Ithaca, N.Y.: Cornell University Press, 1977).

43 L. D. Bel'kind, *Pavel Nikolaevich Iablochkov* (Moscow, 1962); M. A. Shatelin, *Russkie elektrotekhniki XIX veka* (Moscow, 1955).

44 Jonathan Coopersmith, "The Role of the Military in the Electrification of Russia, 1870–1890," in Everett Mendelsohn, Merritt Roe Smith, and Peter Weingart (eds.), *Science, Technology and the Military*, vol. II (Dordrecht, Holland: Kluwer Academic Publishers, 1988).

45 Antony Sutton, *Western Technology and Soviet Economic Development, 1917–
    1965*, 3 vols. (Stanford, Calif.: Hoover Institution Press, 1968, 1971, 1973),
    esp. Vol. II, 1971, pp. 363–72.

46 Alexander Gerschenkron, *Economic Backwardness in Historical Perspective*
    (Cambridge, Mass.: Harvard University Press, 1962).

47 Gregory Crowe, Department of the History of Science, Harvard University,
    is writing a history of Soviet computing. I am grateful to him for assistance
    on this section.

48 On the Soviet space program, see James E. Oberg, *Red Star in Orbit* (New
    York: Random House, 1981); Nicholas Daniloff, *The Kremlin and the Cosmos*
    (New York: Alfred A. Knopf, 1972); Walter A. McDougall, *The Heavens and
    the Earth: A Political History of the Space Age* (New York: Basic Books, 1985);
    Leonid Vladimirov, *Russian Space Bluff* (London: Tom Stacey, 1971); and
    Uspekhi SSSR v issledovanii kosmicheskogo prostranstva (Moscow: Nauka, 1968).

49 Oberg, *Red Star in Orbit*, pp. 74–7; Vladimirov, *Russian Space Bluff* and
    Uspekhi SSSR v issledovanii kosmicheskogo prostranstva.

50 Sutton has given the most thorough analysis, although he is particularly
    unwilling to grant the Soviets any independent achievements.

51 Robert W. Campbell, *The Economics of Soviet Oil and Gas* (Baltimore: The
    Johns Hopkins Press, 1968), pp. 101–20; *Trends in the Soviet Oil and Gas
    Industry* (Baltimore: The Johns Hopkins Press, 1976), pp. 20–2.

52 Loren R. Graham, "Gorbachev's Great Experiment," *Issues in Science and
    Technology* (Winter 1988): 30.

53 Kendall E. Bailes, *Technology and Society under Lenin and Stalin: Origins of the
    Soviet Technical Intelligentsia, 1917–1941* (Princeton, N.J.: Princeton Univer-
    sity Press, 1978), p. 386.

Bibliographic essay

Although many of the references in the notes are to Russian language sources, since this volume is directed toward an English-reading audience, I will restrict this discursive bibliography to English-language books and articles.

An earlier version of this bibliography appeared in the series *Teaching the History of Science: Resources and Strategies,* published under the auspices of the Committee on Education by the History of Science Society.

INTRODUCTION

The history of science and technology in Russia and the Soviet Union is a field of study that is underdeveloped in the West, and good books on the subject in English or other West European languages are correspondingly rare. Nonetheless, a number of sources exist, as the following bibliography illustrates. Because of the youth of the field and the difficulty in gaining access to archives, at least until quite recently, the quality of existing works is uneven and the coverage spotty. In recent years this situation has begun to improve. There is a small but perceptible growth of interest in the history of Russian and Soviet science and technology in research universities. At present a handful of American universities – the Massachusetts Institute of Technology, Harvard, the University of Pennsylvania, Northwestern, Georgetown, Columbia, Arizona, Oregon State – offer occasional courses on the subject, but as yet no more than three or four senior American historians are working full-time in the field.

Counterbalancing the scarcity of good books on the history of Russian and Soviet science in Western languages is a large body of literature on the subject published in the Soviet Union in recent decades. The center of this research is the Institute of the History of Science and Technology of the Russian Academy of Sciences, located in Moscow, with a branch in Leningrad (now St. Petersburg). In just one series of monographs titled Scientific-Biographical Series (*Nauchno-Biograficheskaia Seriia*) this institute sponsored several hundred biographies of Russian and Soviet scientists, almost all of them written in Russian. The Academy of Sciences also published an overview in English of work in Russian: *The History of Science: Soviet Research,* 2 vols. (Moscow, 1985). Although this literature can be used profitably by the Western scholar who knows the Russian language, much of it is flawed by being written from an internalistic and nationalistic point of view. Some of the best Soviet pieces in English on the history of Soviet science can be found in the *Dictionary of Scientific Biography*, where significant deceased Russian and Soviet scientists are described. Users of this source should be sure to check the supplementary volumes for articles written after the editors decided to drop the rule that only Soviet authors could write about Soviet scientists.

I. GENERAL HISTORIES

The best overview in English of the history of science in Russia before 1917 is Alexander Vucinich's two-volume study *Science in Russian Culture* (Stanford, Calif.: Stanford University Press, 1963, 1970). A treatment of the Soviet period, somewhat incomplete in its coverage, is Zhores Medvedev, *Soviet Science* (New York: Norton, 1978). A topic in the history of Soviet science that touches on almost all scientific fields is the role of Marxism. For the 1920s and early 1930s the basic work on this topic is David Joravsky, *Soviet Marxism and Natural Science* (New York: Columbia University Press, 1961). For the role of Marxism in later periods see Loren R. Graham, *Science and Philosophy in the Soviet Union* (New York: Knopf, 1972). The latter book has been expanded and updated to cover events up to the middle 1980s in my *Science, Philosophy, and Human Behavior in the Soviet Union* (New York: Columbia University Press, 1987).

The most important single institution in Russian and Soviet science has been the Academy of Sciences, founded in 1725 by Tsar Peter the Great. Since the Academy traditionally encompassed all fields of knowledge, including both the natural and social sciences, histories of the Academy are virtually general histories of science in Russia, although they do not give much attention to university or industrial research. Two sources treating the early history of the Academy in the tsarist period are Alexander Lipski, "The Foundation of the Russian Academy of Sciences," *Isis* 44 (1953): 349–54; and Ludmilla Schulze, "The Russification of the St. Petersburg Academy of Sciences and Arts in the Eighteenth Century," *British Journal of the History of Science* 18 (1985): 305–35. A work that provides much general information on the Soviet period is Alexander Vucinich, *Empire of Knowledge: The Academy of Sciences of the USSR (1917–1970)* (Berkeley: University of California Press, 1984).

A crucial time for the Academy came in the late 1920s, when it was thoroughly restructured by Soviet authorities. This episode is described in Loren R. Graham, *The Soviet Academy of Sciences and the Communist Party, 1927–1932* (Princeton, N.J.: Princeton University Press, 1967); and in Aleksey E. Levin, "Expedient Catastrophe: A Reconsideration of the 1929 Crisis at the Soviet Academy of Science," *Slavic Review* 47, no. 2 (1988): 261–79. I described later, much less traumatic, reform of the Academy in "Reorganization of the USSR Academy of Sciences," in *Soviet Policy-Making*, edited by Peter Juviler and Henry Morton (New York: Praeger, 1967), pp. 133–62.

Although the Academy dominated Russian and Soviet science, other scientific societies also existed. A description of the fate of the prerevolutionary societies is James M. Swanson's "The Bolshevization of Scientific Societies in the Soviet Union: An Historical Analysis of the Character, Function and Legal Position of Scientific and Scientific-Technical Societies in the USSR, 1929–1936," unpublished Ph.D. dissertation, Department of History, Indiana University, Blooming-ton, Ind., 1968.

One of the most distinctive features of Soviet science is the organization of research in the institute system, an innovation adopted in the 1920s. The estab-lishment and early history of this system are described in Mark Adams, "Sci-ence, Ideology, and Structure: The Kol'tsov Institute, 1900–1970," in *The Social Context of Soviet Science*, edited by Linda Lubrano and Susan Gross Solomon (Boulder, Colo., Westview Press, 1980), pp. 173–204; Paul Josephson, "The Ioffe Physico-Technical Institute and the Birth of Soviet Physics," unpublished Ph.D. dissertation, Massachusetts Institute of Technology, Cambridge, Mass., 1986; and Loren R. Graham, "The Formation of Soviet Research Institutes: A Combina-tion of Revolutionary Innovation and International Borrowing," in *Russian and Slavic History*, edited by Don Karl Rowney and G. Edward Orchard (Columbus: Slavica, 1977), pp. 49–75. The work of an important institute with origins long before the Soviet government arose is described in Stanwyn G. Shefler, *The Komarov Institute: 250 years of Russian Research* (Washington, D. C.: Smithsonian Institution Press, 1967).

Another striking feature of the history of Russian science, at least in the nineteenth century, is the role of women. Russian women were among the first in the world to receive doctorates in mathematics, physiology, zoology, chemis-try, and other fields, as discussed in Ann Hibner Koblitz, "Science, Women and the Russian Intelligentsia: The Generation of the 1860's," *Isis* 79 (1988): 208–26.

A recent book that concentrates more on the impact of science and technology on Soviet society than it does on the history of science per se is Loren R. Graham (ed.), *Science and the Soviet Social Order* (Cambridge, Mass.: Harvard University Press, 1990).

II. HISTORIOGRAPHY

The best overview of Western works on the history and social study of science and technology in the Soviet Union is Susan Solomon, "Reflections on Western Studies of Soviet Science," in *The Social Context of Soviet Science* (edited by Lubrano and Solomon, see Section I of this bibliography), pp. 1–29. Articles describing the evolution of Soviet interpretations of the history of science are David Joravsky, "Soviet Views on the History of Science," *Isis* 46 (1955): 3–13; and Alexander Vucinich, "Soviet Marxism and the History of Science," *Russian Review* 41 (1982): 123–42.

One Soviet contribution to historiography – not in fact itself on the history of Russian or Soviet science – caused a great controversy over methodology and interpretation in the field as a whole, as described in the text. This was Boris Hessen's "The Social and Economic Roots of Newton's *Principia*," in N. I. Bukha-rin et al. (eds.), *Science at the Crossroads* (London: Kniga, 1931; reprinted, London: F. Cass, 1971). The context of this important work is explored in Loren Graham, "The Socio-Political Roots of Boris Hessen: Soviet Marxism and the History of Science," *Social Studies of Science* 15 (1985): 705–22.

The social study of science and technology is usually termed *naukovedenie*, or "science studies." The nature and evolution of this field are explored in Linda

Lubrano, *Soviet Sociology of Science* (Columbus: American Association for the Advancement of Slavic Studies, 1976); and Yakov Rabkin, "*Naukovedenie:* The Study of Scientific Research in the Soviet Union," *Minerva* 14 (1976): 61–78.

III. SPECIAL SUBJECTS

Mathematics. As described in the text, it is in mathematics that Russia and the Soviet Union made the greatest contributions. Unfortunately, the importance of Russian and Soviet mathematics is poorly reflected in English-language sources. Not even Lobachevskii, the creator of non-Euclidean geometry, is the subject of a full biography in English. V. F. Kagan's *N. Lobachevsky and His Contribution to Science* (Moscow: Foreign Languages Publishing House, 1957) is perhaps the source most often cited, but is clearly inadequate. Alexander Vucinich has explored some of the nontechnical aspects of Lobachevskii's life in his "Nikolai Ivanovich Lobachevskii: The Man behind the First Non-Euclidean Geometry," *Isis* 53 (1962): 465–81. The best source on the circumstances of the creation of Lobachevskian geometry is an unpublished senior thesis by Gregory Crowe, "The Life and Work of Nikolai Ivanovich Lobachevsky: A Study of the Factors Leading to the Discovery and Acceptance of the First Non-Euclidean Geometry," Harvard University, 1986.

A happy exception to the dearth of English-language materials on the history of Russian and Soviet mathematics is Anne Hibner Koblitz's biography of the first significant woman mathematician of modern times, *A Convergence of Lives: Sofiia Kovalevskaia: Scientist, Writer, Revolutionary* (Boston: Birkhäuser, 1983). Biographical material is also available on Nikolai Luzin, a founder of the twentieth-century "Moscow School" of mathematics, in three articles: Aleksei E. Levin, "Anatomy of a Public Campaign: 'Academician Luzin's Case' in Soviet Political History," *Slavic Review* (Spring 1990): 90–108; Esther R. Phillips, "Nicolai Nicolaevich Luzin and the Moscow School of the Functions," *Historia Mathematica* 5 (1978): 275–305; and Allen Shields, "Years Ago: Luzin and Egorov," *Mathematical Intelligencer* 9, no. 4 (1987): 24–7. An interesting analysis of the Luzin affair is Aleksey E. Levin, "Anatomy of a Public Campaign: 'Academician Luzin's Case' in Soviet History," *Slavic Review* (Spring 1990): 90–108.

Biological sciences. The Russian and Soviet contribution in biology is less significant than that in mathematics, but the available materials are, somewhat paradoxically, more numerous. Topics that have attracted the attention of Western historians are the reception of Darwinism in Russia in the nineteenth century and the Lysenko affair in the twentieth; these both relate to a third topic that has attracted scholars: genetics.

On Darwinism, the most discriminating work is Daniel Todes, *Darwin without Malthus: The "Struggle for Existence" and Russian Evolutionary Thought in the Nineteenth Century* (Oxford: Oxford University Press, 1989). Some of Todes's main ideas can be found in his "Darwin's Malthusian Metaphor and Russian Evolutionary Thought, 1859–1917," *Isis* 78 (1987): 537–51; and his "V. O. Kovalevskii: The Genesis, Content, and Reception of His Paleontological Work," *Studies in the History of Biology* 2 (1978): 99–165. The profound influence of the Russian tradition of morphology on the formulation of Soviet Darwinism is the subject of Mark Adams, "Severtsov and Schmalhausen: Russian Morphology and the Evolutionary Synthesis," in *The Evolutionary Synthesis: Perspectives on the Unification of Biology,* edited by Ernst Mayr and William Provine (Cambridge, Mass.: Harvard University Press, 1980), pp. 193–225. A more general treatment of Darwinism in Russia is Alexander Vucinich, *Darwin in Russian Thought* (Berkeley/Los Angeles: University of California Press, 1989). James Allen Rogers has three articles on the subject: "The Reception of Darwin's *Origin of Species* by Russian Scientists," *Isis* 64 (1973): 484–503; "Charles Darwin and Russian Scientists," *Russian Review* 19, no. 4 (1960): 371–83; and "Russian Opposition to Darwinism in the Nineteenth Century," *Isis* 65 (1974): 487–505. Another source on this topic is Sarah Swinburne White, "The Reception in Russia of Darwinian Doctrines Concerning Evolution," unpublished Ph.D. dissertation, University of London, 1968. Also useful is George L. Kline, "Darwinism and the Russian Orthodox Church," in *Continuity and Change in Russian and Soviet Thought,* edited by Ernest J. Simmons (New York: Russell and Russell, 1967), pp. 307–28.

One proponent of a characteristically Russian modification of evolutionary theory is Peter Kropotkin, known for his work on "mutual aid" within species. There is no biography of Kropotkin that takes full account of both his biological and political interests, but his political views and activities are analyzed in Martin Allen Miller, *Kropotkin* (Chicago: University of Chicago Press, 1976). Also worthy of examination is James Allen Rogers, "Prince Peter Kropotkin, Scientist and Anarchist: A Biographical Study of Science and Politics in Russian History," unpublished Ph.D. dissertation, Harvard University, Cambridge, Mass., 1957.

A more widely known – indeed, notorious – figure in Russian biology is T. D. Lysenko. The most complete work on Lysenko is David Joravsky, *The Lysenko Affair* (Cambridge, Mass.: Harvard University Press, 1979). Another important work, written by a Soviet biologist who became an opponent of Lysenko, is Zhores Medvedev, *The Rise and Fall of T. D. Lysenko* (New York: Columbia University Press, 1969). More idiosyncratic is Dominique Lecourt, *Proletarian Science? The Case of Lysenko,* translated by Ben Brewster (London: NLB, 1988). For a study of late Lysenkoism see Mark Adams, "Genetics and Molecular Biology in Khrushchev's Russia," unpublished Ph.D. dissertation, Harvard University, Cambridge, Mass., 1973, a source that also contains much information on biochemistry.

That the field of population genetics was largely established in Soviet Russia before Lysenko's rise to power is the significant conclusion of a series of articles by Mark Adams. Two were published in the *Journal of the History of Biology:* "The Founding of Population Genetics: Contributions of the Chetverikov School, 1924–1934," in vol. 1, no. 1 (1968): 23–39; and "Towards a Synthesis: Population Concepts in Russian Evolutionary Thought 1925–1935," in vol. 3, no. 1 (1970): 107–29. Adams discusses how the term "gene pool" derives from a Russian concept in "From Gene Fund to Gene Pool: On the Evolution of Evolutionary Language," *Studies in the History of Biology* 3 (1979): 241–85. And Adams summarizes the significance of the Russian strength in population biology in "Sergei Chetverikov, the Kol'tsov Institute, and the Evolutionary Synthesis," in *The Evolutionary Synthesis* (edited by Mayr and Provine, cited earlier), pp. 242–78. Adams is currently working with Soviet scholars on a joint edition of the letters and papers of Theodosius Dobzhansky, the prominent Soviet geneticist who emigrated to the United States.

The related subject of eugenics is the topic of Loren R. Graham, "Science and Values: The Eugenics Movement in Germany and Russia in the 1920's," *The American Historical Review* 82, no. 5 (1977), 1133–64, an article that shows that interest in eugenics was not tied uniquely to right-wing political movements. Further comparative analysis of eugenics can be found in Mark Adams (ed.), *The Well-Born Science: Eugenics in Germany, France, Brazil, and Russia* (New York: Oxford University Press, 1990).

Before the advent of industrialization there was a strong school of ecology and conservation in the Soviet Union, one with important prerevolutionary roots. Douglas Weiner has explored this topic in a number of articles, including "The Historical Origins of Soviet Environmentalism," *Environmental Review* 6, no. 2 (1982), 42–62; and "Community Ecology in Stalin's Russia: 'Socialist' and 'Bourgeois' Science," *Isis* 75 (1984): 684–96. Weiner has also written an important book analyzing the early history of Soviet conservation and identifying the roots of Lysenkoism: *Models of Nature: Conservation and Ecology in the Soviet Union, 1917–1935* (Bloomington: Indiana University Press, 1987).

Biomedical sciences: Physiology, medicine, and public health. Russia has a rich tradition in physiology and physiological psychology, as the names of I. M. Sechenov, V. M. Bekhterev, and Ivan Pavlov illustrate. David Joravsky's *Russian Psychology: A Critical History* (Oxford: Blackwell, 1989) treats this subject. Other sources are Daniel Todes, "Biological Psychology and the Tsarist Censor: The Dilemma of Scientific Development," *Bulletin of the History of Medicine* 58 (1984), 529–44; B. P. Babkin, *Pavlov* (Chicago: University of Chicago Press, 1949); Y. P. Frolov, *Pavlov and His School* (London: K. Paul, Trench, Trubner, 1937); Todes, "From Radicalism to Scientific Convention: Biological Psychology in Russia from Sechenov to Pavlov," unpublished Ph.D. dissertation, University of Pennsylvania, Philadelphia, 1981; James E. Brett, "Materialist Philosophy in 19th Century Russia: The Physiological Psychology of I. M. Sechenov," unpublished Ph.D. dissertation, University of California at Los Angeles, 1975; Alex Kozulin, *Psychology in Utopia: Toward a Social History of Soviet Psychology* (Cambridge, Mass.: MIT Press, 1984); Kozulin, *Vygotsky's Psychology: A Biography of Ideas* (Cambridge, Mass.: Harvard University Press, 1990); James Wertsch, *Vygotsky and the Social Formation of Mind* (Cambridge, Mass.: Harvard University Press, 1985); A. R. Luria, *The Making of Mind: A Personal Account of Soviet Psychology,* edited by Michael Cole and Sheila Cole (Cambridge, Mass.: Harvard University Press, 1979); René van der Veer and Jaan Valsiner, *Lev Vygotsky: His Life and Work* (Oxford: Blackwell, 1991); and Maury David Shenk, "Ivan Pavlov and the Soviet Psychology of the Individual: The Teplov-Nebylitsyn School of Differential Psychophysiology," unpublished senior thesis, Harvard University, Cambridge, Mass., 1988. Daniel Todes of Johns Hopkins University is currently writing a biography of Pavlov.

Two recent works examine physicians' organizations. The medical society founded in the name of Nikolai Pirogov, one of Russia's great anatomists, played an important role in late nineteenth- and early twentieth-century Russian medicine; Nancy Mandelker Frieden has examined its history in *Russian Physicians in an Era of Reform and Revolution, 1856–1905* (Princeton, N.J.: Princeton University Press, 1981). John F. Hutchinson has written on professionalism among Russian doctors in the nineteenth century in "Society, Corporation or Union? Russian Physicians and the Struggle for Professional Unity (1890–1913)," *Jahrbücher für Geschichte Osteuropas* 30, no. 1 (1982): 37–53. Also see his *Politics and Public Health in Revolutionary Russia, 1890–1918* (Baltimore: The Johns Hopkins University Press, 1990). Doctors themselves – or at least those who were women – are the the focus of Jeanette E. Tuve, *The First Russian Women Physicians* (Newtonville, Mass.: Oriental Research Partners, 1984).

Interest among Western scholars in the history of Russian and Soviet public health has grown considerably in the last two decades. Significant new books are Susan Gross Solomon and John F. Hutchinson (eds.), *Health and Society in Revolutionary Russia* (Bloomington: Indiana University Press, 1990); and John F. Hutchinson, *Politics and Public Health in Revolutionary Russia, 1890–1918* (Baltimore: Johns Hopkins University Press, 1990). Solomon's forthcoming *Caring for the Body Politic* will discuss the fate of social hygiene in the Soviet Union. Books that examine public health in earlier periods in tsarist Russia include John T. Alexander, *Bubonic Plague in Early Modern Russia: Public Health Urban Disaster* (Baltimore: Johns Hopkins University Press, 1980); Charles de Mertens, *An Account of the Plague That Raged at Moscow, 1771* (Newtonville, Mass.: Oriental Research Partners, 1977); and Roderick McGrew, *Russia and the Cholera, 1823–1832* (Madison: University of Wisconsin Press, 1965). Social historians are interested in health-care delivery in nineteenth-century rural Russia. Among recent works are Peter Krug, "The Debate over the Delivery of Health Care in Rural Russia: The Moscow Zemstvo, 1864–1878," *Bulletin of the History of Medicine* 50 (1976): 226–41; and two articles by Samuel Ramer: "Who Was the Russian Feldsher?" *Bulletin of the History of Medicine* 50 (1976): 213–35, and "Childbirth and Culture: Midwifery in the Nineteenth Century Russian Countryside," in *The Family in Imperial Russia: New Lines of Historical Research*, edited by David L. Ransel (Urbana: University of Illinois Press, 1978), p. 218–35.

Early studies of Soviet medicine tended to be largely laudatory. See, for example, Henry E. Sigerist, *Socialized Medicine in the Soviet Union* (New York: Norton, 1937). A more critical and well-researched approach is found in Mark G. Field, *Doctor and Patient in Soviet Russia* (Cambridge, Mass.: Harvard University Press, 1957); and in his *Soviet Socialized Medicine: An Introduction* (New York: Free Press, 1967). A book that includes considerable history is Gordon Hyde, *The Soviet Health Service: An Historical and Comparative Study* (London: Lawrence and Wishart, 1974). A good historical study of early Soviet public health is Christopher Davis, "Economic Problems of the Soviet Health Service: 1917–1930," *Soviet Studies* 35, no. 3 (1983), 343–61.

Two other medical topics are covered in Julie V. Brown, "The Professionalization of Russian Psychiatry, 1857–1911," unpublished Ph.D. dissertation, University of Pennsylvania, Philadelphia, 1981, which looks at the history of Russian psychiatry from a sociological viewpoint; and John F. Hutchinson, "Tsarist Russia and the Bacteriological Revolution," *Journal of the History of Medicine and Allied Sciences* 4, no. 4 (1985): 420–39, on intellectual, professional, and political resistance to an acceptance of bacteriology up to the 1905 Revolution.

Chemistry. Chemistry has a strong tradition in Russia, dating back to the first significant Russian scientist, Mikhail Lomonosov, continuing through A. M. Butlerov and D. I. Mendeleev in the nineteenth century, and persisting into the Soviet period, when N. N. Semenov won the Nobel Prize for his work on the kinetics of chemical reactions. This history, like that of much of Russian science, is poorly covered in the English-language literature. A well-known biography of Lomonosov is B. N. Menshutkin's *Russia's Lomonosov, Chemist, Courtier, Physicist, Poet* (Princeton, N.J.: Princeton University Press, 1952). Unfortunately, Menshutkin's biography contains serious errors, such as the contention that Lomonosov did not believe in phlogiston. A Soviet biography of Lomonosov has recently been translated into English: Galina E. Pavlova and Aleksandr S. Fedorov, *Mikhail Vasil'evich Lomonosov: His Life and Work,* translated by Arthur Aksenov (Moscow: Mir Publishers, 1984). Another source is a collection edited by Henry M. Leicester, *Mikhail Vasil'evich Lomonosov and the Corpuscular Theory* (Cambridge, Mass.: Harvard University Press, 1970). Despite the numerous publications about Lomonosov, especially in Russian, a critical evaluation of his place in the history of science has not yet been written.

Chemistry was established as a profession in Russia by the mid-nineteenth century. This development is analyzed in a valuable dissertation written in 1988 in the history department of Columbia University by Nathan Marc Brooks titled "The Formation of a Community of Chemists in Russia: 1700–1870." Chemistry was particularly strong at Kazan' University, where A. M. Butlerov worked. Relevant sources on chemistry at Kazan' and on Butlerov, one of the founders of structural chemistry, are S. N. Vinogradov, "Chemistry at Kazan University in the Nineteenth Century: A Case History of Intellectual Lineage," *Isis* 56 (1985) 168–73; and Henry M. Leicester, "Alexander Mikhailovich Butlerov," *Journal of Chemical Education* 17 (May 1940): 208–9. On the controversial question of the relative contributions of August Kekulé and Butlerov to the origins of structural chemistry, a useful article is Alan J. Rocke's "Kekulé, Butlerov, and the Historiography of the Theory of Chemical Structure," *British Journal for the History of Science* 14, no. 46 (1981): 27–55.

No adequate biography of the great chemist Dmitrii Ivanovich Mendeleev exists in any language, not even Russian. Indeed, many of the existing treatments of Mendeleev are filled with errors. On the positive side, the Soviet historian and philosopher B. M. Kedrov wrote an excellent description of the discovery of the table of the elements titled *The Day of One Great Discovery* (Moscow: Nauka, 1958). Unfortunately, this book has not been translated into English, but it is summarized in Kedrov's article on Mendeleev in the *Dictionary of Scientific Biography.* Another useful source is Bernadette Bensaude-Vincent, "Mendeleev's Periodic System of Chemical Elements," *British Journal of the History of Science* 19 (1986): 3–17. For an emphasis on the effect that writing a textbook had on Mendeleev at the time he was developing the periodic table, see Loren R. Graham, "Textbook Writing and Scientific Creativity: The Case of Mendeleev," *National Forum* (Winter 1983): 22–3. Two (unpublished) Ph.D. dissertations on Mendeleev are Beverly Almgren, "Mendeleev: The Third Service, 1834–1882," Brown University, Providence, R.I., 1968; and Francis Stackenwalt, "The Thought and Work of Dmitrii Ivanovich Mendeleev on the Industrialization of Russia, 1867–1907," University of Illinois, Urbana, 1976.

One reason that no adequate biography of Mendeleev has yet been written is that he was as active in politics and social issues as he was in chemistry. The future biographer faces a mountain of archival material, most of it collected in the Mendeleev Museum in St. Petersburg. A useful article on Mendeleev's social opinions is "Mendeleev's Views on Science and Society," *Isis* 58 (1967): 242–51.

A third chemist of this generation is better known as a composer of symphonies and opera, but Aleksandr Borodin made his living as a professor of chemistry at the St. Petersburg Academy of Medicine and Surgery. A helpful, but far from complete biography is Nikolai I. Figurovskii and Yurii I. Solov'ev's *Aleksandr Profir'evich Borodin: A Chemist's Biography*, translated by Charlene Steinberg and George Kauffman (New York: Springer Verlag, 1988). See also George Sarton, "Borodin. 1833–87," *Osiris* 7 (1939): 224–51.

A memoir by an important Soviet chemist who emigrated to the United States, V. N. Ipatieff, is *Life of a Chemist* (Stanford, Calif.: Stanford University Press, 1946).

Soviet historians wrote a great deal on the history of chemistry, but as with other topics in this bibliography, I will not attempt to describe the Russian-language literature. For an overview of the evolution of Soviet interpretations of the history of chemistry readers can, however, turn to Yakov M. Rabkin, "Trends and Forces in the Soviet History of Chemistry," *Isis* 67 (1976): 257–73.

Physics and astronomy. The Soviet Union was very strong in the theoretical foundations of physics. Unfortunately, there is very little scholarly work in English on the history of Soviet physics. Good articles on the early history are Paul Josephson, "The Early Years of Soviet Nuclear Physics," *Bulletin of the Atomic Scientists* 43, no. 10 (1987): 36–9; Josephson, "Physics, Stalinist Politics of Science and Cultural Revolution," *Soviet Studies* 40, no. 2 (1988): 245–65; and Josephson, "Physics and Soviet–Western Relations in the 1920s and 1930s," *Physics Today* 41, no. 9 (1988): 54–61. The founder of Soviet work in solid state physics was A. F. Ioffe, a major figure in the history of Soviet science. For a history of his institute and its wider context see Josephson, *Physics and Politics in Revolutionary Russia* (Berkeley: University of California Press, 1991). Peter Kapitsa, once the Soviet Union's best-known physicist because of his capture by Stalin while on home leave from England in 1934, has been the subject of several popular biographies, which have utilized only a small portion of the available literature. Lawrence Badash has included some of the correspondence between Kapitsa and his wife, Anna, in his *Kapitza, Rutherford, and the Kremlin* (New Haven, Conn.: Yale University Press, 1985). Also see J. W. Boag, P. E. Rubinin, and D. Shoenberg (eds.), *Kapitza in Cambridge and Moscow: Life and Letters of a Russian Physicist* (New York: Elsevier Science Publishers, 1990). Kapitsa's collection of articles *Experiment, Theory, Practice* has been published in English (Dordrecht/Boston: D. Reidel, 1980). Also useful is the collection of *Peter Kapitsa on Life and Science: Addresses and Essays*, edited and translated by Albert Parry (New York: Macmillan, 1968). For an interesting exchange of letters between Kapitsa and Stalin, see "Peter Kapitsa: The Scientist Who Talked Back to Stalin," *Bulletin of Atomic Scientists* (April 1990): 26–33.

An aspect of Soviet physics that attracted attention in the West was atomic weapons and atomic energy. See David Holloway's *Entering the Nuclear Arms Race: The Soviet Decision to Build the Atomic Bomb, 1939–1945*, Working Paper 9,

International Security Studies Program (Washington, D.C.: The Wilson Center, 1979); and his book *The Soviet Union and the Arms Race* (New Haven, Conn.: Yale University Press, 1985). An old book still of some value is Arnold Kramish's *Atomic Energy in the Soviet Union* (Stanford, Calif.: Stanford University Press, 1959). I. Golovin's Soviet biography of Igor Kurchatov, the head of the Soviet atomic bomb project, has been translated into English by William H. Dougherty: *I. V. Kurchatov: A Socialist-Realist Biography of the Soviet Nuclear Scientist* (Bloomington, Ind.: Selbstverlag Press, 1968). After the accident of the Chernobyl nuclear power reactor in the spring of 1986, many Westerners became interested in Soviet policies toward atomic energy. Sources on this topic include Paul Josephson, "The Historical Roots of the Chernobyl Disaster," *Soviet Union/Union Sovietique* 13, no. 3 (1986): 275–99; David R. Marples, *Chernobyl and Nuclear Power in the USSR* (New York: Macmillan, 1987); Marples, *The Social Impact of the Chernobyl Disaster* (New York: St. Martin's Press, 1988); and Zhores Medvedev, *Legacy of Chernobyl* (London: Basil Blackwell, 1990).

In recent years the most famous Soviet physicist was the late Andrei Sakharov. Although no complete biography of him exists, he wrote a remarkable memoir before his death: Andrei Sakharov, *Memoirs*, translated by Richard Lourie (New York: Knopf, 1990). A number of collections of Sakharov's writings and of writings about him also exist, including *Sakharov Speaks*, edited by Harrison E. Salisbury (New York: Vintage Books, 1974); *On Sakharov*, edited by Alexander Babyonyshev and translated by Guy Daniels (New York: Knopf, 1982); and Sakharov's *My Country and the World*, translated by Guy Daniels (New York: Vintage Books, 1975).

An interesting attempt to compare Soviet and Western research in high-energy physics is John Irvine and Ben R. Martin, "Basic Research in the East and West: A Comparison of the Scientific Performance of High-Energy Physics Accelerators," *Social Studies of Science* 15, no. 2 (1985): 293–341.

Astronomy is a field in which Russia achieved eminence long before the Revolution of 1917. Unfortunately, there are few English-language works on the history of Russian and Soviet astronomy. An exception is Otto Struve, "The Poulkovo. Observatory (1839–1941), *Sky and Telescope* 1, no. 4 (1941): 3–14, 19.

A history of repression in Soviet astronomy during the purges is Robert McCutcheon, "The Purge of Soviet Astronomy: 1936–1937, with a Discussion of Its Background and Aftermath," unpublished master's thesis, Georgetown University, Washington, D.C., 1985. An excellent summary is his "The 1936–1937 Purge of Soviet Astronomers," *Slavic Review* (Spring 1991): 100–17.

Space exploration. A Russian pioneer in space research, somewhat similar to Robert Goddard in the United States, was Konstantin Tsiolkovskii (1857–1935), who is the subject of an uncritical Soviet biography in English: A. A. Kosmodemianskii, *Konstantin Tsiolkovsky: His Life and Work* (Moscow: Foreign Languages Publishing House, 1956). Tsiolkovskii's collected works were translated into English by NASA: *The Collected Works of K. E. Tsiolkovskiy*, edited by B. N. Iur'ev and A. A. Blagonravov (Moscow, 1951–1959: NASA TT F-236, 237, 238, Washington, D.C., 1965). An interpretation of the image of Tsiolkovskii in Soviet literature is Rita DeDomenico's "The Official Image of Konstantin Tsiolkovsky in the Soviet Union, 1959–1970," unpublished senior thesis, Harvard University, Cambridge, Mass., 1986.

Two popular works on Soviet space exploration are James Oberg, *Red Star in Orbit* (New York: Random House, 1981); and Leonid Vladimirov, *The Russian Space Bluff: The Inside Story of the Soviet Drive to the Moon* (New York: Dial Press, 1973). A comparison of the Soviet and American space programs that won a Pulitzer Prize is Walter McDougall, *The Heavens and the Earth* (New York: Basic Books, 1985). McDougall's work is much stronger in its use of English sources than of Russian ones.

Geology. Although the Soviet Union produced more geologists than any other country in the world, very little exists in English on the history of Russian and Soviet geology. One noteworthy work is the biography of Academician Vladimir Vernadsky by Kendall Bailes, titled *Science and Russian Culture in an Age of Revolution: Vernadsky and His Scientific School, 1863–1945* (Bloomington: Indiana University Press, 1989).

The Soviet Union was slow in adapting to the revolution in geology brought about by plate tectonics. Robert M. Wood gives some of the reasons for this lag in his "Geology vs. Dogma: The Russian Rift," *New Scientist* (June 12, 1980): 234–7.

Technology. Relatively little has been written in the West on the history of technology in Russia and the Soviet Union, but interest in the subject is beginning to grow. An overview attempting to explain the difficulty for Russia and the Soviet Union to catch up with Western countries in technology is Loren R. Graham, "The Fits and Starts of Russian and Soviet Technology," in James P. Scanlon (ed.), *Technology, Culture and Development: The Experience of the Soviet Model* (Armonk, N.Y.: M. E. Sharpe, 1992), pp. 3–24. Studies of the early metallurgy industry are Arcadius Kahan, "Entrepreneurship in the Early Development of Iron Manufacturing in Russia," *Economic Development and Cultural Change* 10 (1962): 395–412; and Ian Blanchard, *Russia's Age of Silver: Precious-Metal Production and Economic Growth in the Eighteenth Century* (London: Routledge, 1989). The casting of bells was an important technology related to the casting of cannons; its history in Russia is explored in Edward V. Williams, *The Bells of Russia: History and Technology* (Princeton, N.J.: Princeton University Press, 1985). An early account of the Tula arms factory is in Iosif Gamel, *Description of the Tula Weapon Factory in Regard to Historical and Technical Aspects* (New Delhi: Amerind, 1988). An interesting example of American influence on prerevolutionary Russian technology is Joseph Bradley, *Guns for the Tsar: American Technology and the Small Arms Industry in Nineteenth-Century Russia* (DeKalb: Northern Illinois University Press, 1990). William Blackwell provides an introduction to Russian industrialization in *The Beginnings of Russian Industrialization, 1800–1860* (Princeton, N.J.: Princeton University Press, 1968). The early history of railroads in Russia is explored in Richard M. Haywood, *The Beginnings of Railway Development in Russia and the Reign of Nicholas I, 1835–1842* (Durham, N.C.: Duke University Press, 1969). Jonathan Coopersmith has worked on the history of electrification and has published *The Electrification of Russia, 1880–1926* (Ithaca: Cornell University Press, 1992).

Engineers have also proved a fruitful area of research. Harley Balzer has written a valuable study of prerevolutionary technical education: "Educating

Engineers: Economic Politics and Technical Training in Tsarist Russia," unpublished Ph.D. dissertation, University of Pennsylvania, 1980). Also, see Balzer's edited volume *Professions in Russia at the End of the Old Regime* (Ithaca, N.Y.: Cornell University Press, 1991). An outstanding history of the role of technology and engineers in the political and social development of the Soviet Union is Kendall Bailes, *Technology and Society under Lenin and Stalin: Origins of the Soviet Technical Intelligentsia, 1917–1941* (Princeton, N.J.: Princeton University Press, 1978). Studies of Taylorism in Soviet Russia include Bailes, "Alexei Gastev and the Soviet Controversy over Taylorism, 1918–1924," *Soviet Studies* 29, no. 3 (1977): 373–94; and Zenovia Sochor, "Soviet Taylorism Revisited," *Soviet Studies* 33, no. 2 (1981): 2446–64. Another useful work is Nicholas Lampert, *The Technical Intelligentsia and the Soviet State* (New York: Macmillan, 1979). Other works with much valuable historical materials on Soviet industrialization include Bruce Parrott, *Politics and Technology in the Soviet Union* (Cambridge, Mass.: MIT Press, 1983); Robert Lewis, *Science and Industrialization in the USSR* (New York: Macmillan, 1979); Lewis A. Siegelbaum, *Stakhanovism and the Politics of Productivity in the USSR, 1935–1941* (Cambridge: Cambridge University Press, 1988); and Hiroaki Kuromiya, *Stalin's Industrial Revolution* (Cambridge: Cambridge University Press, 1988). Also see Loren Graham, *The Ghost of the Executed Engineer* (Cambridge, Mass.: Harvard University Press, 1993).

The influence of Western technology is the focus of two studies. A massive work whose author is unwilling to grant independent industrial achievements to the Soviet Union is Antony Sutton, *Western Technology and Soviet Economic Development*, 3 vols. (Stanford, Calif.: Stanford University Press, 1968–1973). Mark Kuchment wrote an article detailing the birth of the Soviet microelectronics industry and the role played in it by two American engineers: "Active Technology Transfer and the Development of Soviet Microelectronics," in *Selling the Rope to Hang Capitalism?* edited by Charles Perry and Robert Pfaltzgraff, Jr. (Washington, D.C.: Pergamon-Brassey, 1987), pp. 60–9.

Policy studies. Because of the long rivalry between the United States and the Soviet Union in international relations, a competition necessarily involving science and technology, a large literature exists on science and technology policy in the Soviet Union. Few of these works are of interest to historians, but I will mention some that might be useful.

I took a historical look at Soviet science policy in "The Development of Science Policy in the Soviet Union," in *Science Policies of Industrial Nations*, edited by T. Dixon Long and Christopher Wright (New York: Praeger, 1975), pp. 12–58. An updated version includes Gorbachev's reforms in science: Loren R. Graham, "Science and Technology Trends in the Soviet Union," in *Framework for Interaction: Technical Structures in Selected Countries outside the European Community*, edited by Herbert Fusfeld (Troy, N.Y.: Rensselaer Polytechnic Institute, 1987), pp. II-D-1 to II-D-44. Harley Balzer also wrote on science under Gorbachev: "Is Less More? Soviet Science in the Gorbachev Era," *Issues in Science and Technology* vol 1, no. 4 (1985): 29–46. Also see Balzer's *Soviet Science on the Edge of Reform* (Boulder, Colo.: Westview Press, 1989). For a critical view of Soviet science written by a prominent researcher and administrator in the USSR, see Roald Sagdeev, "Science and Perestroika: A Long Way to Go," *Issues in Science and Technology* 4, no. 4 (1988): 48–52.

Paul Josephson discusses early Soviet science policy in "Science Policy in the Soviet Union, 1917–1927," *Minerva* 26, no. 3 (1988): 342–69. Soviet science policy in the period 1945–1975 is treated in Mark Adams, "Biology after Stalin: A Case Study," *Survey* 102 (Winter 1977–78): 53–80. Works that describe the main institutions making science policy in the Soviet Union include E. Zaleski et al., *Science Policy in the USSR* (Paris: Organization for Economic Cooperation and Development, 1969); Paul Cocks, *Science Policy USA–USSR*, vol. II (Washington, D.C.: National Science Foundation, 1980); and John Thomas and Ursula Kruse-Vaucienne's edited volume *Soviet Science and Technology* (Washington, D.C.: George Washington University Press, 1976).

An excellent analysis of the strengths and weaknesses of fundamental science in the USSR is Thane Gustafson, "Why Doesn't Soviet Science Do Better Than It Does?" in Linda Lubrano and Susan Gross Solomon (eds.), *The Social Context of Soviet Science* (Boulder, Colo.: Westview Press, 1980), pp. 31–68. This volume also contains articles by Bruce Parrott (on the organization of Soviet applied research), Linda Lubrano (on Soviet scientific collectives), Kendall Bailes (on the social backgrounds of technical specialists), and me (on genetic engineering).

Important topics of discussion among Soviet science policy specialists were the place of science in Marxist ideology and the role of the "STR" (the "scientific-technical revolution") in Soviet society. Helpful sources on these topics include Paul Josephson, "Science and Ideology in the Soviet Union: The Transformation of Science into a Direct Productive Force," *Soviet Union* 8, no. 2 (1981): 159–85; Julian Cooper, "The Scientific and Technical Revolution in Soviet Theory," in *Technology and Communist Culture*, edited by Frederic J. Fleron (New York: Praeger, 1977); Robert Miller, "The Scientific-Technical Revolution and the Soviet Administrative Battle," in *The Dynamics of Soviet Politics*, edited by Paul Cocks et al. (Cambridge, Mass.: Harvard University Press, 1976), pp. 137–155; and Erik Hoffmann, "Soviet Views of 'The Scientific-Technological Revolution,' " *World Politics* (July 1978): 615–44.

An interesting article on the growth of scientific personnel in the USSR, portraying the Soviet overtaking of the United States in the number of research workers, is Louvan Nolting and Murray Feshbach's "R and D Employment in the USSR," *Science* 207 (February 1, 1980): 493–503. Nolting has also published a series of reports (Foreign Economic Reports, Department of Commerce) on the structure and organization of Soviet science and technology.

A valuable analysis of the political role of Soviet science by Stephen Fortescue is *The Communist Party and Soviet Science* (London: Macmillan, 1987). Another book treating some of the same issues is Peter Kneen's *Soviet Scientists and the State* (Albany: State University of New York (SUNY) Press, 1984). Works written by emigrés who previously worked in the Soviet science establishment provide special insights; these include Mark Azbel, *Refusenik: Trapped in the Soviet Union* (Boston: Houghton Mifflin, 1981); Mark Popovsky, *Manipulated Science* (Garden City, N.Y.: Doubleday, 1979); and Vladimir Kresin, "Soviet Science in Practice: An Insider's View," in *The Soviet Union Today*, edited by James Cracraft (Chicago: Bulletin of Atomic Scientists, 1983).

Four works treating Soviet industrial research from economic and political standpoints are Joseph Berliner, *The Innovation Decision in Soviet Industry* (Cambridge, Mass.: Harvard University Press, 1976); Bruce Parrott, *Politics and Technol-*

ogy in the Soviet Union (Cambridge, Mass.: MIT Press, 1983); Erik Hoffmann and Robbin Laird, *Technocratic Socialism: The Soviet Union in the Advanced Industrial Era* (Durham, N.C.: Duke University Press, 1985); and Raymond Hutchings, *Soviet Science, Technology and Design: Interaction and Convergence* (London: Oxford University Press, 1976).

Mark Beissinger wrote an important study of the pattern of reform and reaction in Soviet management and technical research in his *Scientific Management, Socialist Discipline, and Soviet Power* (Cambridge, Mass.: Harvard University Press, 1988).

A topic of particular interest to American scholars who may wish to do research in the former Soviet Union, no matter what the field, is the history of scholarly exchanges between the United States and the USSR. The most thoughtful analysis of the subject is by Linda Lubrano, "National and International Politics in US–USSR Scientific Cooperation," *Social Studies of Science* 11 (1987): pp. 451–80. Also see *Review of US–USSR Interacademy Exchanges and Relations*, Report of the National Academy of Sciences (Washington, D.C., 1977); and Yale Richmond, *U.S.–Soviet Cultural Exchanges, 1958–1986: Who Wins?* (Boulder, Colo.: Westview Press, 1987).

Index

absolute zero, 23
Académie Royal de Sciences, Paris, 84, 97
Academy of Agricultural Sciences, 132
Academy of Sciences, Bukharin and, 142,
    157, 158, 168; Count Tolstoi and, 38;
    Cultural Revolution and, 94, 95, 96,
    97, 98, 122; early members of, 29, 31,
    35, 36; financial support for, 87; first
    elected president of, 82, 83, 84, 85, 86;
    N. Fuss and, 42, 43, 47; history of sci-
    ence and, 137, 139; Lysenko and, 128,
    131–4; mathematics and, 215, 217;
    Mendeleev and, 49, 80; organization
    of, 180, 181, 182, 183, 188; I. Pavlov
    and, 240; post-Communist reforms,
    190, 191; protests against, 189; resis-
    tance to the Communist party, 273–4;
    Russian Revolution and, 81; Sakharov
    and, 170, 171, 174, 177, 178; sources
    on, 294
Academy of Sciences of Ukraine, 128
Accelerator technology, 212
Adams, M., 240, 282, 290, 295, 297, 298, 305
Adzhubei, A., 278
Aepinus, F., 28
Afanasieva, T., 208
Afghanistan, invasion of, 170, 171
Agassiz, L., 71
Agol, I., 117
agriculture, 166, 167
airplanes, production of, 254, 260
Akhundov, M., 278, 282
Aksakov, S., 264
Aksenov, A., 263
Albats, E., 278, 279
Aleksandrov, A. D., 117, 212, 214, 216, 287
Aleksandrov, A. P., 209
Aleksandrov, D. A., 154, 278, 282
Aleksandrov, P. S., 215, 216
Alexander, J., 245, 290, 299
Alexander I, 32, 33, 34, 35, 36, 39, 40, 41,
    47, 264
Alexander II, 32, 36, 37, 38, 57, 247

Alexei Mikhailovich, tsar, 16
Alikhanov, A., 209
Allilueva, S., 131
Almgren, B., 300
Alvin Clark and Sons, 222
Amalrik, A., 79, 268
Ambartsumian, V., 117, 222
American Association for the Advance-
    ment of Science, 85
Andreev, K., 214
Andrusov, N., 233
Anokhin, P., 117
anthropology, 93
anti-Semitism, 147, 168, 186, 192, 212, 214,
    219
Apothecary Bureau, 245
Arabic numerals, 16, 30
Arabic science, 10ff.
Arbuzov, A., 227, 228
arc light, 254
Archimedes, 11
architecture, 16
Arctic, 21, 232
Aristotle, 11, 14, 16
Arkhangel'skii, A., 232
Armenia, 10, 222, 244
Arnold, V., 217
artillery, 17
Artsimovich, L., 208, 209
astronomy, 220–4, 302
astrophysics, 212
atomic weapons, 159, 201, 209, 212, 301
atomic weights, 53
automobile, production of, 254, 256
aviation, 5, 245, 260
Avicenna (Ibn Sina), 10
Avogadro's principle, 48
Azbel, M., 305
Azerbaidzhan, 124, 244
Azrael, J., 282, 283

Babkin, B., 289, 298
Babyonyshev, A., 302